THERMODYNAMICS OF NONEQUILIBRIUM PROCESSES

THERMODYNAMICS OF NONEQUILIBRIUM PROCESSES

by

S. WIŚNIEWSKI, B. STANISZEWSKI, R. SZYMANIK

D. REIDEL PUBLISHING COMPANY
DORDRECHT – HOLLAND/BOSTON – U.S.A.

PWN–POLISH SCIENTIFIC PUBLISHERS
WARSZAWA

Library of Congress Cataloging in Publication Data

Wiśniewski, Stefan.
Thermodynamics of nonequilibrium processes.

Translation of Termodynamika procesów nierównowagowych.
Bibliography: p.
Includes index.
1. Irreversible processes. 2. Thermoelectricity.
3. Thermomagnetism. I. Staniszewski, Bogumił, joint author. II. Szymanik, Ryszard, joint author.
III. Title.
QC318. 17W5713 536'.7 75-41486
ISBN 90-277-0331-0

TERMODYNAMIKA PROCESÓW NIERÓWNOWAGOWYCH
First published in 1973 by PWN—Polish Scientific Publishers, Warszawa
Translated from the Polish by Eugene Lepa

Distributors for Albania, Bulgaria, Chinese People's Republic, Czechoslovakia, Cuba, German Democratic Republic, Hungary, Korean People's Democratic Republic, Mongolia, Poland, Rumania, Democratic Republic of Vietnam, the U.S.S.R. and Yugoslavia
ARS POLONA—RUCH
Krakowskie Przedmieście 7, 00-068 Warszawa, Poland

Distributors for the U.S.A., Canada and Mexico
D. REIDEL PUBLISHING COMPANY, INC.
Lincoln Building, 160 Old Derby Street, Hingham, Mass. 02043, U.S.A.

Distributors for all other countries
D. REIDEL PUBLISHING COMPANY
P.O. Box 17, Dordrecht, Holland

Printed in Poland by DUAM

TABLE OF CONTENTS

PREFACE

This book presents the fundamentals of nonequilibrium thermodynamics and its principal engineering applications. It has grown out of the teaching experience of the authors and is intended as a textbook for technical universities, being addressed to students of those departments which give courses in nonequilibrium thermodynamics, and also for doctoral students.

The authors believe this book may also be useful to those engineers who come in contact with problems pertaining to nonequilibrium thermodynamics in the course of their work or take an interest in these problems.

To use this book it is sufficient to have a knowledge of classical phenomenological thermodynamics to the extent taught in technical universities and a background in vector calculus and some elements of tensor calculus. It has been the intention of the authors to present nonequilibrium thermodynamics from the point of view of the engineer's needs, interests, and way of thinking, but it is also their hope that specialists from other fields will also find some useful information in it.

Chapters 1, 2, 3, 5 and 7 have been written by S. Wiśniewski, Chapter 4 by B. Staniszewski, and Chapter 6 by R. Szymanik.

PRINCIPAL NOTATION

a	thermal diffusivity
A	area; Richardson–Dushman constant
A_j	chemical affinity of reaction j
$\boldsymbol{B}$	magnetic field strength (historically: magnetic induction)
c_i	molar concentration of component i
c_p	specific heat at constant pressure
c_v	specific heat at constant volume
D	diffusion coefficient
$\boldsymbol{D}$	electric displacement
D'	thermal diffusion coefficient
D''	Dufour coefficient
e	electric charge per unit mass
e_e	electric charge per unit number of electrons
e_t	total specific energy
$\boldsymbol{E}$	electric field strength
$\boldsymbol{F}$	mass force
$\boldsymbol{g}$	gravitational acceleration
h	specific enthalpy; heat transfer coefficient
$\tilde{h}_e$	partial specific enthalpy of electrons
$\tilde{h}_i$	partial specific enthalpy of component i
$\boldsymbol{H}$	magnetic displacement (historically: magnetic field strength)
I	electric current
$\boldsymbol{j}_i$	diffusion flux of component i
$\boldsymbol{j}_e$	electron flux
$\boldsymbol{J}_e$	electric current density
$\boldsymbol{J}_{es}$	saturation current density
$\boldsymbol{J}_i$	flux of component i
$\boldsymbol{J}_j$	rate of the chemical reaction j
$\boldsymbol{J}_q'$	heat flux in the energy balance equation
$\boldsymbol{J}_q''$	heat flux in the entropy balance equation
$\boldsymbol{J}_s$	entropy flux
$\boldsymbol{J}_u$	internal energy flux
k	number of components; Boltzmann's constant
k_T	thermal diffusion ratio
l	length
l_{ab}	phenomenological coefficient stemming from entropy source strength

L_{ab}	phenomenological coefficient stemming from the dissipation function
m	mass
M	molar mass
N_A	Avogadro's number
p	pressure
P	power
q	heat per unit quantity of substance
Q	heat
$Q_e''^*$	heat of transport of electrons
$Q_i''^*$	heat of transport of component i
$\dot{Q}$	heat flow
$\dot{Q}_P$	Peltier heat flow
$\dot{Q}_T$	Thomson heat flow
r	number of chemical reactions; radius
R	gas constant
R_M	universal gas constant
s	specific entropy
$\tilde{s}_e$	partial specific entropy of electrons
$\tilde{s}_i$	partial specific entropy of component i
s_T	Soret coefficient
S	entropy
S_e^*	entropy of transport of electrons
S_i^*	entropy of transport of component i
t	time
T	absolute temperature
u	specific internal energy
$\tilde{u}_i$	partial specific internal energy of component i
U	internal energy
U_e^*	energy of transport of electrons
v	specific volume
$\mathbf{v}$	centre-of-mass or barycentric velocity
$\mathbf{v}_a$	reference velocity
$\mathbf{v}_M$	mean molar velocity
V	volume
V_i	ionization potential
V_M	molar volume
$\tilde{V}_{Mi}$	partial molar volume of component i
x_i	mass fraction of component i
X_a	thermodynamic force
y	Planck's function
y'	generalized Planck's function
Z_A	figure omfteri
z_i	molar fraction of component i

α	thermal diffusion factor
$\boldsymbol{\delta}$	unit tensor
ε	relative emissivity
$\in$	electron charge
ε_0	permittivity of free space
ε_{AB}	Seebeck coefficient for pair of conductors A and B
η	shear (or ordinary) viscosity
η_v	bulk (or volume) viscosity
λ	thermal conductivity
μ_a	chemical potential of atoms
μ_e	chemical potential of electrons
μ_i	chemical potential of component i
μ_i'	total potential of component i
ν	stoichiometric coefficient
Π_{AB}	Peltier coefficient for pair of conductors A and B
ρ	density; charge density; resistivity
ρ_i	component density of component i
σ	electrical conductivity
$\boldsymbol{\sigma}$	stress tensor
σ_T	Thomson coefficient
τ	$\frac{1}{3}$ trace of viscosity part of stress tensor; Thomson coefficient
$\boldsymbol{\tau}$	viscosity part of stress tensor
φ	electric potential
ϕ	work function
Φ_s	entropy source strength
Φ_v	Rayleigh dissipation function
ψ	specific potential energy; electrostatic potential
Ψ	dissipation function
$\boldsymbol{\omega}$	angular velocity

CHAPTER I

THE FUNDAMENTAL CONCEPTS AND LAWS OF CLASSICAL THERMODYNAMICS

1. The Scope and Methods of Classical Thermodynamics

Thermodynamics is an area of physics which is concerned primarily with problems associated with internal energy, and particularly with its conversion into other forms and transfer between a system and its environment. The second concept which is also especially characteristic of thermodynamics is that of entropy. Both of these concepts occur directly or indirectly in all problems considered by thermodynamics.

Classical thermodynamics, which came into being in the mid-19th century, deals in principle with changes of state which can be viewed as a set of successive equilibrium states. For this reason, by analogy with electrostatics or mechanical statics, it should be called thermostatics. In keeping with custom, the appellation *classical thermodynamics* will be used for thermostatics in this book. Phenomena which do not pass through successive equilibrium states are considered by *nonequilibrium thermodynamics*, also called the *thermodynamics of irreversible processes*.

Classical thermodynamics is based on several fundamental laws which are the outcome of many years of physical observations and are known as the *laws of thermodynamics*. These laws are experimentally-verified axioms from which a large number of other laws and relations can be derived by theoretical means, hence building up large areas of this science. In this chapter we shall consider only the most general fundamental concepts, laws, and relations of classical thermodynamics from the physical point of view. Thermodynamics is often adapted so as to suit specific needs and then has special names, such as engineering thermodynamics, chemical thermodynamics, etc.

The range of topics considered by thermodynamics is quite extensive and in many cases reaches out into other areas of physics. The thermodynamics of electric and magnetic phenomena has been particularly developed.

Thermodynamics operates with fundamental concepts which should be understood unambiguously by author and reader alike. Although the development of classical thermodynamics may be regarded as having been completed in respect of formulation of new laws and relations, only in recent years have attempts been made to afford a logical treatment to the axiomatic foundations of thermodynamics in university textbooks. Endeavours along these lines have been made in particular by American thermodynamic scientists: Keenan, Callen, Tribus, Giles, and others. The reader who would like to recall or get a deeper insight into classical thermodynamics is referred to textbooks with the most modern treatments ([1]–[8]). The present chapter will take up only those aspects of classical thermodynamics which are indispensable for an understanding of nonequilibrium thermodynamics or for bringing out the differences between nonequilibrium and classical thermodynamics. Considerations of space prevent any detailed discussion here of the latest trends in axiomatization of the foundations of thermodynamics; instead, the fundamental concepts and laws can merely be given an orderly treatment that differs little from the traditional.

A primary concept in thermodynamics, one taken from philosophy, is that of matter. Matter occurs in the form of substance and in the form of a field. Substance has a corpuscular structure; it consists of molecules, atoms, and constituent parts of atoms, i.e. elementary particles. Field matter also occurs in the form of various kinds of force fields, such as gravitational field, and electromagnetic field.

A physical body consists of a substance which has strictly defined boundaries, real or abstract. Air, water, and steel are examples of substances. The air contained in a room, the water in a vessel, and a steel rod are examples of physical bodies.

In the study of physical phenomena, it is often the custom to isolate in space a region, called a physical system, which constitutes the object under consideration; in thermodynamic discussions this is termed a *thermodynamic system*. A thermodynamic system may contain a single body or many, as well as a field which, in the form of a gravitational field, always accompanies a substance. Everything that lies beyond the system boundary and has a direct bearing on its behaviour is called the *environment* of the system.

Walls separating a system from its environment may be penetrable to various kinds of interactions between the system and the environment. These interactions may cause changes in the state of the system. Different appellations are used for various systems, depending on the interactions between system and environment which occur at the system boundary.

An isolated system is separated from the environment by a wall which prevents the flow of substance and the penetration of energy. In conformity with the definition, both the quantity of energy and the quantity of substance do not change in an isolated system.

In respect of substance being able to flow across the system boundary, systems are classified as closed or open. A flow system is an open system if substance flows into or from it across its boundary.

A system is said to be *adiabatic* if its state can be changed only by means of operations called work. An adiabatic system is sealed off from its environment by an adiabatic wall. A diathermal wall allows conduction of heat, that is, permits the environment to exert an action other than work.

Once a thermodynamic system or physical body has been separated out from the environment, we can begin to describe it in terms of its properties or such properties of the environment as enable the behaviour of the system to be considered. For this purpose, thermodynamics draws upon two methods: phenomenological and statistical. Accordingly, from the point of view of method used to consider phenomena, a distinction is made between phenomenological thermodynamics and statistical thermodynamics, and if the laws of quantum mechanics are taken into account, quantum thermodynamics.

The *phenomenological method* consists in providing a macroscopic description of the systems considered and the phenomena occurring in them, without regard for the molecular and atomic structure of the substance, in terms of physical properties which are often perceptible and directly measurable. Physical phenomenological theories associate physical quantities by general, fundamental relations which lend themselves to experimental verification. The inability to present a phenomenological description of the inner structure of a substance necessitates experimental determination of such individual physical properties of the substance as specific heat and heat conductivity. The phenomenological method is most convenient for concrete engineering applications and consequently we shall henceforth use it almost exclusively.

The *statistical method* consists in considering ensembles of many molecules constituting the given system. To simplify the microscopic description, the space under consideration is divided into regions in which the properties of the individual molecules can be averaged and statistical concepts introduced, such as pressure, temperature, internal energy, and entropy, which have an exact phenomenological meaning; then, the macroscopic phenomenological method can be employed without going into what happens with the individual molecules. In this way, these two methods which in principle are essentially different can be linked together. One may speak of phenomenological concepts with an exact statistical interpretation only if the motion of the molecules is perfectly chaotic, that is if the individual events are independent and if averaging may be carried out over a space containing a sufficiently large number of molecules for the principles of statistics and probability theory to be applicable. At the same time, the averaging space should be small enough to be viewed, in the macroscopic treatment, as a point particle to which such intensive parameters as temperature and pressure could be assigned.

The principal advantage of phenomenological theories is that the laws are general and experimentally verifiable.

The principal advantage of statistical theories, on the other hand, is that they provide a more profound treatment of concepts and laws, and make it possible to calculate the physical properties of a substance from basic data concerning the molecular structure and to determine by computation physical constants which can be found only empirically when phenomenological methods are employed.

2. Determination of a Thermodynamic State

The thermodynamic state of a body is determined uniquely if the body can be reproduced from the same substance in another place in a manner sufficiently accurate for thermodynamic consideration. Thus, in order to determine a state one need not know the individual physical properties which depend only on the species of substance and the geometric shape of the body, which is usually not essential as far as thermodynamics is concerned. The thermodynamic state of a substance, body, or system is specified by a simultaneous set of values of physical quantities, called *state parameters*, which are capable of change. The values of state parameters do not depend on how the body arrived at the state under consideration, or – as is said – on its 'history'. The thermodynamic state changes if at least one of the state parameters changes.

It is known from experiments that the parameters of a given equilibrium state are not independent but are associated with each other by mathematical relations, called *equations of state*, which enable some parameters to be used to calculate the others. The form the equation of state takes depends on the properties of the substance under consideration. To determine the state of a body or substance, therefore, it is sufficient to know only some parameters, treated as independent variables. The other parameters, viewed as dependent variables, are sometimes called *state functions.*

The state of so-called simple bodies can be specified in an especially straightforward manner, viz. by means of two parameters, hence by a point in a plane. Simple bodies are bodies which are at rest, homogeneous (that is, not subject to field forces, concentrated forces, and surface tension), isotropic, and in thermodynamic equilibrium.

An isolated system is characterized by the absence of interactions with the environment. Such a system spontaneously attains a state of thermodynamic equilibrium. Kinds of thermodynamic equilibrium are distinguished just as are kinds of mechanical equilibrium (Fig. 1.1).

A system in a state of stable equilibrium changes arbitrarily little under an arbitrarily small action by the environment. A system in a state of neutral equilibrium may experience a finite change without any change in the environment. A system in a state of unstable (or labile) equilibrium changes considerably under the influence of an

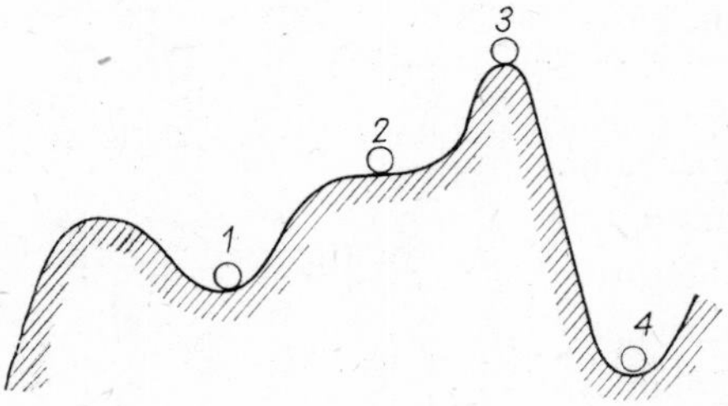

Fig. 1.1 Kinds of equilibrium: 1 meta-stable, 2 neutral, 3 unstable (labile), 4 stable

arbitrarily small action by the environment. Under appropriately small stimuli, a state of metastable equilibrium has the same properties as a state of stable equilibrium, whereas under stimuli greater than the limiting value its properties are like those of a state of unstable equilibrium.

Extensive parameters are determined for the entire volume of a body; they depend on the amount of substance in the body and hence cannot be the same for a body as a whole and for its individual parts. Volume, internal energy, and entropy are instances of extensive parameters. Extensive parameters are additive and their value for an entire system is calculated by summing the values of these parameters for the individual parts of the system. The extensive parameters of homogeneous bodies are proportional to the amount of substance constituting the body. Upper-case letters are customarily used to indicate extensive parameters.

Intensive parameters may be the same for an entire body and its individual parts; temperature and pressure are examples. These parameters may be assigned to particular points of the body, thus forming the corresponding fields, e.g. temperature field. If the values of a given parameter do not vary in time at any point in the space under consideration, the field is said to be steady; otherwise it is called unsteady. A body with identical values of intensive parameters at all points is a homogeneous body. Intensive parameters are in general denoted by lower-case letters.

Specific parameters for homogeneous substances, especially intensive parameters, are formed as the ratios of extensive parameters to the amount of substance. Specific parameters on a mass basis are indicated by lower-case letters, by analogy with the corresponding notation for extensive parameters.

Examples of parameter notation: volume V, mass m, specific volume $v = V/m$.

Specific quantities on a kilomole basis will be called molar quantities and designated just as the corresponding extensive quantities, but with the index M. An example of a molar quantity: molar volume $V_M = V/n = vM$, where n is the number of kilomoles and M is the conversion factor for kilomoles to kilograms, equal numerically to the molecular mass and having a denomination of kg/kmole.

There also exist extensive quantities which are not thermodynamic parameters, e.g. work, from which specific quantities can be formed in a similar manner.

3. The Zeroth Law of Thermodynamics

Thermodynamics is an area of physics which in principle is taught after a course in mechanics. The reader, therefore, is assumed to be well acquainted with the concepts of mechanics. All mechanical quantities are definable in terms of fundamental quantities such as length, time, and mass, or length, time, and force. The transition to thermodynamic quantities in the phenomenological approach requires the introduction of an auxiliary quantity, that of temperature or entropy. At present, the practice in thermodynamics often is first to introduce the concept of entropy and then to define temperature in terms of entropy. Inasmuch as the discussion of the fundamentals of thermodynamics is to be kept to a minimum here, we shall resort to the classical treatment in which the temperature concept is introduced before that of entropy.

Temperature specifies the extent to which a body has been heated, and hence enables warmer bodies to be distinguished from cooler ones. When treated as a fundamental concept, it is not defined in terms of other physical quantities.

In order to determine the properties of temperature, consider two bodies *1* and *2* (Fig. 1.2) which are in contact with body *3* by means of nondeformable diathermal walls. The other walls of bodies *1*, *2*, and *3* which delimit them from the environment are adiabatic and nondeformable. A system of this kind tends spontaneously to

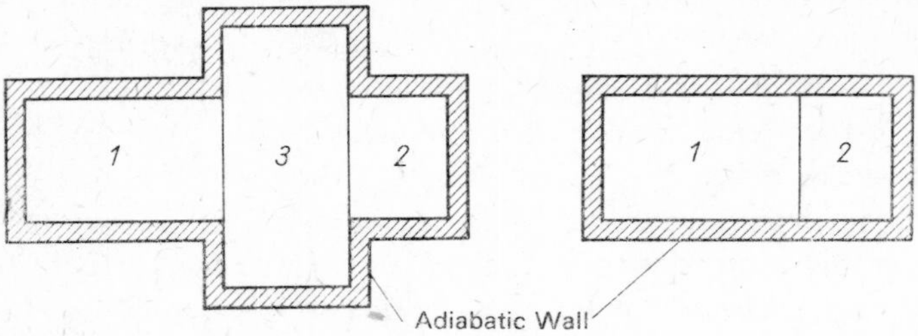

Fig. 1.2 Transitivity of thermal equilibrium

thermal equilibrium, that is, to an equalization of temperatures, body *3* being in equilibrium with bodies *1* and *2*. When body *3* is removed and bodies *1* and *2* are put in direct contact through their diathermal walls, their temperatures remain unchanged, and hence these bodies are in thermal equilibrium.

The foregoing conclusion may be formulated briefly as follows: two bodies in thermal equilibrium with a third body are also in thermal equilibrium with each other. This is the law of transitivity of thermal equilibrium which is so important in determining the temperature property that Fowler and many thermodynamic scientists regarded it as the *zeroth law of thermodynamics*.

4. The First Law of Thermodynamics

Interactions between a system and its environment lead to a change in the state of the system. The phenomenon of a change in the state of a system is described in classical thermodynamics by a thermodynamic process, a set of consecutive states of a system. A process is an equilibrium (quasi-static) process if it takes place very slowly so that it consists of states of stable equilibrium. In the equilibrium case a process can be represented by a plot in which parameters necessary and sufficient for determining the stable state of the system are laid off along the axes. Phenomena which cannot be considered as a set of stable equilibrium states of a substance are called *nonequilibrium processes*.

A process is reversible if it is possible to return from its final state to its initial state in a manner so that the environment is restored to its initial state. In the special case, the inverse to the reversible process may be carried out, running through the same intermediate states in the opposite direction, the interaction between the system and the environment being of the same magnitude but of opposite sign.

In order for a process to be reversible it is a necessary but not a sufficient condition that it run over successive equilibrium states, this being so because some resistances do not vanish no matter how slow the phenomenon. The main reasons why phenomena are irreversible are: the process is nonstatic, thus causing an increase in the work input and a decrease in the work output of the system; nonstatic processes equalize pressures, temperatures, concentrations, etc., without performing external work; and work or energy is dissipated owing to resistances. All of these involve losses of work and, from the thermodynamic point of view, give the same result, that is an increase in entropy.

The action of the system on the environment may be called work if the result of that action can be reduced only to a change in the position of a weight, outside the system, relative to a reference level. It is often assumed in physical or chemical thermodynamics that work added to a system is positive, and work done by the system is negative. In engineering thermodynamics, the convention concerning signs is reversed. Since the present book is addressed to engineers, the latter convention will be employed here.

The work of an equilibrium process between states 1 and 2 can be calculated by means of an integral of the form

$$W_{1,2} = \int_{\pi 1,2} y \, \mathrm{d}X = \int_{X_1}^{X_2} y_\pi(X) \, \mathrm{d}X , \tag{1.1}$$

where y is the generalized force, that is the quantity causing work to be performed, whereas $\mathrm{d}X$ is an elementary generalized displacement or the differential of the quantity which must undergo change in order for the work to be non-zero. If the pressure p, or the negative of the force $(-F)$, surface tension $(-\sigma)$, magnetic displacement $(-H)$, or electric field strength $(-E)$ is taken for the generalized force, then correspondingly the volume V, length (path) l, area A, magnetic moment M, or the polarization of the entire volume of dielectric (VP) is the quantity X. Thus, the work of a change in volume V is

$$W_{1,2} = \int_{\pi 1,2} p \, \mathrm{d}V . \tag{1.2}$$

Expression (1.1) can be integrated if the path of the process π between states 1 and 2 is known (first integral) or if the functional dependence of the generalized force y on the quantity X, characteristic of the given process π, is known (second integral).

Work is not a state parameter but a functional of the path of the process. If the state does not change, the work is equal to zero. Except for work in a potential field of forces, the work depends on the path of the process. In view of this, the integrand in (1.1) is not a total (also called exact) differential but is a linear differential expression, which is often denoted in thermodynamics by đ so as to distinguish it from the total differential:

$$\text{đ}W = y \, \mathrm{d}X , \tag{1.3}$$

and if $y = p$, and $X = V$,

$$\text{đ}W = p \, \mathrm{d}V . \tag{1.4}$$

For irreversible processes a distinction should be made between the work of the process as defined by (1.1) and the external work done on the system by macroscopic external forces. Owing to the dissipation of work by various kinds of resistance, especially friction, the external work of an irreversible process always satisfies the inequality

$$W_{e1,2} \leqslant \int_{\pi 1,2} y \, dX . \tag{1.5}$$

The effect of the environment on a closed adiabatic system is confined to the external work of the process. As has been found by repeated observations of adiabatic systems (Fig. 1.3), the external work of an adiabatic process does not depend on how it is realized between the extreme states. The foregoing assertion is the empirical basis of the first law of thermodynamics. Work is independent of the path in a potential field of forces and then the work is equal to the potential difference. The conclusion suggests itself that the external work of an adiabatic process is equal to the difference between two values of the state function, which is the potential for the adiabatic system, in the extreme states of the transformation. For an adiabatic system, in which changes in kinetic and potential energy may be neglected, this function is the internal energy of the system, and is denoted by the symbol U. The external work of an adiabatic process thus is equal to the drop in the internal energy of the system:

$$W_{e\,ad1,2} = U_1 - U_2 . \tag{1.6}$$

The sign of the change in the internal energy follows from the convention that work removed from a system, that is done at the expense of a drop in internal energy, is

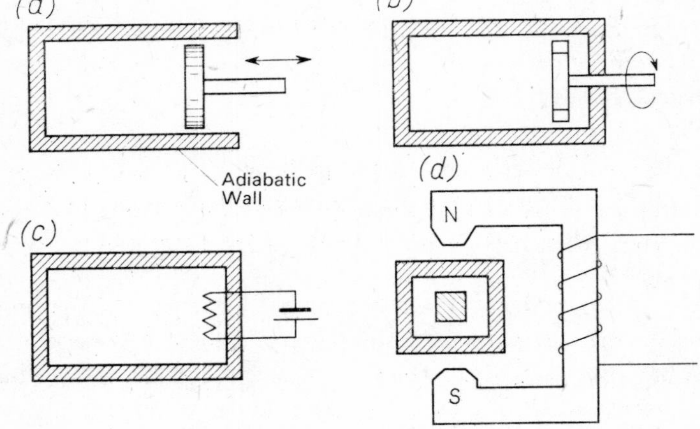

Fig. 1.3 Various work modes of an adiabatic system: (a) work of change of volume, (b) work of friction, (c) work of electric current and (d) work of magnetization

positive. Thermodynamics as a rule considers increments in internal energy or relative values above a reference state. In statistical thermodynamics, the internal energy of a body is defined as the sum of the energies of the molecules, relative to the centre of mass, and the energy of the interaction. Like work, internal energy is an extensive quantity and a specific internal energy per unit mass of substance can be formed.

When a closed system is not adiabatic, the work depends on the path of the process. The difference between the work of the process under consideration and the work of the adiabatic process between the same states is equal to the heat of the process:

$$Q_{1,2}=W_{1,2}-W_{\mathrm{ad}1,2}=U_2-U_1+W_{1,2}\,. \tag{1.7}$$

The external heat $Q_{e1,2}$ of the process is defined in similar manner on the basis of the external work, that is

$$Q_{e1,2}=W_{e1,2}-W_{e\,\mathrm{ad}1,2}=U_2-U_1+W_{e1,2}\,. \tag{1.8}$$

The difference between the heat $Q_{1,2}$ and the external heat $Q_{e1,2}$ is the 'heat' arising from dissipation of work inside the system (in the particular case, frictional heat):

$$Q_{i1,2}=Q_{1,2}-Q_{e1,2}=W_{i1,2}=W_{1,2}-W_{e1,2}\,. \tag{1.9}$$

Equation (1.8), defining the external heat of a closed system, expresses the first law of thermodynamics. This equation states that a change in the internal energy of a closed system is equal to the difference between the external heat and the external work of the same process:

$$U_2-U_1=Q_{e1,2}-W_{e1,2}\,. \tag{1.10}$$

It is easily remarked that the convention concerning the sign of heat is the reverse of that for the sign of work. Heat entering a system, that is increasing its internal energy, is treated as positive. Heat arising from the dissipation of work is positive for irreversible processes, and zero for reversible processes. As in the case with work, heat depends on the path of the process; hence the differential notation of equation (1.7), which expresses the first law of thermodynamics, is of the form

$$đQ=\mathrm{d}U+đW\,. \tag{1.11}$$

For equilibrium processes, the linear differential expression $đQ$, used to calculate the heat of the process, may be rewritten as

$$đQ=đQ_e+đQ_i=\mathrm{d}U+\sum_{i=1}^{n} y_i\,\mathrm{d}X_i \tag{1.12}$$

or, on introduction of a state function called *generalized enthalpy* which is defined as

$$H=U+\sum_{i=1}^{n} y_i X_i\,, \tag{1.13}$$

it may also be recast in the form

$$đQ=\mathrm{d}H-\sum_{i=1}^{n} X_i\,\mathrm{d}y_i\,. \tag{1.14}$$

If $y=p$ and $X=V$, then enthalpy may be defined as

$$H=U+pV, \tag{1.15}$$

and the linear differential expression used to evaluate the heat of the process assumes the form:

$$đQ=\mathrm{d}U+p\,\mathrm{d}V=\mathrm{d}H-V\,\mathrm{d}p. \tag{1.16}$$

The ratio of the heat of the process to the amount of substance participating in the process and to the temperature increment during this process is called the *specific heat* of the process. The specific heat is called the *mean specific heat* if the temperature increment is finite, and the *specific heat in the given state* if the temperature increment is infinitesimal.

The specific heat as constant volume or the specific heat at constant pressure occur most often in thermodynamic considerations.

For real gases, by equation (1.16) the specific heat in a given state at constant volume is

$$c_v=\frac{đQ_v}{m\,\mathrm{d}T}=\left(\frac{\partial u}{\partial T}\right)_v, \tag{1.17}$$

whereas the specific heat at constant pressure is

$$c_p=\frac{đQ_p}{m\,\mathrm{d}T}=\left(\frac{\partial h}{\partial T}\right)_p. \tag{1.18}$$

5. The Second Law of Thermodynamics

The differential expression $đQ$, expressing the elementary heat of an equilibrium process, always allows itself to be reduced to a linear differential expression known as a *Pfaff differential expression* or *Pfaffian*:

$$đ\Pi=\sum_{i=1}^{n} y_i\,\mathrm{d}x_i, \tag{1.19}$$

where x_i are independent variables, and y_i are dependent variables.

In some cases, Pfaffians are exact differentials of a function $F(x_i)$ of the independent variables x_i $(i=1, 2, 3, \dots, n)$, and then

$$đ\Pi=\mathrm{d}F=\sum_{i=1}^{n}\frac{\partial F}{\partial x_i}\,\mathrm{d}x_i, \tag{1.20}$$

that is

$$y_i=\frac{\partial F}{\partial x_i}. \tag{1.21}$$

If the second-order partial derivatives of the function $F(x_i)$ are continuous, then they are equal to each other and independent of the order of differentiation:

$$\frac{\partial y_i}{\partial x_k}=\frac{\partial y_k}{\partial x_i}. \tag{1.22}$$

If a Pfaff differential expression is not an exact differential of some function $F(x_i)$, then at least in one case

$$\frac{\partial y_i}{\partial x_k} \neq \frac{\partial y_k}{\partial x_i}. \qquad (1.23)$$

In some cases, Pfaffians may be exact differentials of $F(x_i)$ only upon multiplication by an integrating factor $\psi(x_i)$ which, in the general case, is also a function of the independent variables x_i, and then

$$\mathrm{d}F = \psi \sum_{i=1}^{n} y_i \, \mathrm{d}x_i. \qquad (1.24)$$

Pfaffians of this kind are said to be integrable. All others are not integrable.

It is easily shown that a Pfaff differential expression with two independent variables

$$đ\Pi = y_1 \, \mathrm{d}x_1 + y_2 \mathrm{d}x_2 \qquad (1.25)$$

always has an integrating factor.

A Pfaffian with more than two independent variables has an integrating factor only if certain conditions are met. By the principle of Carathéodory, if in the neighbourhood of any point of a simply connected region, containing no singular points, there are points which are not accessible from that point along the path specified by the Pfaff differential equation $đ\Pi = 0$, then the Pfaff differential expresion $đ\Pi$ has an integrating factor. If at least one integrating factor exists, then an infinite number of such factors exists.

Any change in the state of a simple body is characterized by two independent variables. In this case the differential expression determining the heat $đQ$ is always integrable. The inverse absolute temperature is one of the integrating factors of the expression $đQ$.

The differential of the entropy for reversible process of simple bodies is defined by the relationship

$$\mathrm{d}S = \frac{đQ}{T} = \frac{\mathrm{d}U + y\,\mathrm{d}X}{T} = \frac{\mathrm{d}H - X\,\mathrm{d}y}{T}. \qquad (1.26)$$

In the special case when $y = p$ and $X = V$,

$$\mathrm{d}S = \frac{đQ}{T} = \frac{\mathrm{d}U + p\,\mathrm{d}V}{T} = \frac{\mathrm{d}H - V\,\mathrm{d}p}{T}. \qquad (1.27)$$

To integrate these equations one needs to know an equation of state relating the state function to the state parameters. Since $\mathrm{d}S$ is an exact differential of the state function S, called *entropy*, the result of the integration does not depend on the path of the process or on how it is realized, provided that both the initial and final states are stable equilibrium states.

If there are more than two independent variables, it is necessary to adopt a postulate based on empirical observations. This is one of the verbal formulations of the second law of thermodynamics (given by Carathéodory): in the neighbourhood of every state there are states which are not accessible by an adiabatic process, that is along the path $đQ=0$. Accordingly, the expression $đQ$ has an integrating factor, by the Carathéodory principle. In thermodynamics it is more convenient to use an integrating denominator which is the inverse of the integrating factor. When the differential expression $đQ$ is divided by the integrating denominator of the thermodynamic temperature T, the result is the total differential of the entropy S.

Especially interesting conclusions are obtained upon considering the changes in entropy during irreversible phenomena in an adiabatic system. As an example of such a phenomenon, we shall consider the spontaneous equalization of temperatures in an adiabatic system. Two bodies forming an adiabatic system may be at different temperatures, T_I and T_{II}, only if they are separated by an adiabatic wall. When this wall is replaced by a diathermal wall, there is a spontaneous equalization of temperatures with no external work being performed. This is a typically irreversible phenomenon; no spontaneous return of such bodies to their initial temperatures has been observed. If the temperature T_I of body I is greater than the temperature T_{II} of body II, then the heat flows from body I to body II. The elementary heat of the process, $đQ$, is positive for body II and negative for body I and amounts to $-đQ$. The sum of the entropy increments of the bodies taking part in the phenomenon

$$\sum dS = \frac{đQ}{T_{II}} - \frac{|đQ|}{T_I} = \frac{T_I - T_{II}}{T_I T_{II}} |đQ| > 0 \tag{1.28}$$

is positive if $T_I > T_{II}$. Otherwise, that is if $T_{II} > T_I$, the sign of the heat changes and the result of the reasoning remains the same.

The irreversible phenomenon of temperature equalization considered above entails a loss of external work which could be obtained with the given temperature difference, and, if the system is adiabatic, also with an increase in entropy. By considering this example and many other cases of irreversible processes in adiabatic systems one may arrive at the principle of entropy increase: the entropy of a closed adiabatic system increases during irreversible processes, but does not during reversible processes. This principle is one of the verbal formulations of the second law of thermodynamics for processes.

Any thermodynamic system plus the environment constitute an isolated system which is a special case of an adiabatic system if all external interactions, and not just heat exchange, are absent. This case can also have the principle of entropy increase applied to it in the form: the sum of the entropies of all bodies taking part in a phenomenon increases during irreversible processes, and remains unchanged during reversible processes.

6. The Third Law of Thermodynamics

It has been found experimentally that at absolute temperatures approaching zero isothermal processes between initial and final equilibrium states proceed without change in entropy. On this basis M. Planck formulated the third law of thermodynamics: as the absolute temperature approaches zero, the entropy of every chemically homogeneous body of finite density and in a state of stable equilibrium approaches a zero limiting value, regardless of the pressure, state of aggregation, and chemical modification of the substance.

The third law of thermodynamics is now formulated more generally: the entropy of any system in a state of stable equilibrium tends to a finite value as the absolute temperature tends to zero.

The third law of thermodynamics is of major importance not only in the range of very low temperatures, but also for the wide range of intermediate and high temperatures. This makes it possible to eliminate from the equations the constant values of entropy at absolute zero, taken as the reference temperature, and reduces the amount of data required for thermodynamic calculations when there are chemical reactions, and processes of dissociation or ionization. If the chemical composition of a substance does not change, for purposes of calculation it is sufficient to know the relative values of the entropy above an arbitrary reference state.

7. The Properties of Multicomponent Fluids

The properties of multicomponent fluids depend directly on their chemical composition, and this is specified in terms of the fractions of the components. The fraction of a component is the ratio of the amount of that component to the amount of the entire solution. Quantities concerning individual components will be labelled with an index $i=1, 2, 3, \ldots, k$ corresponding to the label of the component. Various kinds of fraction are formed, depending on how the amount of substance is specified.

The mass fraction of component i,

$$x_i=\frac{m_i}{m}, \tag{1.29}$$

is the ratio of the mass m_i of that component to the mass m of all components in the given volume.

The molar fraction of component i,

$$z_i=\frac{n_i}{n}, \tag{1.30}$$

is the ratio of the number n_i of kilomoles of that component to the number n of kilomoles of the solution, which is equal to the sum of kilomoles of all components:

$$n=\sum_{i=1}^{k} n_i. \tag{1.31}$$

Since the sum of the parts is equal to the whole, the sum of the fractions of all k components of a solution is equal to unity:

$$\sum_{i=1}^{k} x_i=1, \qquad \sum_{i=1}^{k} z_i=1. \tag{1.32}$$

The concepts of component quantities and partial quantities are introduced when multicomponent fluids are considered. Component quantities are defined for the substance of a component occurring throughout the entire given volume V of fluid at the same temperature T. Component quantities are additive.

The obvious statement that the sum of the masses of all components is equal to the mass of the multicomponent fluid

$$m = \sum_{i=1}^{k} m_i, \tag{1.33}$$

and division of both sides of the equation by the volume of the whole fluid yield the result that the density ρ of the multicomponent fluid is equal to the sum of the component densities ρ_i of all the components:

$$\rho = \sum_{i=1}^{k} \rho_i. \tag{1.34}$$

Likewise, on dividing both sides of (1.31) by the volume of the whole fluid, we obtain

$$c = \sum_{i=1}^{k} c_i, \tag{1.35}$$

where $c = n/V$ denotes the concentration of the fluid, and $c_i = n_i/V$ denotes the concentration of component i.

If the numerator and denominator of equations (1.29) and (1.30) which define the mass fraction and molar fraction, respectively, are divided by the fluid volume, it follows that

$$x_i = \frac{\rho_i}{\rho}, \qquad z_i = \frac{c_i}{c}. \tag{1.36}$$

The component pressure[1] p_i is the pressure which component i would exert if it itself occupied the entire volume of the fluid and were at the same temperature as the fluid. Solutions of ideal gases satisfy Dalton's law:

$$\sum_{i=1}^{k} p_i = p \tag{1.37}$$

– the sum of component pressures is equal to the pressure of the multicomponent fluid.

For ideal gases the number of kilomoles in the numerator and the denominator of the defining equation (1.30) of the molar fraction can be calculated from the Clapeyron equation of state, and then

$$z_i = \frac{R_M T}{pV} \frac{p_i V}{R_M T} = \frac{p_i}{p}. \tag{1.38}$$

[1] Usually referred to as partial pressure.

Partial specific quantities $\tilde{z}_i$ concerning individual components are defined by means of the partial derivative of the given extensive quantity Z with respect to the amount of substance m_i, the amounts of the other components remaining constant and at constant pressure and temperature:

$$\tilde{z}_i = \left(\frac{\partial Z}{\partial m_i}\right)_{p,\,T,\,m_{j \neq i}} . \tag{1.39}$$

Specific quantities concerning the fluid as a whole are calculated from the partial specific quantities by the weighted-sum law:

$$z = \sum_{i=1}^{k} x_i \tilde{z}_i , \tag{1.40}$$

and per unit volume, by relation (1.36) as

$$\rho z = \sum_{i=1}^{k} \rho_i \tilde{z}_i . \tag{1.41}$$

In the special case, a relationship is obtained between the partial specific volumes $\tilde{v}_i$ and the component densities ρ_i:

$$\sum_{i=1}^{k} \rho_i \tilde{v}_i = \rho v = 1 . \tag{1.42}$$

In addition to partial specific quantities, defined by (1.39), use is also made of partial molar quantities formed if kilomoles are taken for the units of the amount of substance:

$$\tilde{Z}_{Mi} = \left(\frac{\partial Z}{\partial n_i}\right)_{p,\,T,\,n_{j \neq i}} . \tag{1.43}$$

The molar quantity of the solution is obtained from the partial molar quantities $\tilde{Z}_{Mi}$

$$Z_M = \frac{Z}{n} = \sum_{i=1}^{k} z_i \tilde{Z}_{Mi} . \tag{1.44}$$

The conversion factor M for kilomoles to kilograms for the solution can be determined from the conversion factors M_i for the individual components by starting from the obvious relation

$$nM = m = \sum_{i=1}^{k} m_i = \sum_{i=1}^{k} n_i M_i , \tag{1.45}$$

and it is found to be

$$M = \sum_{i=1}^{k} z_i M_i . \tag{1.46}$$

The foregoing notation and relations permit formulae to be derived for converting the fractions of any substance:

$$x_i = \frac{m_i}{m} = \left(\frac{\rho_i}{\rho}\right)_{V,T} = \frac{z_i M_i}{\sum_{i=1}^{k} z_i M_i}, \tag{1.47}$$

$$z_i = \frac{n_i}{n} = \frac{c_i}{c} = \frac{x_i}{M_i \sum_{i=1}^{k} \frac{x_i}{M_i}}. \tag{1.48}$$

If a multicomponent system is being considered, the entropy S of the system is a function not only of the internal energy U and the volume V but also the quantity of substance of the individual components, in terms of the mass m_i (or number of kilomoles):

$$S = S(U, V, m_i) \quad (i = 1, 2, 3, \ldots, k).$$

In this case the internal energy can be expressed as a function of the entropy, volume, and quantity of substance of individual components:

$$U = U(S, V, m_i) \quad (i = 1, 2, 3, \ldots, k).$$

The exact differential of the internal energy thus is of the form:

$$\mathrm{d}U = \left(\frac{\partial U}{\partial S}\right)_{V, m_i} \mathrm{d}S + \left(\frac{\partial U}{\partial V}\right)_{S, m_i} \mathrm{d}V + \sum_{i=1}^{k} \left(\frac{\partial U}{\partial m_i}\right)_{S, V, m_{j \neq i}} \mathrm{d}m_i. \tag{1.49}$$

By equation (1.27)

$$\left(\frac{\partial U}{\partial S}\right)_{V, m_i} = T, \tag{1.50}$$

$$\left(\frac{\partial U}{\partial V}\right)_{S, m_i} = -p \tag{1.51}$$

whereas the derivatives

$$\left(\frac{\partial U}{\partial m_i}\right)_{S, V, m_{j \neq i}} = \mu_i \tag{1.52}$$

are called the *chemical potentials* of the components.

When equations (1.50) to (1.52) are taken into account, equation (1.49) yields the Gibbs relation:

$$\mathrm{d}U = T\,\mathrm{d}S - p\,\mathrm{d}V + \sum_{i=1}^{k} \mu_i \,\mathrm{d}m_i. \tag{1.53}$$

As follows from equation (1.15) defining enthalpy and from the Gibbs relation (1.53),

$$\mathrm{d}H = T\,\mathrm{d}S + V\,\mathrm{d}p + \sum_{i=1}^{k} \mu_i \,\mathrm{d}m_i. \tag{1.54}$$

In addition to internal energy and enthalpy, the following are used in thermodynamics as a state function: *free energy*

$$F = U - TS \tag{1.55}$$

and *free enthalpy*

$$G = H - TS, \tag{1.56}$$

whose total differentials are given by

$$\mathrm{d}F = -S\,\mathrm{d}T - p\,\mathrm{d}V + \sum_{i=1}^{k} \mu_i \mathrm{d}m_i \tag{1.57}$$

$$\mathrm{d}G = -S\,\mathrm{d}T + V\,\mathrm{d}p + \sum_{i=1}^{k} \mu_i \mathrm{d}m_i . \tag{1.58}$$

In view of the equations above, other equivalent definitions can be given for the chemical potential

$$\mu_i = \left(\frac{\partial H}{\partial m_i}\right)_{S,\,p,\,m_{j\neq i}} = \left(\frac{\partial F}{\partial m_i}\right)_{T,\,V,\,m_{j\neq i}} = \left(\frac{\partial G}{\partial m_i}\right)_{T,\,p,\,m_{j\neq i}}$$

$$= \tilde{g}_i = \tilde{u}_i + p\tilde{v}_i - T\tilde{s}_i . \tag{1.59}$$

The chemical potential is equal to the partial free enthalpy of the component.

For ideal gases, in the light of

$$z_i = \left(\frac{p_i}{p}\right)_{V,\,T} = \frac{\tilde{v}_i}{v_i} \tag{1.60}$$

we have

$$\mu_i = \tilde{u}_i + p_i v_i - T\tilde{s}_i . \tag{1.61}$$

The chemical potential of a component of a solution of ideal gases can be written as

$$\mu_i = \mu_{0i}(T_0, p_0) + \int_{T_0}^{T} c_{pi}(T)\,\mathrm{d}T - T\int_{T_0}^{T} c_{pi}(T)\frac{\mathrm{d}T}{T} + R_i T \ln\frac{p_i}{p_0}, \tag{1.62}$$

where μ_{0i} denotes the chemical potential of component i in the reference state at temperature T_0 and pressure p_0, p_i is the component pressure, and R_i is the gas constant of the component. Terms in equation (1.62) which are constant or which depend only on the temperature can be written in terms of a function $f_i(T)$; hence, on introduction of the molar fraction of ideal gases, given by equation (1.60), the chemical potential of a component of a solution of ideal gases is defined by

$$\mu_i = f_i(T) + R_i T \ln p z_i . \tag{1.63}$$

In view of the Gibbs relation (1.53), the exact differential of entropy for a multicomponent system is defined by the relation

$$\mathrm{d}S = \frac{1}{T}\left(\mathrm{d}U + p\,\mathrm{d}V - \sum_{i=1}^{k} \mu_i \mathrm{d}m_i\right). \tag{1.64}$$

The reasoning above can easily be extended to still other generalized forces and displacements.

Integration of the Pfaffian on the right-hand side of the Gibbs relation (1.53) is no simple matter since the intensive quantities T, p, and μ_i are functions of all the independent state parameters of the system. To this end use should be made of Euler's relation for homogeneous equations.

The equation $F(x_i)=0$ is a homogeneous equation of degree n if

$$F(ax_i)=a^n F(x_i). \tag{1.65}$$

This means that if each of the independent variables is multiplied by a, the result is the same as when the function F is multiplied by a to the nth power.

If the function $F(ax_i)$ is differentiated with respect to a while the x_i remain constant, then

$$\frac{\partial F(ax_i)}{\partial a}=\sum_{i=1}^{k} x_i\left(\frac{\partial F}{\partial ax_i}\right)_{axj\neq i}, \tag{1.66}$$

which, in accordance with equation (1.65), is equal to

$$\frac{\partial a^n F(x_i)}{\partial a}=na^{n-1}F(x_i). \tag{1.67}$$

If $a=1$, comparison of expressions (1.66) and (1.67) yields Euler's homogeneous relation

$$nF=\sum_{i=1}^{k} x_i\left(\frac{\partial F}{\partial x_i}\right)_{xj\neq i}. \tag{1.68}$$

For homogeneous bodies the extensive thermodynamic parameters are first-degree homogeneous functions of the amount of substance, $n=1$. If the amount of substance of the body is increased a-fold, every extensive parameter also increases a-fold.

By Euler's relation (1.68) the internal energy $U(S, V, m_i)$ can be written as

$$U=S\left(\frac{\partial U}{\partial S}\right)_{V,m_i}+V\left(\frac{\partial U}{\partial V}\right)_{S,m_i}+\sum_{i=1}^{k} m_i\left(\frac{\partial U}{\partial m_i}\right)_{S,V,m_{j\neq i}}, \tag{1.69}$$

whence, on invoking relations (1.50)–(1.52), we obtain an integral form of the Gibbs relation:

$$U=TS-pV+\sum_{i=1}^{k}\mu_i m_i. \tag{1.70}$$

Differentiation of this equation yields

$$\mathrm{d}U=T\,\mathrm{d}S+S\,\mathrm{d}T-p\,\mathrm{d}V-V\,\mathrm{d}p+\sum_{i=1}^{k}\mu_i\,\mathrm{d}m_i+\sum_{i=1}^{k}m_i\,\mathrm{d}\mu_i. \tag{1.71}$$

If the Gibbs relation (1.53) is subtracted from this equation, we obtain the Gibbs–

Duhem equation

$$S\,\mathrm{d}T - V\,\mathrm{d}p + \sum_{i=1}^{k} m_i\,d\mu_i = 0, \tag{1.72}$$

which is so important for nonequilibrium thermodynamics; per unit mass of a multicomponent fluid this is

$$s\,\mathrm{d}T - v\,\mathrm{d}p + \sum_{i=1}^{k} x_i\,\mathrm{d}\mu_i = 0. \tag{1.73}$$

CHAPTER II

THE GENERAL FOUNDATIONS OF NONEQUILIBRIUM THERMODYNAMICS

1. Nonequilibrium Thermodynamics and Classical Thermodynamics

Classical thermodynamics in principle considers equilibrium processes which consist of successive states of stable equilibrium. In the case of processes which are nonequilibrium but do take place between extreme states of equilibrium, it permits differences of state functions to be calculated since these differences do not depend on how the phenomenon proceeds between the states considered. Giving the equilibrium conditions makes it possible to determine the direction of spontaneous nonequilibrium phenomena. Instead of thermodynamic equations for equilibrium processes, thermodynamic inequalities are obtained for nonequilibrium processes. Classical thermodynamics cannot consider processes which involve a lack of equilibrium since this is bound up with nonhomogeneity of the substance even within a single phase. In this case, a phase cannot be defined as a homogeneous part of a system, whereas the region occupied by the phase is specified by discontinuities of the intensive parameters and the properties of the substance at the phase boundary.

Equilibrium processes are merely a theoretical model. The equilibrium state is rarely met with in nature and of relatively little interest to engineering. Stationary states, not varying in time, occur most frequently in industrial processes. In the case of stationary states flows between a system and its environment come into equilibrium, allowing technological processes to be carried out continuously. The state of equilibrium is an extreme case of stationary state when the flows between system and environment vanish. In the thermodynamic respect, a characteristic feature of a system in equilibrium is that there are no effects involving dissipation of work or energy, as occur in the stationary state during actual irreversible processes.

Considerations in science usually begin with the simplest problems, with idealized systems which are easier to analyse. As it developed, thermodynamics went on from consideration of equilibrium states to consideration of processes which take place with no equilibrium, particular account being taken of stationary states. The most important area of applications for nonequilibrium thermodynamics is that of studying phenomena evoked by several causes, called *thermodynamic forces*. In these cases there is interference of phenomena caused by various forces. For instance, if temperature and concentration gradients are present, then in addition to the ordinary, Fourier-law conduction of heat under the influence of the temperature gradient there is a flow of heat arising from the concentration gradient, this being known as the *Dufour effect*. When a multicomponent fluid has temperature gradients, thermal diffusion, or the *Soret effect*, occurs in addition to ordinary diffusion caused by concentration gradients.

Cross-effects, produced when nonequilibrium processes interfere with each other, are described mathematically by adjoining extra terms to phenomenological laws which are valid for individual nonequilibrium processes. Thus, for example, if a term proportional to the temperature gradient is added to the right-hand side of Fick's first law for diffusion in a binary fluid, account can be taken of the fact that diffusion of a substance is also due to the concentration gradient (ordinary diffusion), as well as the temperature gradient (thermal diffusion). If a term dependent on the concentration gradient is added to the right-hand side of Fourier's equation, heat flow is found to be brought about by the temperature gradient (ordinary heat conduction), as well as the concentration gradient (Dufour effect).

The first attempts to construct a theory of interfering nonequilibrium phenomena were undertaken by Thomson for the thermoelectric effect in 1854. However, he could not theoretically validate the method he used. In 1850 Clausius introduced the concept of noncompensated heat as a measure of irreversibility. The rate of local entropy change for nonhomogeneous systems was calculated by W. Natanson (1896), P. Duhem (1911), G. Jaumann (1911 and 1918), and E. Lohr (1917 and 1924). Jaumann introduced the concept of entropy production and entropy flux.

The commutative law which Lars Onsager formulated in 1931, now also known as *Onsager's principle of microscopic reversibility*, has been of fundamental importance for the development of modern nonequilibrium thermodynamics. This law was supplemented by H.B.G. Casimir in 1945. The Onsager principle, just as other postulates of nonequilibrium thermodynamics, cannot be substantiated theoretically by phenomenological methods but merely verified experimentally. In this respect, it displays a similarity to the laws of classical thermodynamics. Some authors have proposed calling Onsager's principle the fourth law of thermodynamics.

In most cases nonequilibrium phenomena of a vectorial nature are described with great accuracy by linear phenomenological equations such as Fourier's law, Fick's laws, Ohm's law, etc. The proportionality factors in these laws were determined theoretically or empirically. Phenomena of a scalar character, such as relaxation phenomena and reaction rates, are describable by linear laws only in the vicinity of an equilibrium state.

A well-knit theory of the phenomenological thermodynamics of nonequilibrium processes, within the domain of validity of linear phenomenological equations, was worked out in the period from 1940 to 1962 by German scientists (C. Eckart, J. Meixner, R. Haase), Belgian scientists (I. Prigogine, P. Glandsorff), Dutch scientists (S.R. de Groot, P. Mazur), and American scientists. Fundamental monographs on this field of the science [15, 21, 22, 25] were written during this period.

Nonlinear theories of the mechanics of continuous media were constructed by C. Truesdell and W. Noll, and subsequently extended to thermodynamics by B. D. Coleman [19, 31, 37]. These theories take a small number of thermodynamic quantities and deformation gradients as their starting point, but assume that these quantities depend on the history of the phenomenon. It is further assumed that a fully defined entropy exists for arbitrarily large deviations from the equilibrium state.

A general characterization of the state of a substance is ascertained if it is known how that state arose from the equilibrium state. The state of a substance, with due account for its individual properties, is specified by the constitutive equations (*Materialgleichungen* in German).

The further growth of nonequilibrium thermodynamics is also bound up with a deepening of its foundations from the molecular-statistical point of view. This method makes it possible theoretically to validate the postulates of nonequilibrium thermodynamics, particularly Onsager's principle, and is aimed at calculating those coefficients which previously had to be determined experimentally. Nonequilibrium processes are also considered by means of quantum-physical methods.

Nonequilibrium thermodynamics is an indispensable tool for the continued development of other physical and engineering sciences such as hydrodynamics, aerodynamics, plasma physics, magnetohydrodynamics, solid state physics, engineering applications of thermoelectricity, electronics, biophysics, biochemistry, electrochemistry, and so forth. It may also be employed for a theoretical description of processes of combustion, isotope separation, complex chemical reactions, etc. Of a number of topics dealt with by nonequilibrium thermodynamics, we shall discuss only those which enable the methods used by the science to be understood and are linked directly with engineering applications. Since this book is addressed to engineers, the phenomenological method has been chosen.

2. The Postulate of Local Thermodynamic Equilibrium

Nonequilibrium thermodynamics considers systems which are not in thermodynamic equilibrium. These are nonhomogeneous systems in which at least some of the intensive parameters are functions of time and position. The state of a substance forming such a system is specified by introducing the concept of local thermodynamic state.

A local thermodynamic state is determined for a substance at individual points of a system which is not in equilibrium. Elementary volumes of substance surrounding the given points are considered to this end. These volumes must be so small that the substance in them can be treated as homogeneous. At the same time they should contain a sufficient number of molecules for the principles of statistics and the methods of phenomenological thermodynamics to be applicable to them.

Extensive quantities are not directly suitable for specifying the state of a substance. They should be taken per unit mass of substance or unit volume. For nonhomogeneous substances, not in equilibrium, these quantities are formed and denoted as follows.

Specific quantities per unit mass of substance are set up as the ratios of extensive quantities, denoted by the upper-case letter Z, to the mass m as the mass tends to zero, and are denoted by the corresponding lower-case letter z:

$$z = \lim_{m \to 0} \frac{Z}{m} = \frac{dZ}{dm} \cdot \tag{2.1}$$

Example: specific volume

$$v = \frac{dV}{dm}.$$

Specific quantities per amount of substance given by the number of kilomoles n will be called *molar quantities* and labelled by the index M:

$$Z_M = \lim_{n \to 0} \frac{Z}{n} = \frac{dZ}{dn} = zM, \tag{2.2}$$

where M is the conversion factor for kilomoles to kilograms, numerically equal to the molecular mass but having a denomination of kg/kmole. Example: molar volume

$$V_M = \frac{dV}{dn} = vM.$$

The ratios of extensive quantities to the volume may be called the *densities* of those quantities and indexed by v:

$$Z_v = \lim_{V \to 0} \frac{Z}{V} = \frac{dZ}{dV} = \frac{z}{v} = \rho z. \tag{2.3}$$

Example: internal energy density

$$U_v = \frac{dU}{dV} = \frac{u}{v} = \rho u.$$

Remark: The notation above does not apply to mass density, briefly called density:

$$\rho = \frac{dm}{dV} = \frac{1}{v}.$$

The ratio of the number of kilomoles to the volume, that is the inverse molar volume, is called the *molar concentration*:

$$c = \frac{dn}{dV} = \frac{1}{V_M}. \tag{2.4}$$

In many cases the postulate of local thermodynamic equilibrium is applicable to substances not in thermodynamic equilibrium.

Postulate I: Although a thermodynamic system as a whole may not be in equilibrium, arbitrarily small elements of its volume are in local thermodynamic equilibrium and have state functions which depend on state parameters through the same relationships as in the case of equilibrium states in classical thermodynamics.

For a system to which the postulate of local thermodynamic equilibrium is applicable, specific entropy and specific internal energy may be determined at every point in the same way as for substances in equilibrium, and then the thermodynamic temperature and thermodynamic pressure are definable as

$$T=\left(\frac{\partial u}{\partial s}\right)_v, \qquad p=-\left(\frac{\partial u}{\partial v}\right)_s. \tag{2.5}$$

In this case, the following equations hold at every point: the classical-thermodynamic equations of state, relating the thermodynamic functions to the state parameters, e.g. $f(v, p, T)=0$ or $f(u, v, T)=0$, and the Gibbs and Gibbs–Duhem relations, used when considering multicomponent media. When there is local thermodynamic equilibrium the equations of state do not contain gradients of the state parameters.

The range of validity of the postulate of local thermodynamic equilibrium is determinable phenomenologically only by experiment. The further the system is from the equilibrium state, the more the assumption of local thermodynamic equilibrium is at variance with reality. As shown by experiments, the postulate of local thermodynamic equilibrium is valid if the gradients of intensive thermodynamic functions which are decisive in regard to equilibrium in the system under consideration are small and their local values vary relatively slowly in comparison with the relaxation time of the local state of the substance. The first condition corresponds to the requirement that the change in an intensive parameter be small over the molecular mean free path. The second condition stems from the fact that the entropy concept is introduced in statistical thermodynamics in considering a substance over a period of time that is long compared to the relaxation time of its local state, that is for a substance in a state of statistical equilibrium. It may be assumed that the postulate of local thermodynamic equilibrium holds only if energy dissipation rapidly damps large deviations from the equilibrium state. The assumption of local thermodynamic equilibrium is not true for very rarefied gases wherein molecular collisions are infrequent or for phenomena occurring at very low temperatures at which energy is dissipated very slowly.

The state of a system not in equilibrium is specified by the set of local states of the substance at all points of the system. It is, therefore, determined by the fields of the intensive parameters of state, necessary and sufficient for specifying the local state of the substance. In using the appellation *nonequilibrium thermodynamics*, one does of course take into account the absence of an equilibrium state by the system. A nonequilibrium process is described uniquely if the set of consecutive states of the system during realization of the process is known.

3. Local Formulation of the Second Law of Thermodynamics

A change in the entropy of an elementary volume of substance,

$$\mathrm{d}S=\mathrm{d}S_e+\mathrm{d}S_i, \tag{2.6}$$

may be due to two causes: entropy inflow from the environment, and entropy production inside the volume under consideration. The entropy inflow $\mathrm{d}S_e$ from the environment is due to the conduction flux of internal energy, inflow of entropy along with substance in the form of a convection entropy flux transported along with the macroscopic flow of the substance as a whole, and the resultant entropy flux caused by diffusion of the individual components; hence

$$dS_e \gtrless 0 \tag{2.7}$$

may be positive or negative, and in the special case, equal to zero. For closed, thermally homogeneous systems

$$dS_e = \frac{đQ_e}{T}, \tag{2.8}$$

where $đQ_e$ is the elementary heat due to thermal interaction between the system and the environment.

Entropy production inside the elementary volume considered caused by irreversible phenomena is the local value of the sum of entropy increments, calculated in accordance with the laws of classical thermodynamics. By the second law of thermodynamics, entropy production

$$dS_i \geqslant 0 \tag{2.9}$$

is always positive for irreversible phenomena and zero for reversible phenomena.

If entropy production is taken per unit time, the result is the entropy production rate:

$$P = \frac{dS_i}{dt} \geqslant 0. \tag{2.10}$$

The quantity

$$\Phi_s = \frac{dP}{dV} = \frac{d(\rho s_i)}{dt} \geqslant 0, \tag{2.11}$$

that is, the entropy produced inside the given volume by irreversible phenomena, is called the *entropy source strength* when taken per unit volume and unit time. This quantity is positive for all real phenomena and zero only for ideal reversible phenomena.

The product of the entropy source strength and the absolute temperature,

$$\Psi = T\Phi_s \geqslant 0, \tag{2.12}$$

called the *dissipation function*, is also a non-negative quantity.

When phenomena at the interface between two phases are considered, the amount of entropy produced is taken per unit surface area.

The principle of entropy increase, used in classical thermodynamics, asserts that a change in entropy as a result of irreversible phenomena inside a closed adiabatic system is always positive. This principle does, however, admit a situation such that the entropy may decrease at some place in the system owing to irreversible phenomena, providing that entropy production at another place in the system compensates this loss.

The classical statement of the second law of thermodynamics for any system and its environment requires that the sum of the entropy increments of all bodies taking part in an irreversible phenómenon be positive. With this formulation, owing to irreversible phenomena inside the system, the entropy of the system may decrease if concomitant irreversible phenomena in the environment give rise to a sufficiently large entropy increment.

The local formulation of the second law of thermodynamics in the form (2.9) allows irreversible phenomena, called *coupled phenomena,* which entail a decrease in entropy, to occur at some place provided that occurring concurrently at that same place are irreversible phenomena, called *coupling phenomena,* which result in such a considerable production of entropy that the ultimate overall entropy increment is positive.

The second law of thermodynamics can be given a local formulation as follows: If coupling phenomena involving an increase in entropy occur at any place, entropy-decreasing phenomena coupled to them may also occur at that same place, but the overall effect of all the irreversible phenomena at a given place in space must always yield an increase in entropy.

In actual fact the local formulation of the second law of thermodynamics follows from the application of the classical formulations to an elementary volume. As will be shown by examples further on in the book, quantities occurring in the formula for the entropy source strength may be considered to be thermodynamic forces when they are treated as causes of nonequilibrium processes, or they may be considered to be generalized flows when they are treated as the effects of thermodynamic forces. This classification is not unique but to a degree is a matter of convention. Generalized flows are called *strengths* if they are scalars, and *fluxes* if they are vectors. Forces and generalized flows having the same index are said to be conjugated. They are tensors of the same rank.

The entropy source strength or dissipation function, which is the product of the entropy source strength and the absolute temperature, may be written as a sum of the products of thermodynamic forces and the generalized flows conjugate to them. The entropy source strength is a sum of products of the form

$$\Phi_s = \sum_{a=1}^{n} J'_a X'_a \geqslant 0. \tag{2.13}$$

The dissipation function is similar in form:

$$\Psi = T\Phi_s = \sum_{a=1}^{n} J_a X_a \geqslant 0. \tag{2.14}$$

In these formulae J_a and J'_a are generalized flows, and X_a and X'_a are thermodynamic forces.

The products of conjugated thermodynamic forces and the effects of their action must be scalars, and hence are the products of two scalars, the dot products of two vectors, or double dot products of two tensors of rank two:

$$\Psi = \sum_{a=1}^{n_0} J_a X_a + \sum_{a=1}^{n_1} J_a \cdot X_a + \sum_{a=1}^{n_2} \mathbf{J}_a : \mathbf{X}_a, \tag{2.15}$$

where n_0 is the number of scalar, n_1 the number of vectorial, and n_2 the number of tensorial (rank two) thermodynamic forces.

Since Ψ and Φ_s can be transformed in various ways, and the products factored in different ways, both the forces and the generalized flows are not uniquely defined by the form of these expressions. The choice of thermodynamic forces must, however, be made so that in the equilibrium state when the thermodynamic forces vanish ($X_a=0$), the entropy source strength must also be zero. The entropy source strength must furthermore be invariant under the Galilean transformation since the irreversibility or reversibility of the phenomenon must be invariant under that transformation.

Finally, it must be noted that in contradistinction to entropy, the entropy source strength and the dissipation function are not state functions since they depend on the mode of change between given states.

Example 2.1. Determine the entropy source strength, dissipation function, thermodynamic forces, and generalized flows for conduction of heat by an isotropic solid.

Solution. To begin with, consider one-dimensional conduction of heat by a rod segment of length $\mathrm{d}x$ and constant cross-sectional area A (Fig. 2.1). The lateral surfaces of the rod are insulated so that they may be considered adiabatic. The increment of the entropy of the rod element under consideration is

$$\mathrm{d}S = \frac{Q + \text{đ}Q}{T + \mathrm{d}T} - \frac{Q}{T} \approx -\frac{Q\,\mathrm{d}T}{T^2} + \frac{\text{đ}Q}{T}. \tag{1}$$

The entropy increment above can be split up into a reversible change in entropy

$$\mathrm{d}S_e = \frac{\text{đ}Q}{T} \gtrless 0 \tag{2}$$

and an irreversible entropy production

$$\mathrm{d}S_i = -\frac{Q\,\mathrm{d}T}{T^2} > 0. \tag{3}$$

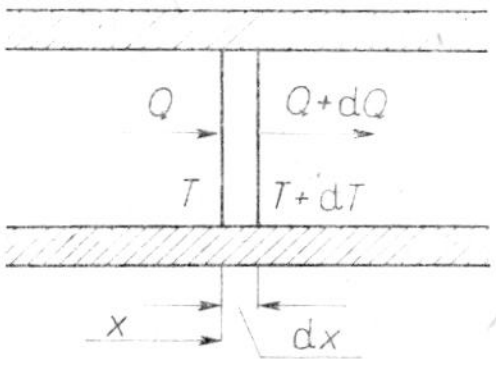

Fig. 2.1 One-dimensional conduction of heat by a rod

The rate of entropy production is obtained by taking the entropy production per unit time

$$P = \frac{\mathrm{d}S_i}{\mathrm{d}t} = -\frac{\dot{Q}\,\mathrm{d}T}{T^2}. \tag{4}$$

The entropy source strength is obtained by taking the entropy production rate per unit volume

$$\Phi_s = \frac{dS_i}{dt\,dV} = -\frac{J_q}{T^2}\frac{dT}{dx}, \tag{5}$$

where

$$J_q = \frac{dQ}{A\,dt} \tag{6}$$

is the conducted heat flux.

For three-dimensional conduction of heat in isotropic solids, the heat flux and the temperature gradient become vectors and the entropy production intensity is given by the expression

$$\Phi_s = -\boldsymbol{J}_q \cdot \frac{1}{T^2}\nabla T = \boldsymbol{J}_q \cdot \nabla\left(\frac{1}{T}\right) = -\boldsymbol{J}_s \cdot \nabla(\ln T), \tag{7}$$

where the entropy flux

$$\boldsymbol{J}_s = \frac{\boldsymbol{J}_q}{T} \tag{8}$$

has made its appearance.

Proceeding directly from the entropy source strength, we can take the inverse temperature gradient $\nabla(1/T)$ for the thermodynamic force and then the heat flux $\boldsymbol{J}_q$ is the generalized flow, or the gradient of the natural logarithm of the temperature, $\nabla(\ln T)$, can be adopted as the force and then the entropy flux taken with a minus sign, $-\boldsymbol{J}_s$, is the generalized flow.

If a dissipation function of the form

$$\Psi = -\Phi_s = \boldsymbol{T}\frac{\boldsymbol{J}_q}{T}\cdot\nabla T = -\boldsymbol{J}_q\cdot\nabla(\ln T) = -\boldsymbol{J}_s\cdot\nabla T \tag{9}$$

is taken as the starting point, the gradient of the natural logarithm of the temperature, $\nabla(\ln T)$, may be the thermodynamic force, whereas $-\boldsymbol{J}_q$, the heat flux taken with the opposite sign, is the generalized flow; or the temperature gradient ∇T may be the force and $-\boldsymbol{J}_s$, the entropy flux taken with the opposite sign, is the generalized flow.

4. Linear Phenomenological Equations

Generalized flows, or the effects due to thermodynamic forces, are in the general case functions of all n forces

$$J_a(X_b) \quad (b = 1, 2, 3, \ldots, n).$$

Using a Maclaurin-series expansion, we can write

$$J_a = J_a(X_b = 0) + \sum_{b=1}^{n} \frac{\partial J_a}{\partial X_b} X_b + \frac{1}{2!}\sum_{b,c=1}^{n} \frac{\partial^2 J_a}{\partial X_b \partial X_c} X_b X_c + \ldots. \tag{2.16}$$

If the thermodynamic forces become zero, the system goes over into an equilibrium state in which the generalized flows become zero, and hence

$$J_a(X_b = 0) = 0.$$

In the first approximation, that is for states not very far from the equilibrium state, we can confine ourselves to linear relationships called *linear phenomenological equations*:

$$J_a = \sum_{b=1}^{n} L_{ab} X_b . \tag{2.17}$$

Example: When there are three thermodynamic forces, the linear phenomenological equation is of the form

$$J_a = L_{a1} X_1 + L_{a2} X_2 + L_{a3} X_3 .$$

Postulate II: The generalized flows J_a in the first approximation depend on all the thermodynamic forces X_b. The proportionality factors, called *phenomenological coefficients*, are defined as follows:

$$L_{ab} = \left(\frac{\partial J_a}{\partial X_b} \right)_{X_c} = \left(\frac{J_a}{X_b} \right)_{X_c = 0} \qquad (c \neq b) . \tag{2.18}$$

The thermodynamic forces and the generalized flows are quantities determined by analysis of the expression for entropy source strength or dissipation function.

The physical meaning of the phenomenological coefficients can be elucidated by considering only specific topics (cf. Example 2.2). At this juncture it can be said in general terms that these coefficients are not functions of thermodynamic forces, hence also are not functions of the effects of their action in the form of generalized flows; on the other hand, they can be functions of the parameters of the local state of a substance and depend on the kind of substance constituting the system under consideration. Coefficients with the same indices, L_{aa}, relate the conjugated forces and generalized flows, e.g. the heat flux and the inverse temperature gradient. Phenomenological coefficients L_{ab} with different indices ($a \neq b$) concern interfering cross-phenomena.

The admissibility of the postulate concerning the linear dependence of generalized flows on the thermodynamic forces must be verified experimentally and accordingly has been considered a postulate of linear nonequilibrium thermodynamics. This postulate is valid for most cases involving flows of substance, heat, or electric current. The greatest deviations from this postulate appear for chemical reactions and relaxation phenomena. Most chemical reactions take place so far from the equilibrium state that even as a first approximation the reaction rate cannot be assumed to be linearly dependent on the chemical affinity.

If the linear phenomenological equation (2.17) is introduced into expression (2.14) for the dissipation function

$$\Psi = \sum_{a,b=1}^{n} L_{ab} X_a X_b \geqslant 0 , \tag{2.19}$$

it is seen that the dissipation function is a quadratic form in all forces. The same conclusion holds for the entropy source strength:

$$\Phi_s = \sum_{a,b=1}^{n} l_{ab} X'_a X'_b \geqslant 0 , \tag{2.20}$$

where, for clarity, the phenomenological coefficients have been denoted by a lower-case letter. If it is assumed that $X_a = TX'_a$, the coefficients L_{ab} and l_{ab} stemming from reasoning based on the dissipation function and on the entropy source strength, respectively, are found to be associated by the obvious relationship

$$l_{ab} = TL_{ab}\,. \tag{2.21}$$

The quadratic form (2.19) or (2.20) may be written as a matrix:

$$\Psi = \sum_{a,b=1}^{n} L_{ab} X_a X_b = [X_1\, X_2 \dots X_n] \begin{bmatrix} L_{11} & L_{12} & \dots & L_{1n} \\ L_{21} & L_{22} & \dots & L_{2n} \\ \vdots & \vdots & \vdots & \vdots \\ L_{n1} & L_{n2} & \dots & L_{nn} \end{bmatrix} \begin{bmatrix} X_1 \\ X_2 \\ \vdots \\ X_n \end{bmatrix} \geqslant 0\,.$$

A necessary and sufficient condition for the quadratic form above to be non-negative is that all its principal minors be non-negative and this may be written as

$$\begin{vmatrix} L_{aa} & L_{ab} \\ L_{ba} & L_{bb} \end{vmatrix} = L_{aa} L_{bb} - L_{ab} L_{ba} \geqslant 0 \quad (a \neq b\,;\; a\,, b = 1\,, 2\,, 3\,, \dots\,, n)\,. \tag{2.22}$$

Recasting the dissipation function in the form

$$\Psi = \sum_{a=1}^{n} L_{aa} X_a^2 + \sum_{a,b=1}^{n} \frac{L_{ab} + L_{ba}}{2} X_a X_b \geqslant 0 \qquad (a \neq b)$$

yields a symmetric matrix of coefficients and principal minors

$$\begin{vmatrix} L_{aa} & \dfrac{L_{ab} + L_{ba}}{2} \\ \dfrac{L_{ab} + L_{ba}}{2} & L_{bb} \end{vmatrix} = L_{aa} L_{bb} - \tfrac{1}{4}(L_{ab} + L_{ba})^2 \geqslant 0 \tag{2.23}$$

$$(a \neq b\,;\; a,\; b = 1\,, 2\,, 3\,, \dots\,, n)\,.$$

Since the dissipation function cannot be negative if only single forces occur,

$$\Psi = L_{aa} X_a^2 \geqslant 0\,,$$

all phenomenological coefficients with the same indices cannot be negative,

$$L_{aa} \geqslant 0 \quad (a = 1\,, 2\,, 3\,, \dots\,, n)\,, \tag{2.24}$$

and phenomenological coefficients with different indices must satisfy the condition

$$L_{aa} L_{bb} \geqslant \tfrac{1}{4}(L_{ab} + L_{ba})^2 \quad (a \neq b\,;\; a,\; b = 1\,, 2\,, 3\,, \dots\,, n)\,. \tag{2.25}$$

If in the system under consideration there is no metastable equilibrium (for which $J_a = 0$ and $\Psi = 0$, although $X_b \neq 0$) and all thermodynamic forces and generalized flows are independent, then the equality signs in expressions (2.24) and (2.25) drop out. All real phenomena are described in this case by equations with phenomenological coefficients which satisfy the inequalities given above.

In the general case, generalized flows are functions of all forces of different tensorial character. For an isotropic medium there are restrictions which follow from the Curie symmetry principle (also known as Curie's theorem).

The *Curie principle*: quantities whose tensorial characters differ by an odd number of ranks cannot interact in an isotropic medium.

The Curie principle can be substantiated as follows. If the thermodynamic forces X_b and generalized flows J_a are tensors of the same rank, the phenomenological coefficients L_{ab} are scalars which depend on the local state of the substance but in an isotropic medium do not depend on the gradients of intensive quantities. If the J_a and X_b are tensors of different ranks, the phenomenological coefficients are tensors of rank equal to the difference between the ranks of these tensors. A phenomenological coefficient which is a tensor of even rank can exist in an isotropic medium, whereas one of odd rank causes the medium to be nonisotropic and must thus become zero in an isotropic medium.

The Curie symmetry principle implies that in an isotropic medium there can be no cross effects between scalar and vector phenomena. Scalar parts are separated from quantities which are tensors of rank two so as to leave tensors of zero trace (cf. Section III.7). For an isotropic medium, therefore, the dissipation function or entropy source strength can be decomposed into three non-negative parts

$$\Psi = \Psi_0 + \Psi_1 + \Psi_2 . \tag{2.26}$$

These are, respectively, the sums of the products of the thermodynamic forces, and generalized flows with the character of scalars, vectors and tensors of rank two:

$$\left.\begin{aligned} \Psi_0 &= \sum_{a=1}^{n_0} J_a X_a \geqslant 0, \\ \Psi_1 &= \sum_{a=1}^{n_1} \mathbf{J}_a \cdot \mathbf{X}_a \geqslant 0, \\ \Psi_2 &= \sum_{a=1}^{n_2} \mathsf{J}_a : \mathsf{X}_a \geqslant 0, \end{aligned}\right\} \tag{2.27}$$

where n_0 is the number of scalar, n_1 the number of vectorial, and n_2 the numbe of tensorial (rank two) thermodynamic forces.

In view of the foregoing, the phenomenological equations for an isotropic medium take the form:

$$\left.\begin{aligned} J_a &= \sum_{b=1}^{n_0} L_{ab} X_b , \\ \mathbf{J}_a &= \sum_{b=1}^{n_1} L_{ab} \mathbf{X}_b , \\ \mathsf{J}_a &= \sum_{b=1}^{n_2} L_{ab} \mathsf{X}_b . \end{aligned}\right\} \tag{2.28}$$

It should be noted that the thermodynamic forces are also functions of the generalized flows:

$$X_a(J_b) \quad (b=1, 2, 3, \ldots, n).$$

Using the Maclaurin expansion and confining ourselves in the first approximation to linear relationships, we obtain linear phenomenological equations of the form

$$X_a = \sum_{b=1}^{n} K_{ab} J_b, \tag{2.29}$$

where the phenomenological coefficients are defined as

$$K_{ab} = \left(\frac{\partial X_a}{\partial J_b}\right)_{J_c} = \left(\frac{X_a}{J_b}\right)_{J_c=0} \quad (c \neq b). \tag{2.30}$$

Further considerations concerning the phenomenological equations (2.29) and phenomenological coefficients (2.30) are analogous to those previously carried out for those quantities concerning generalized flows. In particular, the phenomenological coefficients K_{ab} satisfy the inequalities which follow from the positive value of the dissipation function and state that all coefficients of the same indices are positive

$$K_{aa} > 0 \quad (a=1, 2, 3, \ldots, n), \tag{2.31}$$

and all coefficients of different indices must satisfy the condition

$$K_{aa} K_{bb} > \tfrac{1}{4}(K_{ab} + K_{ba})^2 \quad (a \neq b;\ a,\ b=1, 2, 3, \ldots, n). \tag{2.32}$$

The phenomenological coefficients L_{ab} and K_{ab} are related by an equation of the form

$$K_{ab} = L_{ba}^{-1}, \tag{2.33}$$

where L_{ba}^{-1} is the inverse of the matrix L_{ba}.

The relation above can also be written as

$$\sum_{b=1}^{n} K_{ab} L_{bc} = \delta_{ac} \quad (b, c=1, 2, 3, \ldots, n). \tag{2.34}$$

Example 2.2. Determine the relationship between the phenomenological coefficients L_{ab} of the equations for generalized flows and the phenomenological coefficients K_{ab} of the equations for thermodynamic forces, if there are two forces and two generalized flows.

Solution. The phenomenological equations for generalized flows are of the form:

$$J_1 = L_{11} X_1 + L_{12} X_2, \tag{1}$$

$$J_2 = L_{21} X_1 + L_{22} X_2. \tag{2}$$

Solving these equations for the thermodynamic forces yields

$$X_1 = \frac{L_{22}}{L_{11} L_{22} - L_{12} L_{21}} J_1 - \frac{L_{12}}{L_{11} L_{22} - L_{12} L_{21}} J_2, \tag{3}$$

$$X_2 = -\frac{L_{21}}{L_{11} L_{22} - L_{12} L_{21}} J_1 + \frac{L_{11}}{L_{11} L_{22} - L_{12} L_{21}} J_2. \tag{4}$$

In view of the phenomenological equations

$$X_1 = K_{11} J_1 + K_{12} J_2, \tag{5}$$

$$X_2 = K_{21} J_1 + K_{22} J_2, \tag{6}$$

relations linking the phenomenological coefficients are obtained:

$$K_{11} = \frac{L_{22}}{L_{11} L_{22} - L_{12} L_{21}}, \tag{7}$$

$$K_{12} = -\frac{L_{12}}{L_{11} L_{22} - L_{12} L_{21}}, \tag{8}$$

$$K_{21} = -\frac{L_{21}}{L_{11} L_{22} - L_{12} L_{21}}, \tag{9}$$

$$K_{22} = \frac{L_{11}}{L_{11} L_{22} - L_{12} L_{21}}. \tag{10}$$

Example 2.3. Find the phenomenological coefficients for heat conduction.

Solution. Heat conduction (Example 2.1) can be described by means of thermodynamic forces $-\nabla(\ln T)$ or $\nabla(1/T)$ and generalized flow in the form of the heat flux $\boldsymbol{J}_q$. The first force follows from the dissipation function, and the second, from the entropy source strength. The phenomenological coefficients emerging from considerations based on the dissipation function have been denoted by the upper-case letter L and those based on entropy source strength, by the lower-case letter l, whereupon

$$-\boldsymbol{J}_q = L_{qq} \nabla (\ln T) = \frac{L_{qq}}{T} \nabla T, \tag{1}$$

or

$$-\boldsymbol{J}_q = -l_{qq} \nabla \left(\frac{1}{T}\right) = \frac{l_{qq}}{T^2} \nabla T. \tag{2}$$

Heat conduction is also described by Fourier's empirical equation:

$$\boldsymbol{J}_q = -\lambda \nabla T, \tag{3}$$

where λ is the heat conductivity.

Phenomenological coefficients associated with heat conduction thus depend as follows on the heat conductivity:

$$L_{qq} = \lambda T, \tag{4}$$

$$l_{qq} = \lambda T^2 \tag{5}$$

and

$$l_{qq} = T L_{qq}. \tag{6}$$

The heat conductivity λ is a function of the temperature in the general case. The phenomenological coefficient L_{qq} resulting from reasoning based on the dissipation function in general depends on the temperature to a lesser degree than does the coefficient l_{qq} obtained by reasoning based on the entropy source strength; indeed, it may be treated as a constant quantity over a broad temperature range.

5. The Onsager Reciprocal Relations

The microphysical equations of motion for individual molecules of a system are usually invariant under time reversal $t \to -t$, that is, if the sign of the time is changed, all the molecules traverse their paths in the opposite direction. This is not so if the forces causing the deviations from the equilibrium state are forces determined by a vector product (Coriolis forces in rotating systems, Lorentz forces in magnetic fields). In the latter case the signs of both vectors must be changed if there is to be a reversal of the phenomenon of molecular motion when the sign of the time is changed.

In a magnetic field a molecule traverses its path in the opposite direction only if the velocity of the molecule and the magnetic displacement have their signs changed. This follows from the form of the expression for the Lorentz force which is proportional to the vector product of the molecular velocity $\boldsymbol{v}_i$ of component i and the magnetic field strength $\boldsymbol{B}$. This force, per unit mass of component i, is

$$\boldsymbol{F}_i = e_i \left[\boldsymbol{E} + \frac{1}{c} (\boldsymbol{v}_i \times \boldsymbol{B}) \right], \tag{2.35}$$

where e_i is the electric charge per unit mass of component, and $\boldsymbol{E}$ is the electric field strength.

A similar situation arises in rotating systems in which the molecule travels along the path in the opposite direction if the sign of the time is changed and if concurrently the sign of the linear and angular velocity is changed. This is due to the molecule being acted on by a Coriolis force proportional to the vector product of the translational velocity $\boldsymbol{v}_i$ and the angular velocity $\boldsymbol{\omega}$. The force per unit mass of component i in a rotating system is

$$\boldsymbol{F}_i = \omega^2 \boldsymbol{r} + 2(\boldsymbol{v}_i \times \boldsymbol{\omega}), \tag{2.36}$$

where $\boldsymbol{r}$ is the distance from the axis of rotation, $\omega^2 \boldsymbol{r}$ is the centrifugal force, and $2(\boldsymbol{v}_i \times \boldsymbol{\omega})$ is the Coriolis force (both forces are per unit mass).

The foregoing reasoning concerning how theoretically the direction of the molecular motion can be reversed is the essence of the microscopic principle of reversibility which underlies the statistical validation of the Onsager relations between phenomenological coefficients (cf. Appendix II).

The state of a system can be specified by means of type α parameters, which are even functions of molecular velocity (e.g. energy, concentration), or type β parameters, which are odd functions of molecular velocity (e.g. momentum density). Thermodynamic forces arising from the deviation of the values of these parameters from equilibrium values will have similar properties.

Associated with the microscopic principle of reversibility is the Onsager principle which is taken as the next postulate of the phenomenological thermodynamics of nonequilibrium processes.

Postulate III: With a proper choice of thermodynamic forces and generalized flows, the phenomenological coefficients are related by

$$L_{ab} = \varepsilon_a \varepsilon_b L_{ba} \tag{2.37}$$

if there are no forces determined by a vector product, and related by

$$L_{ab}(\boldsymbol{B}, \boldsymbol{\omega}) = \varepsilon_a \varepsilon_b L_{ba}(-\boldsymbol{B}, -\boldsymbol{\omega}) \tag{2.38}$$

if such forces do occur. In the foregoing relations

$$\varepsilon_c = 1 \qquad \text{for type } \alpha \text{ forces},$$

$$\varepsilon_c = -1 \qquad \text{for type } \beta \text{ forces}.$$

Relationships of the type $L_{ab} = L_{ba}$ between phenomenological coefficients are called *Onsager reciprocal relations*, whereas all the others are known as *Onsager–Casimir relations*.

Application of the Onsager relations to inequality (2.22) or (2.25) yields

$$L_{aa} L_{bb} - L_{ab}^2 > 0. \tag{2.39}$$

The three postulates given above are used in the classical entropy balance method of linear nonequilibrium thermodynamics. The entropy source strength or dissipation function is determined from the entropy balance, and this makes it possible to determine the thermodynamic forces on which the generalized flows depend linearly. The Onsager reciprocal relations associate together the phenomenological coefficients pertaining to cross-effects.

6. Transformations of Thermodynamic Forces and Generalized Flows

As shown above, the form of the expression for the entropy source strength or dissipation function does not uniquely determine the thermodynamic forces or generalized flows. Furthermore, other reasoning permits different definitions to be given for these quantities. For open systems containing multicomponent fluids, the heat flux can be defined in several alternative ways (cf. Section III.8). The diffusion flux can be defined in several ways (cf. Section III.3), depending on the choice of reference velocity. Thus, thermodynamic forces and generalized flows can be transformed in various ways.

First of all, consider the case when the individual forces are not interrelated and neither are the generalized flows. In that event, any choice of forces gives the same number of products in the sums forming the expression for the dissipation function

$$\Psi = \sum_{a=1}^{n} J_a X_a = \sum_{a=1}^{n} J'_a X'_a, \tag{2.40}$$

where J_a and X_a denote the old, and J'_a and X'_a denote the new, generalized flows and thermodynamic forces.

In the special case, the new generalized flows may depend linearly on the old flows,

$$J'_a = \sum_{b=1}^{n} k_{ab} J_b \qquad (a = 1, 2, 3, \ldots, n), \tag{2.41}$$

and then, by formula (2.40),

$$\sum_{a=1}^{n} J'_a X'_a = \sum_{b=1}^{n} J_b \sum_{a=1}^{n} k_{ab} X'_a = \sum_{b=1}^{n} J_b X_b.$$

Hence the old and new forces, X_b and X'_a, are also related linearly by

$$X_b = \sum_{a=1}^{n} k_{ab} X'_a \quad (b=1,2,3,\ldots,n). \tag{2.42}$$

The old and new phenomenological coefficients, L_{bc} and L'_{ad}, can be used to write

$$J'_a = \sum_{d=1}^{n} L'_{ad} X'_d = \sum_{b=1}^{n} k_{ab} J_b = \sum_{b,c=1}^{n} L_{bc} k_{ab} X_c = \sum_{b,c,d=1}^{n} L_{bc} k_{ab} k_{dc} X'_d,$$

which yields the relation

$$L'_{ad} = \sum_{b,c=1}^{n} L_{bc} k_{ab} k_{dc} \quad (a, d=1,2,3,\ldots,n). \tag{2.43}$$

Applying the Onsager relations

$$L_{bc} = L_{cb} \quad (b, c=1,2,3,\ldots,n) \tag{2.44}$$

to equation (2.43), we obtain

$$L'_{ad} = \sum_{b,c=1}^{n} L_{cb} k_{ab} k_{dc} = L'_{da} \quad (a, d=1,2,3,\ldots,n). \tag{2.45}$$

If thermodynamic forces and generalized flows, related by phenomenological coefficients which obey the Onsager relations, are subjected to a linear transformation such that the dissipation function or the entropy source strength are not affected, we arrive at phenomenological equations with coefficients which satisfy the Onsager relations.

Next we consider the case when the thermodynamic forces are independent, whereas the generalized flows are linked by the linear relation

$$\sum_{a=1}^{n} k_a J_a = 0. \tag{2.46}$$

If the constant k_n is not equal to zero, the nth generalized flow

$$J_n = -\sum_{a=1}^{n-1} \frac{k_a}{k_n} J_a \tag{2.47}$$

can be calculated from the equation above and inserted into expression (2.14) for the dissipation function, whence

$$\Psi = \sum_{a=1}^{n} J_a X_a = \sum_{a=1}^{n-1} J_a \left(X_a - \frac{k_a}{k_n} X_n \right). \tag{2.48}$$

This expression now contains only $n-1$ independent forces $\left(X_a - \frac{k_a}{k_n} X_n \right)$ and $n-1$ independent generalized flows J_a.

For both cases, i.e. for n and $n-1$ thermodynamic forces, the phenomenological equations take the respective forms

$$J_a = \sum_{c=1}^{n} L_{ac} X_c \quad (a=1,2,3,\ldots,n), \tag{2.49}$$

$$J_a = \sum_{c=1}^{n-1} L'_{ac} \left(X_c - \frac{k_c}{k_n} X_n \right) \quad (a=1,2,3,\ldots,n-1), \tag{2.50}$$

where, in conformity with the Onsager reciprocal relations,

$$L'_{ac} = L'_{ca} \quad (a,\ c=1,2,3,\ldots,n-1). \tag{2.51}$$

Substitution of relation (2.50) into formula (2.47) yields

$$J_n = -\sum_{a,c=1}^{n-1} L'_{ac} \left(\frac{k_c}{k_n} X_a - \frac{k_a k_c}{k_n^2} X_n \right). \tag{2.52}$$

Comparison of the coefficients in expressions (2.49)–(2.52) leads to the relations:

$$L_{ac} = L'_{ac} \quad (a,\ c=1,2,3,\ldots,n-1). \tag{2.53}$$

$$L_{an} = -\sum_{c=1}^{n-1} \frac{k_c}{k_n} L'_{ac} \quad (a=1,2,3,\ldots,n-1), \tag{2.54}$$

$$L_{na} = -\sum_{c=1}^{n-1} \frac{k_c}{k_n} L'_{ac} \quad (a=1,2,3,\ldots,n-1), \tag{2.55}$$

$$L_{nn} = \sum_{a,c=1}^{n-1} \frac{k_a k_c}{k_n^2} L'_{ac}. \tag{2.56}$$

Relations (2.53)–(2.56) imply the validity of the Onsager reciprocal relations for the phenomenological coefficients

$$L_{ac} = L_{ca} \quad (a,\ c=1,2,3,\ldots,n). \tag{2.57}$$

The phenomenological coefficients in the preceding relations are, in accordance with equation (2.46), related by

$$\sum_{c=1}^{n} k_c L_{ac} = 0 \quad (a=1,2,3,\ldots,n), \tag{2.58}$$

$$\sum_{a=1}^{n} k_a L_{ac} = 0 \quad (c=1,2,3,\ldots,n), \tag{2.59}$$

where only $2n-1$ equations are independent since, by relations (2.58) or (2.59),

$$\sum_{a,c=1}^{n} k_a k_c L_{ac} = 0. \tag{2.60}$$

The result of the considerations above may be put into words: The Onsager reciprocal relations remain valid if a homogeneous relationship exists between the generalized flows.

A similar discussion can be carried out if the generalized flows are independent and a linear relationship exists between the thermodynamic forces:

$$\sum_{a=1}^{n} l_a X_a = 0 . \tag{2.61}$$

The Onsager reciprocal relations also hold in this case.

Finally, the linear relation (2.61) between the thermodynamic forces will be posited concurrently with the linear relation (2.46) between the generalized flows. These two relations enable the dissipation function to be written as

$$\Psi = \sum_{a=1}^{n} J_a X_a = \sum_{a=1}^{n-1} J_a \left(X_a + \frac{k_a}{k_n} \sum_{b=1}^{n-1} \frac{l_b}{l_n} X_b \right), \tag{2.62}$$

whence, on interchanging the dummy indices, we have

$$J_a = \sum_{c=1}^{n-1} L'_{ac} \left(X_c + \frac{k_c}{k_n} \sum_{b=1}^{n-1} \frac{l_b}{l_n} X_b \right) = \sum_{c=1}^{n-1} \left(L'_{ac} + \frac{l_c}{l_n} \sum_{b=1}^{n-1} \frac{k_b}{k_n} L'_{ab} \right) X_c \tag{2.63}$$

$$(a = 1, 2, 3, \ldots, n-1) .$$

Combination of this expression with equation (2.47) yields

$$J_n = - \sum_{a,c=1}^{n-1} \frac{k_a}{k_n} \left(L'_{ac} + \frac{l_c}{l_n} \sum_{b=1}^{n-1} \frac{k_b}{k_n} L'_{ab} \right) X_c . \tag{2.64}$$

The generalized flows are expressible only in terms of independent forces, X_n being removed from the phenomenological equation by means of relation (2.61):

$$J_a = \sum_{c=1}^{n} L_{ac} X_c = \sum_{c=1}^{n-1} \left(L_{ac} - \frac{l_c}{l_n} L_{an} \right) X_c . \tag{2.65}$$

Comparison of expressions (2.63) and (2.64) with expression (2.65) gives the following relations between the phenomenological coefficients:

$$L_{ac} - \frac{l_c}{l_n} L_{an} = L'_{ac} + \frac{l_c}{l_n} \sum_{b=1}^{n-1} \frac{k_b}{k_n} L'_{ab} \quad (a, c = 1, 2, 3, \ldots, n-1), \tag{2.66}$$

$$L_{nc} - \frac{l_c}{l_n} L_{nn} = - \sum_{a=1}^{n-1} \frac{k_a}{k_n} \left(L'_{ac} + \frac{l_c}{l_n} \sum_{b=1}^{n-1} \frac{k_b}{k_n} L'_{ab} \right) \quad (c = 1, 2, 3, \ldots, n-1). \tag{2.67}$$

We now have only $n(n-1)$ relations for n^2 coefficients, i.e. there are n degrees of freedom in their choice and their matrix need not be symmetric. At the same time, the phenomenological coefficients can be chosen so that relations (2.53)–(2.56) be satisfied, and then the coefficients L_{ab} statisfy the Onsager reciprocal relations, provided that the L'_{ab} do.

In the choice of relations between the phenomenological coefficients L_{ab} and L'_{cd} which follow from the n degrees of freedom, account is taken of $n-1$ Onsager relations $L'_{ab} = L'_{ba}$ and any other nth relation. It should be borne in mind that there are other restrictions, not taken into account above, implied by the positive value of the dissipation function. Inequalities (2.24) and (2.39) must, therefore, be satisfied. This problem will be considered in greater detail in Example 2.6.

The results of the considerations above can be put into words: If homogeneous linear relations exist between the generalized flows as well as between the forces, the phenomenological coefficients are uniquely determined and the Onsager reciprocal relations need not be obeyed. However, the phenomenological coefficients can always be chosen so that the Onsager reciprocal relations are satisfied.

Since the Onsager reciprocal relations may not be satisfied when relationships exist between the thermodynamic forces and between generalized flows, it is best to pursue considerations of nonequilibrium thermodynamics after eliminating the relations between the forces.

Example 2.4. Consider the transformation of $n=2$ generalized flows by means of a coefficient specified by the matrix

$$k=\begin{bmatrix}1 & 0\\ a & 1\end{bmatrix}.$$

Solution. Since

$$k_{11}=1, \quad k_{12}=0, \quad k_{21}=a, \quad k_{22}=1, \tag{1}$$

the new generalized flows are of the form:

$$J_1'=k_{11}J_1+k_{12}J_2=J_1, \tag{2}$$

$$J_2'=k_{21}J_1+k_{22}J_2=aJ_1+J_2. \tag{3}$$

The phenomenological coefficients L_{ab}' for the new generalized flows can be expressed in terms of the phenomenological coefficients L_{ab} [equation (2.43)] for the old generalized flows as

$$L_{11}'=L_{11}, \tag{4}$$

$$L_{12}'=aL_{11}+L_{12}, \tag{5}$$

$$L_{21}'=aL_{11}+L_{21}, \tag{6}$$

$$L_{22}'=a^2L_{11}+a(L_{12}+L_{21})+L_{22}. \tag{7}$$

If the matrix of the old phenomenological coefficients is symmetric

$$L_{12}=L_{21}, \tag{8}$$

so is the matrix of new phenomenological coefficients:

$$L_{12}'=L_{21}'. \tag{9}$$

Example 2.5. Carry out the transformations of thermodynamic forces and generalized flows if there are two independent forces X_1 and X_2 and two linearly dependent generalized flows,

$$kJ_1+J_2=0, \tag{1}$$

so that only independent flows remain.

Solution. For the case under consideration, the dissipation function can be written as

$$\Psi=J_1X_1+J_2X_2=J_1(X_1-kX_2). \tag{2}$$

In view of this, the phenomenological equations can be put into the form

$$J_1=L_{11}X_1+L_{12}X_2, \tag{3}$$

$$J_2=L_{21}X_1+L_{22}X_2, \tag{4}$$

or

$$J_1' = L(X_1 - kX_2) = L'X'. \tag{5}$$

if

$$X' = X_1 - kX_2 \tag{6}$$

is taken for the sole thermodynamic force. Substitution of relation (5) into formula (1) yields

$$J_2 = -kL'(X_1 - kX_2). \tag{7}$$

When condition (1) is taken into account in equations (3) and (4), relations are obtained between the phenomenological coefficients

$$kL_{11} + L_{21} = 0, \quad kL_{12} + L_{22} = 0. \tag{8}$$

By relations (1) and (8),

$$J_1 = -\frac{1}{k} J_2 = L_{11} X_1 - \frac{L_{22}}{k} X_2. \tag{9}$$

Comparison of relations (9) and (5) results in

$$L' = L_{11} = \frac{L_{22}}{k^2}. \tag{10}$$

In view of relations (8), we also have

$$L_{12} = L_{21} = -kL' = -\frac{L_{22}}{k}, \tag{11}$$

that is to say, the Onsager reciprocal relations are satisfied in phenomenological equations (3) and (4).

Example 2.6. Transform the thermodynamic forces and generalized flows when there are two independent generalized flows and two linearly dependent thermodynamic forces,

$$kX_1 + X_2 = 0. \tag{1}$$

Solution. In the case under consideration the dissipation function can be written as

$$\Psi = J_1 X_1 + J_2 X_2 = (J_1 - kJ_2) X_1. \tag{2}$$

The phenomenological equations solved for the thermodynamic forces are of the form

$$X_1 = K_{11} J_1 + K_{12} J_2, \tag{3}$$

$$X_2 = K_{21} J_1 + K_{22} J_2 \tag{4}$$

or

$$X_1 = K'(J_1 - kJ_2) = K'J', \tag{5}$$

where we have confined ourselves to one generalized flow

$$J' = J_1 - kJ_2. \tag{6}$$

On inserting relation (5) into formula (1), we obtain

$$X_2 = -kK'(J_1 - kJ_2) = 0. \tag{7}$$

When condition (1) is taken into account in equations (3) and (4), the result is the following relation between the phenomenological coefficients:

$$kK_{11} + K_{21} = 0, \quad kK_{12} + K_{22} = 0. \tag{8}$$

On the basis of relations (1) and (8), we have

$$X_1 = -\frac{1}{k} J_2 = K_{11} J_1 - \frac{K_{22}}{k} J_2. \tag{9}$$

Comparison of relations (9) and (5) yields

$$K' = K_{11} = \frac{K_{22}}{k^2}, \tag{10}$$

and, in view of relations (8), also gives

$$K_{12} = K_{21}, \tag{11}$$

that is, the Onsager reciprocal relations are satisfied in the phenomenological equations (3) and (4).

Example 2.7. Perform transformations of the thermodynamic forces and generalized flows when there are two linear dependent thermodynamic forces

$$kX_1 + X_2 = 0 \tag{1}$$

and two linearly dependent generalized flows

$$mJ_1 + J_2 = 0. \tag{2}$$

Solution. The dissipation function can now be recast in the form

$$\Psi = J_1 X_1 + J_2 X_2 = (km+1) J_1 X_1. \tag{3}$$

Thus, the phenomenological equations can be rewritten as

$$J_1 = L_{11} X_1 + L_{12} X_2, \tag{4}$$

$$J_2 = L_{21} X_1 + L_{22} X_2 \tag{5}$$

or

$$J_1 = L' X', \tag{6}$$

where a new thermodynamic force

$$X' = (km+1) X_1 = -\left(m + \frac{1}{k}\right) X_2 \tag{7}$$

and phenomenological coefficient

$$L' = L'_{11} \tag{8}$$

have been introduced.

It follows from equations (2), (6), and (7) that

$$J_1 = (km+1) L' X_1, \tag{9}$$

$$J_2 = -m(km+1) L' X_1, \tag{10}$$

whereas equations (1), (4), and (5) lead to

$$J_1 = (L_{11} - kL_{12}) X_1, \tag{11}$$

$$J_2 = (L_{21} - kL_{22}) X_1. \tag{12}$$

Comparison of these expressions for generalized flows yields the following relations between the phenomenological coefficients:

$$L_{11} - kL_{12} = (km+1) L', \tag{13}$$

$$L_{21} - kL_{22} = -m(km+1) L'. \tag{14}$$

There are $n=2$ degrees of freedom in the choice of phenomenological coefficients in phenomenological equations (4) and (5). Thus, if it is assumed that $n-1=1$ Onsager reciprocal relation

$$L_{12} = L_{21} \tag{15}$$

is valid, and that the coefficients L_{22} and L' are proportional to each other,

$$L_{22}=pL', \tag{16}$$

then relations (13) and (14) reduce to

$$L_{11}=-(k^2m^2-k^2p-1)L', \tag{17}$$

$$L_{12}=L_{21}=-(km^2-kp+m)L'. \tag{18}$$

Next, we consider the restrictions imposed on the value of the proportionality factor p by the positive value of the dissipation function. The dissipation function is positive if phenomenological coefficients with the same indices are positive,

$$L_{11}>0, \quad L_{22}>0, \quad L'=L'_{11}>0, \tag{19}$$

and coefficients with different indices obey the inequality

$$L_{11}L_{22}-L_{12}^2>0. \tag{20}$$

The inequalities $L'>0$ and $L_{22}>0$ restrict the proportionality factor to positive values $p>0$. The inequalities $L'>0$ and $L_{11}>0$ yield the condition

$$p>m^2-\frac{1}{k^2}. \tag{21}$$

Finally, inequality (20) reduces to the condition

$$p>m^2>0. \tag{22}$$

7. Stationary States

Depending on the boundary conditions imposed on the thermodynamic system, states in which the intensive parameters are independent of time may be equilibrium states or nonequilibrium stationary states. In an equilibrium state interactions in the nature of flows between the system and environment vanish and the intensive state parameters are the same for the entire phase. In a stationary state the intensive parameters which specify the local state of the substance at a stationary point in space do not vary with time,

$$\frac{\partial z}{\partial t}=0. \tag{2.68}$$

Extensive parameters Z, determining the state of the system with boundaries which are stationary with respect to the coordinate system, do not change in a stationary system,

$$\frac{\mathrm{d}Z}{\mathrm{d}t}=0. \tag{2.69}$$

It is easily seen that the stationary state of the substance at every point of the system under consideration is related to the stationary state of the system.

In the particular case, equation (2.69) defining the stationary state of the system can be applied to the entropy, whereby

$$\frac{dS}{dt}=0. \tag{2.70}$$

In accordance with the convention adopted earlier (cf. Section II.3), the expression determining the total entropy change in time can be decomposed into two parts

$$\frac{dS}{dt}=\frac{dS_e}{dt}+\frac{dS_i}{dt}, \tag{2.71}$$

where dS_e/dt is the reversible entropy change in time resulting from the existence of an entropy flux between the system and the environment, whereas

$$P=\frac{dS_i}{dt}>0 \tag{2.72}$$

is always a positive entropy change in time, resulting from irreversible phenomena inside the system, and is called the *rate of entropy production.*

By relations (2.70) and (2.71), in the stationary state we have

$$\frac{dS_e}{dt}=-\frac{dS_i}{dt}<0. \tag{2.73}$$

The total entropy produced inside the system is compensated by entropy leaving the system across the boundaries. The time-reversible entropy change of a system in a stationary state is negative. A nonequilibrium stationary state cannot be attained in an isolated system for which the time-reversible entropy change is zero:

$$\frac{dS_e}{dt}=0. \tag{2.74}$$

Another characteristic feature of the equilibrium state, in addition to the absence of an entropy flow across the system boundaries and the invariance of the system entropy, is the absence of entropy production owing to irreversible phenomena:

$$P=\frac{dS_i}{dt}=0. \tag{2.75}$$

As the considerations above imply, we have

$$\frac{dS}{dt}=\frac{dS_e}{dt}=\frac{dS_i}{dt}=0 \tag{2.76}$$

in the equilibrium state of a system, whereas the stationary state must have only

$$\frac{dS}{dt}=\frac{dS_e}{dt}+\frac{dS_i}{dt}=0. \tag{2.77}$$

If a system is to be able to attain a stationary state, the boundary conditions at its boundaries cannot change in time. Below we consider the time variation of the rate of entropy production when a system approaches a stationary state in a way so that the intensive parameters, parameter gradients, and flows caused by those gradients do not vary within the system.

The rate of entropy production inside the given system of volume V is

$$P=\int_V \Phi_s \mathrm{d}V=\int_V \sum_{a=1}^{n} J_a X_a \mathrm{d}V, \tag{2.78}$$

and is hence a function of n thermodynamic forces X_a and n generalized flows J_a which in the general case vary with time, remaining constant at the system boundaries only in the stationary state.

Differentiation of equation (2.78) with respect to time yields

$$\frac{\mathrm{d}P}{\mathrm{d}t}=\int_V \sum_{a=1}^{n} J_a \frac{\partial X_a}{\partial t}\mathrm{d}V+\int_V \sum_{a=1}^{n} \frac{\partial J_a}{\partial t} X_a \mathrm{d}V=\frac{\mathrm{d}_X P}{\mathrm{d}t}+\frac{\mathrm{d}_J P}{\mathrm{d}t}. \tag{2.79}$$

The first term, representing the change in the rate of entropy production as a result of the thermodynamic forces varying in time,

$$\frac{\mathrm{d}_X P}{\mathrm{d}t}=\int_V \frac{\partial_X \Phi_s}{\partial t}\mathrm{d}V=\int_V \sum_{a=1}^{n} J_a \frac{\partial X_a}{\partial t}\mathrm{d}V\leqslant 0, \tag{2.80}$$

cannot be a positive quantity under stationary conditions at the system boundary. This can be shown with concrete examples (cf. Example 2.8).

The second term of equation (2.79), of the form

$$\frac{\mathrm{d}_J P}{\mathrm{d}t}=\int_V \frac{\partial_J \Phi_s}{\partial t}\mathrm{d}V=\int_V \sum_{a=1}^{n} \frac{\partial J_a}{\partial t} X_a \mathrm{d}V\lesseqgtr 0, \tag{2.81}$$

corresponding to a change in the rate of entropy production caused by the generalized flows varying in time, does not have a particular sign in the general case and therefore nothing general can be said about the sign of the change $\mathrm{d}P/\mathrm{d}t$ in the rate of entropy production in time.

It is a different matter when the generalized flows are linearly dependent on the thermodynamic forces,

$$J_a=\sum_{b=1}^{n} l_{ab} X_b \quad (a=1,2,3,\ldots,n) \tag{2.82}$$

and when the phenomenological coefficients are constant, hence also do not vary with time, and are linked by the Onsager reciprocal relations:

$$l_{ab}=l_{ba} \quad (a,\ b=1,2,3,\ldots,n). \tag{2.83}$$

In this case

$$\frac{\mathrm{d}_J P}{\mathrm{d}t} = \int_V \sum_{a=1}^{n} \frac{\partial J_a}{\partial t} X_a \,\mathrm{d}V = \int_V \sum_{a,b=1}^{n} l_{ab} X_a \frac{\partial X_b}{\partial t} \,\mathrm{d}V \tag{2.84}$$

$$= \int_V \sum_{b=1}^{n} J_b \frac{\partial X_b}{\partial t} \,\mathrm{d}V = \frac{\mathrm{d}_X P}{\mathrm{d}t} = \frac{1}{2} \frac{\mathrm{d}P}{\mathrm{d}t} \leqslant 0$$

and thereby

$$\frac{\mathrm{d}P}{\mathrm{d}t} = 2 \frac{\mathrm{d}_X P}{\mathrm{d}t} = 2 \frac{\mathrm{d}_J P}{\mathrm{d}t} \leqslant 0 \,. \tag{2.85}$$

As is seen, with the assumptions made above, the rate of entropy production decreases with time until a minimum is reached in the stationary state. The dissipation function changes in similar fashion.

Relation (2.84) can also be recast in a local form

$$\frac{\partial_X \Phi_s}{\partial t} = \frac{\partial_J \Phi_s}{\partial t} \,. \tag{2.86}$$

A system such that the boundary conditions do not vary with time at its boundaries tends to a stationary state. In the case when the generalized flows are linearly dependent on the thermodynamic forces, and the phenomenological coefficients are constant and obey the Onsager reciprocal relations, the stationary state of the system corresponds to the minimal rate of entropy production and minimal energy dissipation.

If the phenomenological coefficient L depends only on the temperature, it is possible to introduce a function of the temperature, defined by

$$\frac{1}{\theta} = \int_{T_1}^{T} \sqrt{\frac{L}{L_0}} \,\mathrm{d}\left(\frac{1}{T}\right) \tag{2.87}$$

where $T_1 = \text{const}$, and then a constant phenomenological coefficient L_0 appears in the equations. If the phenomenological coefficients are a function only of the temperature and satisfy the Onsager reciprocal relations, the stationary state of the system is bound up with the minimal rate of entropy production.

The foregoing problem can also be considered in the opposite direction, viz. the variational calculus can be used to determine the distribution of thermodynamic parameters which satisfies the condition of minimal entropy production. This topic has been elaborated most fully by Gyarmati [39].

In accordance with the previous considerations, the entropy source strength can be written as

$$\Phi_s = \sum_{a,b=1}^{n} l_{ab} X_a X_b = \sum_{a,b=1}^{n} k_{ab} J_a J_b \,, \tag{2.88}$$

i.e. as a function of the thermodynamic forces or of the generalized flows. These two forms of entropy source strength are completely equivalent.

The entropy production rate

$$P=\int_V \Phi_s \mathrm{d}V=\int_V \sum_{a,b=1}^{n} l_{ab} X_a X_b \mathrm{d}V=\int_V \sum_{a,b=1}^{n} k_{ab} J_a J_b \mathrm{d}V \tag{2.89}$$

is minimal if its variation is equal to zero:

$$\delta P=\delta \int_V \sum_{a,b=1}^{n} l_{ab} X_a X_b \mathrm{d}V=\delta \int_V \sum_{a,b=1}^{n} k_{ab} J_a J_b \mathrm{d}V=0. \tag{2.90}$$

Depending on the form of entropy source strength under consideration, the generalized flows J_a may vary while the forces X_a remain constant (Onsager) or the thermodynamic forces may vary while the generalized flows remain constant (Glansdorff and Prigogine). One may also consider the most general case when both the thermodynamic forces and the generalized flows vary (Gyarmati).

To bring out the differences between the three approaches mentioned above, it is necessary to introduce the potentials

$$\psi=\tfrac{1}{2} \sum_{a,b=1}^{n} l_{ab} X_a X_b \geqslant 0, \tag{2.91}$$

$$\varphi=\tfrac{1}{2} \sum_{a,b=1}^{n} k_{ab} J_a J_b \geqslant 0, \tag{2.92}$$

which are equal to half the value of the entropy source strength when linear phenomenological equations are valid.

As is readily perceived, these potentials have the following properties:

$$\frac{\partial \psi}{\partial X_a}=\sum_{b=1}^{n} l_{ab} X_b=J_a, \tag{2.93}$$

$$\frac{\partial \varphi}{\partial J_a}=\sum_{b=1}^{n} k_{ab} J_b=X_a, \tag{2.94}$$

and

$$\frac{\partial^2 \psi}{\partial X_a \partial X_b}=\frac{\partial J_a}{\partial X_b}=l_{ab}=l_{ba}=\frac{\partial J_b}{\partial X_a}=\frac{\partial^2 \psi}{\partial X_b \partial X_a}, \tag{2.95}$$

$$\frac{\partial^2 \varphi}{\partial J_a \partial J_b}=\frac{\partial X_a}{\partial J_b}=k_{ab}=k_{ba}=\frac{\partial X_b}{\partial J_a}=\frac{\partial^2 \varphi}{\partial J_b \partial J_a}. \tag{2.96}$$

As is seen, the first derivatives of these potentials correspond to linear phenomenological equations, and the second derivatives, to Onsager reciprocal relations.

For the three cases mentioned above, the principle of least dissipation of energy, and hence of minimal entropy production rate, can be given in a local form pertaining to an elementary volume

$$\delta(\Phi_s-\varphi)_{X_a}=0, \quad X_a=\text{const}, \quad \delta X_a=0, \quad \delta J_a\neq 0, \tag{2.97}$$

$$\delta(\Phi_s-\psi)_{J_a}=0, \quad J_a=\text{const}, \quad \delta J_a=0, \quad \delta X_a\neq 0, \tag{2.98}$$

$$\delta[\Phi_s-(\varphi+\psi)]=0, \quad \delta J_a\neq 0, \quad \delta X_a\neq 0, \tag{2.99}$$

or in an integral form pertaining to the entire system under consideration:

$$\delta\int_V(\Phi_s-\varphi)_{X_a}\,\mathrm{d}V=0, \quad X_a=\text{const}, \quad \delta X_a=0, \quad \delta J_a\neq 0, \tag{2.100}$$

$$\delta\int_V(\Phi_s-\psi)_{J_a}\,\mathrm{d}V=0, \quad J_a=\text{const}, \quad \delta J_a=0, \quad \delta X_a\neq 0, \tag{2.101}$$

$$\delta\int_V[\Phi_s-(\varphi+\psi)]\,\mathrm{d}V=0, \quad \delta J_a\neq 0, \quad \delta X_a\neq 0. \tag{2.102}$$

Example 2.8. Find how the rate of entropy production changes with time as a result of heat conduction in a one-component, isotropic body under an invariant temperature distribution at the boundaries of the body.

Solution. By Example 2.1, the entropy source strength for the case under consideration is equal to

$$\Phi_s=\boldsymbol{J}_q\cdot\nabla\left(\frac{1}{T}\right). \tag{1}$$

The time variation of the rate of entropy production owing to the variation of the thermodynamic forces is

$$\frac{\mathrm{d}_X P}{\mathrm{d}t}=\int_V \boldsymbol{J}_q\cdot\frac{\partial}{\partial t}\left[\nabla\left(\frac{1}{T}\right)\right]\mathrm{d}V=\int_A\left[\frac{\partial}{\partial t}\left(\frac{1}{T}\right)\right]\boldsymbol{J}_q\cdot\mathrm{d}\boldsymbol{A}-\int_V\left[\frac{\partial}{\partial t}\left(\frac{1}{T}\right)\right](\nabla\cdot\boldsymbol{J}_q)\,\mathrm{d}V, \tag{2}$$

where integration by parts has been performed. Since the temperature does not change with time at the system boundaries, the surface integral is equal to zero.

The divergence of the heat flux can be determined from an equation expressing the first law of thermodynamics:

$$đq=\mathrm{d}u+p\,\mathrm{d}v. \tag{3}$$

For solids, it may be assumed that

$$\mathrm{d}u=c_v\,\mathrm{d}T \qquad c_v=\text{const}, \quad \mathrm{d}v=0, \tag{4}$$

whereby

$$-\nabla\cdot\boldsymbol{J}_q=\rho\frac{đq}{\mathrm{d}t}=\rho\frac{\partial u}{\partial t}=\rho c_v\frac{\partial T}{\partial t}. \tag{5}$$

When equations (2) and (5) are combined, in view of the fact that $c_v>0$ the result is

$$\frac{\mathrm{d}_X P}{\mathrm{d}t}=-\int_V\frac{\rho c_v}{T^2}\left(\frac{\partial T}{\partial t}\right)^2\mathrm{d}V\leqslant 0. \tag{6}$$

By equation (1) the phenomenological equation for heat conduction is of the form

$$\boldsymbol{J}_q=l_{qq}\nabla\left(\frac{1}{T}\right), \tag{7}$$

and comparison of this form with Fourier's law

$$J_q = -\lambda \nabla T \tag{8}$$

yields a phenomenological coefficient defined as

$$l_{qq} = \lambda T^2 . \tag{9}$$

For bodies with small temperature gradients, the phenomenological coefficient may be assumed to be constant, and then

$$\frac{\mathrm{d}P}{\mathrm{d}t} = 2\frac{\mathrm{d}_X P}{\mathrm{d}t} = -2\int_V \frac{\rho c_v}{T^2}\left(\frac{\partial T}{\partial t}\right)^2 \mathrm{d}V \leqslant 0 . \tag{10}$$

In an isotropic solid the rate of entropy production owing to heat conduction falls off with time until a minimum is reached in an equilibrium state.

Example 2.9. Find the temperature distribution in a one-component, isotropic solid in which there is heat conduction corresponding to the minimal rate of entropy production.

Solution. Equations (1) and (7) of Example 2.8 can be used to express the entropy production rate as

$$P = \int_V \Phi_s \,\mathrm{d}V = \int_V J_q \cdot \nabla\left(\frac{1}{T}\right) \mathrm{d}V = l_{qq}\int_V \left[\nabla\left(\frac{1}{T}\right)\right]^2 \mathrm{d}V, \tag{1}$$

where, as in the preceding example, the phenomenological coefficient l_{qq} has been assumed to be constant, this assumption being valid only for small temperature differences within the solid under consideration.

The minimal entropy production rate is determined by the condition that the first variation vanish:

$$\delta P = \delta\left\{l_{qq}\int_V \left[\nabla\left(\frac{1}{T}\right)\right]^2 \mathrm{d}V\right\} = 0 . \tag{2}$$

If the temperature distribution at the boundary of the space under consideration is independent of time, the temperature variations δT there are equal to zero.

The solution of this variational problem for a region of the substance specified in the x, y, z-coordinate system should be sought in the form of a surface of area A in the x, y-plane, onto which the given region is projected.

The functional

$$v\,[z(x, y)] = \int_A F(x, y, z, p, q)\,\mathrm{d}x\,\mathrm{d}y = \int_A \left[\left(\frac{\partial T^{-1}}{\partial x}\right)^2 + \left(\frac{\partial T^{-1}}{\partial y}\right)^2\right] \mathrm{d}x\,\mathrm{d}y \tag{3}$$

depends on the function z which in turn is a function of two variables, x and y. The notation

$$F(x, y, z, p, q) = \left(\frac{\partial T^{-1}}{\partial x}\right)^2 + \left(\frac{\partial T^{-1}}{\partial y}\right)^2 ,$$

$$z = T^{-1}, \quad p = \frac{\partial z}{\partial x} = \frac{\partial T^{-1}}{\partial x}, \qquad q = \frac{\partial z}{\partial y} = \frac{\partial T^{-1}}{\partial y},$$

has been used in expression (3).

On the periphery of surface A we have

$$F = \left(\frac{\partial z}{\partial x}\right)^2 + \left(\frac{\partial z}{\partial y}\right)^2 = p^2 + q^2 , \tag{4}$$

that is, F is a function of p and q, and not of x, y, z.

Euler's equation for the variational problem above is of the form

$$F_z-\frac{\partial}{\partial x}\{F_p\}-\frac{\partial}{\partial y}\{F_q\}=0, \tag{5}$$

where

$$\frac{\partial}{\partial x}\{F_p\}=F_{px}+F_{pz}\frac{\partial z}{\partial x}+F_{pp}\frac{\partial p}{\partial x}+F_{pq}\frac{\partial q}{\partial x},$$

$$\frac{\partial}{\partial y}\{F_q\}=F_{qy}+F_{qz}\frac{\partial z}{\partial y}+F_{qq}\frac{\partial q}{\partial y}+F_{qp}\frac{\partial p}{\partial y}.$$

In the case under consideration, therefore, we have

$$F_z=\frac{\partial F}{\partial z}=0, \qquad F_{px}=\frac{\partial^2 F}{\partial p\,\partial x}=0, \qquad F_{pz}=0,$$

$$F_{pp}=\frac{\partial^2 F}{\partial p^2}=\frac{\partial^2}{\partial p^2}(p^2+q^2)=2,$$

$$\frac{\partial p}{\partial x}=\frac{\partial}{\partial x}\left(\frac{\partial T^{-1}}{\partial x}\right)=\frac{\partial^2(T^{-1})}{\partial x^2},$$

and hence equation (5) is of the form

$$\frac{\partial^2(T^{-1})}{\partial x^2}+\frac{\partial^2(T^{-1})}{\partial y^2}=0, \tag{6}$$

while for three independent variables it reduces to the familiar form of Laplace's equation

$$\nabla^2\left(\frac{1}{T}\right)=\nabla\cdot\left[\nabla\left(\frac{1}{T}\right)\right]=0. \tag{7}$$

On taking account of the fact that, as in Example 2.8,

$$\boldsymbol{J}_q=l_{qq}\nabla\left(\frac{1}{T}\right), \qquad -\nabla\cdot\boldsymbol{J}_q=\rho c_v\frac{\partial T}{\partial t},$$

we obtain

$$\nabla\cdot\boldsymbol{J}_q=0, \qquad \frac{\partial T}{\partial t}=0. \tag{8}$$

The state of minimal rate of entropy production is, in the given case, a stationary state.

Example 2.10. Use the integral principle of minimal dissipation of energy to derive the differential equation for nonstationary heat conduction in an isotropic solid.

Solution. For an isotropic solid the entropy balance equation is of the form

$$\rho\frac{\partial s}{\partial t}=-\nabla\cdot\boldsymbol{J}_s+\Phi_s, \tag{1}$$

whereas the integral principle of minimum energy dissipation pertaining to the entire system under consideration is defined by relation (2.101), that is after invoking the Gauss–Ostrogradsky theorem

$$\delta\int_V\left[\rho\frac{\partial s}{\partial t}-\psi\right]\mathrm{d}V+\delta\int_A\boldsymbol{J}_s\cdot\mathrm{d}\boldsymbol{A}=0. \tag{2}$$

For isotropic solids

$$\rho\frac{\partial s}{\partial t}=\frac{\rho}{T}\frac{\partial u}{\partial t}=\frac{\rho c_v}{T}\frac{\partial T}{\partial t}, \tag{3}$$

$$J_s = \frac{J_q}{T}, \tag{4}$$

$$\psi = \frac{l_{qq}}{2}\left[\nabla\left(\frac{1}{T}\right)\right]^2. \tag{5}$$

If unchanging conditions are assumed at the boundaries of the system, the variational equation (2) can be written for our case of an isotropic solid in the form

$$\delta \int_V \left[\frac{\rho c_v}{T}\frac{\partial T}{\partial t} - \frac{l_{qq}}{2}\left(\nabla \frac{1}{T}\right)^2\right] \mathrm{d}V = 0. \tag{6}$$

The inverse absolute temperature, $1/T$, is taken as the independent variable subject to variation, whereupon

$$\int_V \left\{\left[\frac{\rho c_v}{T}\frac{\partial T}{\partial t} + \nabla\cdot\left(l_{qq}\nabla\frac{1}{T}\right)\right]\delta\left(\frac{1}{T}\right)\right\} \mathrm{d}V = 0. \tag{7}$$

The foregoing variational equation corresponding to the principle defined by condition (2.101) is equivalent to the existence of a differential heat conduction equation of the form

$$\rho c_v \frac{\partial T}{\partial t} = -\nabla\cdot\left[l_{qq}\nabla\left(\frac{1}{T}\right)\right] = \nabla\cdot(l_{qq}\, T^{-2}\nabla T) \tag{8}$$

The Lagrangian for the variational problem under consideration is

$$\mathscr{L}_q = \frac{\rho c_v}{T}\frac{\partial T}{\partial t} - \frac{l_{qq}}{2}\left[\nabla\left(\frac{1}{T}\right)\right]^2, \tag{9}$$

and the differential equation (8) for nonstationary heat conduction may be considered as an Euler–Lagrange equation

$$\frac{\partial \mathscr{L}_q}{\partial (T^{-1})} - \sum_{\alpha=1}^{3} \frac{\partial}{\partial x_\alpha} \frac{\partial \mathscr{L}_q}{\partial\left[\frac{\partial (T^{-1})}{\partial x_\alpha}\right]} = 0 \tag{10}$$

for the variational problem

$$\delta \int_V \mathscr{L}_q \, \mathrm{d}V = 0. \tag{11}$$

8. The Stability of Thermodynamic Equilibrium

The nonequilibrium processes considered above concerned states which were not very far from states of thermodynamic equilibrium and hence did not encompass all real phenomena. Processes occurring at a great distance from an equilibrium state have recently been of considerable interest. Not all of the postulates formulated above can be used to describe these processes, and furthermore some of the linear relations cease to be valid. It has been observed that instabilities occur during such processes. Methods for the investigation of processes described by nonlinear relations have been presented in the monograph by Glansdorff and Prigogine [40]. The principles adopted there are used to analyse processes for which the postulate of local thermodynamic equilibrium may be considered valid and, consequently, the formal apparatus of classical thermodynamics (particularly the Gibbs relation) can be used to describe the thermodynamic parameters. However, no use is made of the Onsager reciprocal relations and the linear phenomenological equations for generalized flows.

The starting point for our considerations is the entropy balance equation for the entire given system of volume V; this can be written as

$$P=\int_V \Phi_s \, dV=\frac{\partial S}{\partial t}+W \geqslant 0 . \tag{2.103}$$

The foregoing equation states that the rate P of entropy production taking place inside the system in part increases the entropy of the system (the term $\partial S/\partial t$), and in part is carried off across the boundary of the system as an entropy flow

$$W=\int_A (\rho s \boldsymbol{v}+\boldsymbol{J}_s)\cdot d\boldsymbol{A}=\int_V \nabla\cdot(\rho s \boldsymbol{v}+\boldsymbol{J}_s)\, dV . \tag{2.104}$$

The entropy flow is obtained by integrating the convection and conduction entropy fluxes, $\rho s \boldsymbol{v}$ and $\boldsymbol{J}_s$, over the entire external surface of the system of area A.

Examination of the stability of a system requires analysis of the relations describing the deviation of the state of the system from equilibrium. The rate of entropy production in an equilibrium state is zero, that is to say, is of second-order magnitude relative to the deviation from equilibrium. For this reason, it is convenient to expand the terms on the right-hand side of equation (2.103) in a series in the equilibrium values. The entropy S of a nonequilibrium system is expressed by means of the entropy S_{eq} of an equilibrium system up to second-order terms,

$$S=S_{eq}+(\delta S)_{eq}+\tfrac{1}{2}(\delta^2 S)_{eq} . \tag{2.105}$$

Since the entropy S_{eq} of an equilibrium system does not depend on time, differentiation of expression (2.105) with respect to time yields

$$\frac{\partial S}{\partial t}=\frac{\partial(\delta S)_{eq}}{\partial t}+\frac{1}{2}\frac{\partial(\delta^2 S)_{eq}}{\partial t} . \tag{2.106}$$

Similarly, the entropy flux can be expressed with the aid of a first-order term W_{eq}, corresponding to the equilibrium entropy flow, and a second-order term ΔW, that is,

$$W=W_{eq}+\Delta W . \tag{2.107}$$

For example, for heat conduction in solids

$$\frac{1}{T}=\frac{1}{T_{eq}}+\Delta\left(\frac{1}{T}\right), \tag{2.108}$$

$$W_{eq}=\int_A \frac{\boldsymbol{J}_q}{T_{eq}}\cdot d\boldsymbol{A}, \tag{2.109}$$

$$\Delta W=\int_A \boldsymbol{J}_q \Delta\left(\frac{1}{T}\right)\cdot d\boldsymbol{A} . \tag{2.110}$$

In accordance with equations (2.103), (2.106), and (2.107), the relation

$$\frac{\partial(\delta S)_{eq}}{\partial t} = -W_{eq} \tag{2.111}$$

between first-order terms, and the relation

$$\frac{1}{2}\frac{\partial(\delta^2 S)_{eq}}{\partial t} = P - \Delta W \tag{2.112}$$

between second-order terms must be satisfied. The latter equation is valid only in the absence of velocity fluctuations, that is only for solids.

Integration of equation (2.111) with respect to time, beginning from an equilibrium state, yields

$$(\delta S)_{eq} = -\int_0^t W_{eq}\,dt\,. \tag{2.113}$$

For an isolated system the equilibrium entropy flow is equal to zero ($W_{eq}=0$), and hence the classical condition for thermodynamic equilibrium is obtained

$$(\delta S)_{eq} = 0\,. \tag{2.114}$$

On the other hand, small equilibrium changes in the entropy of a diathermal system can be compensated by the entropy flow flowing across the system boundaries. A nonequilibrium process occurs if there is no such compensation.

If the boundary conditions at the system boundaries are assumed to be constant, the second-order term of the entropy flow becomes equal to zero

$$\Delta W = 0\,. \tag{2.115}$$

The condition above holds if changes in both the thermodynamic forces and in the fluxes at the system boundaries vanish. Equation (2.112) is then restricted to the relation

$$\frac{1}{2}\frac{\partial(\delta^2 S)_{eq}}{\partial t} = P \geqslant 0\,. \tag{2.116}$$

The foregoing equation is the thermodynamic criterion of evolution for states close to equilibrium states.

A spontaneous return of a system from a final state f to an equilibrium state eq entails a positive entropy production

$$\int_f^{eq} P\,dt > 0\,, \tag{2.117}$$

whereas the stability condition for thermodynamic equilibrium of a system is

$$\int_{eq}^{f} P\,dt < 0\,. \tag{2.118}$$

A thermodynamic system is stable if every change in state beginning from an unperturbed state is incompatible with the second law of thermodynamics.

Integration of equation (2.116) with respect to time, account being taken of the stability condition (2.118), gives

$$\tfrac{1}{2}(\delta^2 S)_{eq} = \int_0^t P\,dt = \int_0^t dS_i < 0 . \tag{2.119}$$

As follows from formula (2.119), the stability of a system depends solely on the sign of the quantity $(\delta^2 S)_{eq}$ calculated in the equilibrium state. It is not necessary, however, to know the value of the entropy production due to deviations from equilibrium (caused by fluctuations, for instance). The quantity $(\delta^2 S)_{eq}$ characterizes the curvature of the plane describing the entropy close to the equilibrium state.

The stability condition (2.119) for the system can also be given in a local form per unit volume

$$[\delta^2(\rho s)]_{eq} < 0 \tag{2.120}$$

or per unit mass of substance

$$(\delta^2 s)_{eq} < 0 . \tag{2.121}$$

In considering nonequilibrium processes at an arbitrary distance from an equilibrium state, one cannot decompose the balance equation into two separate parts containing first-order or second-order terms since the entropy production is not then a quantity of the second order. In this case inequality (2.116) ceases to be true. Instead, on assuming the postulate of local equilibrium to be valid, one can write the condition for local stability as

$$\delta^2(\rho s) < 0 \tag{2.122}$$

or

$$\delta^2 s < 0 . \tag{2.123}$$

This condition can be presented in integral form for the system as a whole since, by the second law of thermodynamics,

$$\frac{\partial(\delta^2 S)}{\partial t} \geqslant 0 , \tag{2.124}$$

whereas the stability condition gives

$$\delta^2 S < 0 . \tag{2.125}$$

As emerges from the considerations above, phenomenological equations relating generalized flows and thermodynamic forces need not be used in order to determine stability conditions. These conditions may, however, be used when analysing phenomena described by such laws.

The explicit form of the stability conditions for nonequilibrium states is obtained by considering the balance equation for excess entropy production rate. The stability of heat conduction in a solid when the temperature is the sole independent variable will be considered by way of example. In this case,

$$\frac{1}{2}\frac{\partial(\delta^2 s)}{\partial t}=\delta\left(\frac{1}{T}\right)\frac{\partial(\delta u)}{\partial t} \tag{2.126}$$

whereas, by the energy-balance equation

$$\rho\frac{\partial(\delta u)}{\partial t}=-\nabla\cdot(\delta \boldsymbol{J}_q). \tag{2.127}$$

When the equations above are combined, the result is a straightforward example of a balance equation for excess entropy production rate

$$\frac{1}{2}\rho\frac{\partial(\delta^2 s)}{\partial t}=-\delta\left(\frac{1}{T}\right)\nabla\cdot(\delta \boldsymbol{J}_q)=\delta \boldsymbol{J}_q\cdot\delta\left[\nabla\left(\frac{1}{T}\right)\right] -\nabla\cdot\left[\delta \boldsymbol{J}_q\,\delta\left(\frac{1}{T}\right)\right]. \tag{2.128}$$

If boundary conditions of the first kind in the form of temperature distributions

$$\left[\delta\left(\frac{1}{T}\right)\right]_A=0 \tag{2.129}$$

or boundary conditions of the second kind in the form of heat flux distributions

$$(\delta \boldsymbol{J}_q)_A=0 \tag{2.130}$$

do not change with time on the surface of the solid, the stability condition (2.124) of the system becomes

$$\frac{1}{2}\int_V \rho\frac{\mathrm{d}(\delta^2 s)}{\partial t}\,\mathrm{d}V=\int_V \delta \boldsymbol{J}_q\cdot\delta\left[\nabla\left(\frac{1}{T}\right)\right]\mathrm{d}V=P(\delta S)>0, \tag{2.131}$$

where $P(\delta S)$ is the excess entropy production rate which should be distinguished from the variation δP of the entropy production rate occurring, for instance, in equation (2.90).

The positive sign of the excess entropy production rate is not determined by the second law of thermodynamics. If linear phenomenological equations with constant coefficients (l_{qq}=const) are valid, a quadratic form is obtained

$$\frac{1}{2}\int_V \frac{\partial(\delta^2 s)}{\partial t}\,\mathrm{d}V=\int_V l_{qq}\left[\delta\nabla\left(\frac{1}{T}\right)\right]^2\mathrm{d}V>0. \tag{2.132}$$

On the other hand, if they are not constant, the phenomenological coefficients l_{qq} can be replaced by constant phenomenological coefficients l_{qq}^0 by modifying the thermodynamic forces by means of the function

$$\varepsilon^2=\frac{l_{qq}}{l_{qq}^0} \tag{2.133}$$

and the result is

$$J_q = l_{qq} \nabla \left(\frac{1}{T}\right) = l_{qq}^0 \varepsilon^2 \nabla \left(\frac{1}{T}\right). \tag{2.134}$$

The balance equation (2.128) for the excess entropy production rate in this case takes the form

$$\frac{1}{2} \rho \varepsilon^2 \frac{\partial(\delta^2 s)}{\partial t} = \delta J_q \cdot \nabla \left[\varepsilon^2 \delta \left(\frac{1}{T}\right)\right] - \nabla \cdot \left[\varepsilon^2 \delta J_q \delta \left(\frac{1}{T}\right)\right]. \tag{2.135}$$

When the constant boundary conditions (2.129) or (2.130) are taken into account, we obtain a stability condition

$$\frac{1}{2} \int_V \rho \varepsilon^2 \frac{\partial(\delta^2 s)}{\partial t} \, \mathrm{d}V = \int_V \delta J_q \cdot \nabla \left[\varepsilon^2 \delta \left(\frac{1}{T}\right)\right] \mathrm{d}V > 0. \tag{2.136}$$

The heat flux is often specified by means of Fourier's law

$$J_q = -\lambda \nabla T, \tag{2.137}$$

where the heat conductivity λ is related to the phenomenological coefficient by

$$l_{qq} = \lambda T^2. \tag{2.138}$$

If the heat conductivity is constant, $\lambda = \lambda^0$, then

$$\varepsilon^2 = T^2, \tag{2.139}$$

and the stability condition (2.136) takes the form

$$\frac{1}{2} \int_V \rho T^2 \frac{\partial(\delta^2 s)}{\partial t} \, \mathrm{d}V = \lambda^0 \int_V (\delta \nabla T)^2 \mathrm{d}V > 0. \tag{2.140}$$

If the heat conductivity is an arbitrary function of the temperature, $\lambda(T)$, and the body is isotropic, use is made of the transformation

$$\theta = \int_{T_1}^{T} \lambda(T) \, \mathrm{d}T \qquad (T_1 = \text{const}) \tag{2.141}$$

and the function

$$\varepsilon^2 = \lambda T^2. \tag{2.142}$$

Since

$$\nabla \left[\varepsilon^2 \delta \left(\frac{1}{T}\right)\right] = -\nabla(\delta\theta) = -\delta(\nabla\theta) = \delta J_q, \tag{2.143}$$

a stability condition is obtained in the form

$$\frac{1}{2} \int_V \rho \lambda T^2 \frac{\partial(\delta^2 s)}{\partial t} \, \mathrm{d}V = \int_V (\delta J_q)^2 \mathrm{d}V > 0. \tag{2.144}$$

Boundary conditions of the third kind for heat conduction are determined by giving the heat transfer coefficient h and fluid temperature T_f surrounding the surface of a body at a temperature of T_B. These quantities are related to the heat flux on the isothermal surface B of a solid by Newton's law for heat transfer

$$J_{qB}=h(T_B-T_f). \tag{2.145}$$

Boundary conditions of the first kind (2.129) are given for the other parts of the surface. In the simplest case the boundary conditions of the third kind are constant, and hence

$$\delta T_f=0, \quad \delta h=0, \quad \delta J_{qB}=h\,\delta T_B. \tag{2.146}$$

In this case relation (2.135) gives the stability condition

$$\frac{1}{2}\int_V \rho\varepsilon^2 \frac{\partial(\delta^2 s)}{\partial t}\,dV=\int_V \delta \boldsymbol{J}_q\cdot\nabla\left[\varepsilon^2\delta\left(\frac{1}{T}\right)\right]dV+h\int_B\left(\frac{\varepsilon\delta T}{T}\right)^2 dB>0. \tag{2.147}$$

The first term of this expression coincides with the stability condition (2.136) for constant boundary conditions of the first kind on the surface of the body, the second term specifies the stability condition on the surface of the body and can be made a separate inequality in determining necessary and sufficient stability conditions:

$$h\int_B\left(\frac{\varepsilon\delta T}{T}\right)^2 dB>0 \tag{2.148}$$

which means

$$h>0, \tag{2.149}$$

that is, the coefficient of heat transfer must be positive.

If there are variations of fluid temperature, then

$$\delta J_{qB}=h(\delta T_B-\delta T_f) \tag{2.150}$$

and the surface stability condition is of integral form

$$h\int_B\left(\frac{\varepsilon}{T}\right)^2(\delta T_B-\delta T_f)\,\delta T_B\,dB>0 \tag{2.151}$$

or, when relation (2.149) is taken into account, is of local form

$$(\delta T_B-\delta T_f)\,\delta T_B>0, \tag{2.152}$$

that is

$$|\delta T_f|<|\delta T_B| \quad \text{for} \quad \delta T_B\,\delta T_f>0. \tag{2.153}$$

The system is stable if the absolute values of the temperature variations of the fluid are smaller than those of the temperature variations of the body surface.

CHAPTER III

PROCESSES IN FLUID MULTICOMPONENT CONTINUOUS MEDIA

1. The General Form of Balance Equations for Extensive Quantities

Thermodynamic phenomena occurring in fluid media are described by means of balance equations for extensive quantities, set up at first for an elementary volume of fluid and then per unit volume. Intensive quantities specifying the local state of the substance appear in these equations. Accordingly, the balance equations for extensive quantities and all subsequent relationships and conclusions pertaining to classical nonequilibrium thermodynamics are physically meaningful only if the local state parameters of the substance can be determined. A fluid can have its local state parameters determined only if it can be treated as a continuous medium. Intensive parameters described by the macroscopic properties of the medium are concepts pertaining to ensembles of quite considerable numbers of molecules and hence cease to be physically meaningful if the properties of the medium change considerably over distances of the order of the molecular mean free path. Under these conditions, the molecular velocity distribution is not a Maxwell distribution and the postulate of local thermodynamic equilibrium cannot be invoked. In the particular case, the macroscopic phenomenological method cannot be employed to describe phenomena occurring in gases so dilute that the dimensions of the vessel containing the gas or of the body around which the gas flows are of the order of the mean free path. Nor can the phenomenological method be used to describe phenomena inside a shock wave in which, on a length equal to several mean free paths, the macroscopic intensive parameters change in jumps.

At low flow velocities the paths of the fluid particles are gentle and do not intersect. As the velocity increases the flow changes from laminar to turbulent, eddies are set up in the fluid, paths of particles intersect, and the local parameters of the fluid experience fluctuation. These are phenomena occurring in regions which are large in comparison with the molecular mean free path and consequently are macroscopic in character. A fluid in turbulent motion may also be treated as a continuous medium. Since they fluctuate strongly during turbulent flow, local parameters are averaged over an interval of time much longer than the period of the fluctuations. The balance equations for extensive quantities are valid both for the instantaneous values of the local state parameters and for their mean values. As a rule, mean values are used in practice. Fluctuations occurring in turbulent flow deviate relatively little from the local equilibrium state and thus in general the postulate of local thermodynamic equilibrium and the linear phenomenological equations resulting from it can be applied to time-averaged turbulent flow.

Fluid thermodynamics is based directly on the mechanics of continuous media [14]. Determining the thermodynamic state of a fluid requires a mathematical description of the fluid motion. The properties of a fluid can be considered at points which are at rest relative to a reference frame or at points which are moving along with the fluid.

Every intensive parameter in a fluid which is not in equilibrium varies in time and in space. If an intensive quantity $\varphi(t, x, y, z)$ is a function of time and space coordinates, its total differential can be expressed as

$$\mathrm{d}\varphi = \frac{\partial \varphi}{\partial t}\,\mathrm{d}t + \frac{\partial \varphi}{\partial x}\,\mathrm{d}x + \frac{\partial \varphi}{\partial y}\,\mathrm{d}y + \frac{\partial \varphi}{\partial z}\,\mathrm{d}z\,.$$

Dividing this total differential by the time differential yields the total time derivative

$$\frac{\mathrm{d}\varphi}{\mathrm{d}t} = \frac{\partial \varphi}{\partial t} + \frac{\partial \varphi}{\partial x}\frac{\mathrm{d}x}{\mathrm{d}t} + \frac{\partial \varphi}{\partial y}\frac{\mathrm{d}y}{\mathrm{d}t} + \frac{\partial \varphi}{\partial z}\frac{\mathrm{d}z}{\mathrm{d}t}\,.$$

If the point at which the change in the intensive quantity φ occurs, or the observer, moves with a velocity $\boldsymbol{\omega}$ different from the velocity of the fluid, the derivative defined by the equation above can be rewritten as

$$\frac{\mathrm{d}_\omega \varphi}{\mathrm{d}t} = \frac{\partial \varphi}{\partial t} + \boldsymbol{\omega} \cdot \nabla \varphi\,. \tag{3.1}$$

The derivative so defined is applicable, for instance, in considering the propagation of a flame front or detonation wave.

Lagrange's method consists in the time variation of an intensive quantity φ being considered at a point moving along with the substance at its centre-of-mass velocity, that is, being considered by an observer moving along with the fluid. In this case the velocity components $\mathrm{d}x/\mathrm{d}t$, $\mathrm{d}y/\mathrm{d}t$, $\mathrm{d}z/\mathrm{d}t$ of the point coincide with the components v_x, v_y, v_z of the centre-of-mass velocity of the fluid and the equation under consideration takes on a form called a *substantial time derivative*:

$$\frac{\mathrm{d}\varphi}{\mathrm{d}t} = \frac{\partial \varphi}{\partial t} + v_x \frac{\partial \varphi}{\partial x} + v_y \frac{\partial \varphi}{\partial y} + v_z \frac{\partial \varphi}{\partial z} = \frac{\partial \varphi}{\partial t} + v_\alpha \frac{\partial \varphi}{\partial x_\alpha} = \varphi_{,t} + v_\alpha \varphi_{,\alpha}$$

or

$$\frac{\mathrm{d}\varphi}{\mathrm{d}t} = \frac{\partial \varphi}{\partial t} + \boldsymbol{v} \cdot \nabla \varphi\,. \tag{3.2}$$

The substantial derivative thus consists of the local time variation of the intensive quantity in question, as determined by an observer who is at rest relative to the reference frame (first term), and the convection change caused in the quantity by the fluid element under consideration being displaced in space (second term).

In Euler's method (1752), time variations of an intensive quantity are considered at a point at rest, that is, by an observer who is at rest relative to an external coordinate system. In this case

$$\frac{dx}{dt}=\frac{dy}{dt}=\frac{dz}{dt}=0, \qquad \frac{d\varphi}{dt}=\frac{\partial\varphi}{\partial t}$$

and hence, as stated above, the partial time derivative $\partial\varphi/\partial t$ determines the local change in the quantity φ relative to the observer at rest. If this derivative is equal to zero, the field is said to be stationary.

These three alternative ways of considering the properties of a fluid are illustrated by the example of a fisherman observing the 'concentration' of fish in a river. In the first case the fisherman rows his boat on the river; in the second, he sits in his boat as it is carried along by the current; and in the third, he stands on the bank.

The balance equations for the amount of substance, momentum, energy, and entropy are particularly important for nonequilibrium thermodynamics. These extensive quantities, denoted in general by Z, are balanced at first for a volume V enclosed by a control surface A which is at rest and are then taken per unit volume of fluid at rest or moving relative to a coordinate system. Thus, the balance equations for the quantities Z, taken on a per unit volume basis, can be written in terms of the partial time derivative from the point of view of a motionless observer, and in terms of the substantial derivative from the point of view of an observer moving together with the fluid.

An extensive quantity Z for a substance filling a volume V can be calculated as

$$Z=\int_V Z_v \,dV=\int_V \rho z \,dV. \tag{3.3}$$

Since the balancing region is tied firmly to the coordinate system, the partial time derivative of the extensive quantity pertaining to the entire balancing region is equal to the total differential

$$\frac{\partial Z}{\partial t}=\frac{dZ}{dt}=\frac{d}{dt}\int_V \rho z \,dV=\int_V \frac{\partial(\rho z)}{\partial t}\,dV. \tag{3.4}$$

A partial derivative has appeared under the integral sign since ρz is determined per unit volume at rest relative to the coordinate system.

The time variation dZ/dt of the quantity under consideration may be due to:

- addition of Z to the system across the boundary along with substance in the form of a convection flow of Z,
- addition of Z to the system across the boundary without a flow of substance in the form of a conduction flow of Z,
- production of Z inside the region under consideration.

The amount of substance flowing through an elementary surface of area dA (Fig. 3.1) per unit time is $\rho \boldsymbol{v}\cdot d\boldsymbol{A}$ (where $d\boldsymbol{A}$ is a vector whose magnitude is the area dA and whose direction is taken as the outward-drawn normal to the surface); along with the substance there is a convection flux $\rho z \boldsymbol{v}$ of the extensive quantity Z. The amount of extensive quantity Z transported by convection across a surface of area A per unit time is $-\int_A (\rho z \boldsymbol{v})\cdot \mathrm{d}\boldsymbol{A}$.

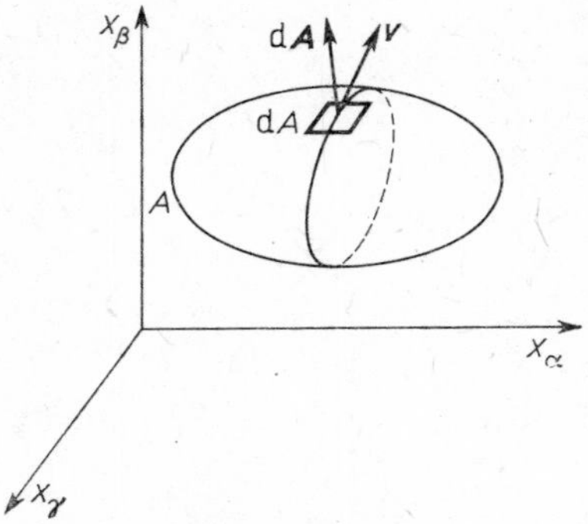

Fig 3.1 Flow of substance across the boundary of a system

The conduction flux $\boldsymbol{J}_z$ of the extensive quantity Z is a vector with the same direction as the flow of this quantity and magnitude equal to the numerical value of the extensive quantity under consideration transported without flow of substance across a unit area of surface perpendicular to the vector in a unit time. The amount of extensive quantity Z transported across a surface of area A per unit time by means of a conduction flux without a flow of substance is $-\int_A \boldsymbol{J}_z\cdot \mathrm{d}\boldsymbol{A}$.

The rate of production Φ_z of an extensive quantity Z at a given point is the ratio of the amount of Z produced inside the elementary volume of substance surrounding that point to the value of that volume and time:

$$\Phi_z = \frac{\mathrm{d}Z}{\mathrm{d}V\mathrm{d}t}\,. \tag{3.5}$$

The amount of extensive quantity Z produced per unit time throughout a space of volume V is $\int_V \Phi_z \mathrm{d}V$.

For the entire space of volume V, at rest relative to the coordinate system, the Z balance equation per unit time is of the form

$$\frac{\mathrm{d}Z}{\mathrm{d}t} = \int_V \frac{\partial(\rho z)}{\partial t}\mathrm{d}V = -\int_A (\rho z \boldsymbol{v})\cdot \mathrm{d}\boldsymbol{A} - \int_A \boldsymbol{J}_z\cdot \mathrm{d}\boldsymbol{A} + \int_V \Phi_z \mathrm{d}V\,.$$

The Gauss–Ostrogradsky theorem can be used to convert integrals over a closed surface of area A into integrals extending over the entire space of volume V contained inside that surface:

$$\int_A (\rho z \boldsymbol{v})\cdot \mathrm{d}\boldsymbol{A} = \int_V [\nabla\cdot(\rho z \boldsymbol{v})]\,\mathrm{d}V\,,$$

$$\int_A \boldsymbol{J}_z\cdot \mathrm{d}\boldsymbol{A} = \int_V (\nabla\cdot \boldsymbol{J}_z)\,\mathrm{d}V\,,$$

and then

$$\frac{dZ}{dt} = \int_V \frac{\partial(\rho z)}{\partial t} dV = -\int_V [\nabla \cdot (\rho z \boldsymbol{v})] dV - \int_V (\nabla \cdot \boldsymbol{J}_z) dV + \int_V \Phi_z dV. \quad (3.6)$$

The balance equation above is valid for the entire arbitrarily chosen volume under consideration, and is thus valid for the elementary volume surrounding a point ($V \to 0$) as well as per unit volume:

$$\frac{\partial(\rho z)}{\partial t} = -\nabla \cdot (\rho z \boldsymbol{v}) - \nabla \cdot \boldsymbol{J}_z + \Phi_z. \quad (3.7)$$

This is the local balance equation set up by an observer at rest for the extensive quantity Z.

The foregoing equation simplifies in special cases. If the given extensive quantity Z obeys a conservation law (as do mass, momentum, and total energy), a change in Z within the given volume can occur only as a result of external interactions. Then the only terms remaining on the right-hand side of equation (3.7) are those constituting the divergences of the convection and conduction fluxes $\rho z \boldsymbol{v}$ and $\boldsymbol{J}_z$, of the extensive quantity Z. The source strength or, briefly, the source term Φ_z, is zero for quantities governed by a conservation law.

The local balance equation for quantities subject to a conservation law is called the *conservation equation* for that quantity and is of the form

$$\frac{\partial(\rho z)}{\partial t} = -\nabla \cdot (\rho z \boldsymbol{v}) - \nabla \cdot \boldsymbol{J}_z. \quad (3.8)$$

If the system is in a stationary state the extensive quantity Z does not vary with time,

$$\frac{dZ}{dt} = 0, \qquad \int_V \Phi_z dV = \int_V [\nabla \cdot (\boldsymbol{J}_z + \rho z \boldsymbol{v})] dV, \quad (3.9)$$

hence, the entire amount of Z produced inside the system must be removed across the system boundary. The divergence of the sum of the conduction and convection fluxes of the extensive quantity governed by a conservation law is equal to zero in the stationary state:

$$\nabla \cdot (\boldsymbol{J}_z + \rho z \boldsymbol{v}) = 0. \quad (3.10)$$

The local balance equations for an observer moving along with the fluid are written in a form requiring use of the substantial derivative. For the amounts of substance of the components equation (3.7) takes the form

$$\frac{\partial \rho}{\partial t} = -\nabla \cdot (\rho \boldsymbol{v}), \quad (3.11)$$

since in this case $z=1$, $\boldsymbol{J}_z=0$ and $\Phi_z=0$.

At this point we combine the defining equation of the substantial derivative (3.2), multiplied by ρ, with the foregoing form of the conservation law for the amount of substance, multiplied by φ, and the familiar mathematical relation

$$\nabla \cdot (\varphi \boldsymbol{v}) = \varphi (\nabla \cdot \boldsymbol{v}) + \boldsymbol{v} \cdot \nabla \varphi \tag{3.12}$$

into account. The result is

$$\rho \frac{\mathrm{d}\varphi}{\mathrm{d}t} = \frac{\partial (\rho \varphi)}{\partial t} + \nabla \cdot (\rho \varphi \boldsymbol{v}) . \tag{3.13}$$

This equation is valid for every local quantity φ which may be a scalar, a component of a vector, or a component of a tensor. In the special case, this equation may be applied to the specific quantity z and combined with equation (3.7); then the balance equation for this quantity for an observer who is moving along with the fluid at its local centre-of-mass velocity $\boldsymbol{v}$ is of the form

$$\rho \frac{\mathrm{d}z}{\mathrm{d}t} = -\nabla \cdot \boldsymbol{J}_z + \Phi_z . \tag{3.14}$$

On the right-hand side of the equation, the divergence of the convection flux of the extensive quantity Z, that is $-\nabla \cdot (\rho z \boldsymbol{v})$, vanishes since the coordinate system is moving along with the fluid. On the left-hand side, $\rho \mathrm{d}z/\mathrm{d}t$—the product of the mass density and the substantial derivative of z, appears instead of $\partial(\rho z)/\partial t$—the partial time derivative of the density of Z.

In the case of a quantity subject to a conservation law, seeing that $\Phi_z = 0$ we get a conservation law of the form

$$\rho \frac{\mathrm{d}z}{\mathrm{d}t} = -\nabla \cdot \boldsymbol{J}_z . \tag{3.15}$$

2. Diffusion Flux

A multicomponent fluid as a whole moves with a macroscopic velocity which is different from the velocities of the individual components. These differences are due to the microscopic motions of the individual molecules and the diffusion resulting from these motions.

The substance flux of component i is a vector whose direction is parallel to the velocity of the component and whose modulus is equal to the amount of substance of the flowing component per unit time and unit surface area perpendicular to the velocity vector of the component. The substance flux can also be calculated as the product of the component density ρ_i and the component velocity $\boldsymbol{v}_i$,

$$\boldsymbol{J}_i = \rho_i \boldsymbol{v}_i . \tag{3.16}$$

In addition to the flow of a component relative to a motionless coordinate system, one can consider the diffusion of the component relative to a coordinate system moving at the reference velocity $\boldsymbol{v}_a$. Depending on the choice of reference velocity, various definitions can be given for the diffusion flux, which in general is written as

$$\boldsymbol{j}_{v_a i}=\rho_i(\boldsymbol{v}_i-\boldsymbol{v}_a). \tag{3.17}$$

The substance and diffusion fluxes, $\boldsymbol{J}_i$ and $\boldsymbol{j}_{v_a i}$, of component i are related by

$$\boldsymbol{J}_i=\boldsymbol{j}_{v_a i}+\rho_i \boldsymbol{v}_a. \tag{3.18}$$

In addition to mass fluxes of substance, molar fluxes also appear in calculations. The molar flux of component i is a vector whose direction is parallel to the velocity vector of the component and whose modulus is equal to the component flow in kilomoles, per unit time and unit surface area. It can also be calculated as the product of the concentration c_i and velocity $\boldsymbol{v}_i$ of the component,

$$\boldsymbol{J}_{Mi}=\frac{1}{M_i}\boldsymbol{J}_i=c_i\boldsymbol{v}_i. \tag{3.19}$$

The molar diffusion flux of component i

$$\boldsymbol{j}_{Mv_a i}=c_i(\boldsymbol{v}_i-\boldsymbol{v}_a) \tag{3.20}$$

is related to the molar density of the component by

$$\boldsymbol{J}_{Mi}=\boldsymbol{j}_{Mv_a i}+c_i\boldsymbol{v}_a. \tag{3.21}$$

Reference velocities defined in a number of various ways are used as the characteristic velocities $\boldsymbol{v}_a$.

The centre-of-mass velocity, also called the *barycentric velocity*, is defined by the relationship

$$\boldsymbol{v}=\frac{1}{\rho}\sum_{i=1}^{k}\rho_i\boldsymbol{v}_i=\sum_{i=1}^{k}x_i\boldsymbol{v}_i. \tag{3.22}$$

The mean molar velocity is

$$\boldsymbol{v}_M=\frac{1}{c}\sum_{i=1}^{k}c_i\boldsymbol{v}_i=\sum_{i=1}^{k}z_i\boldsymbol{v}_i. \tag{3.23}$$

The mean volume velocity is

$$\boldsymbol{v}_v=\frac{1}{V}\sum_{i=1}^{k}\tilde{V}_i\boldsymbol{v}_i=\sum_{i=1}^{k}c_i\tilde{V}_{Mi}\boldsymbol{v}_i \tag{3.24}$$

where $\tilde{V}_{Mi}$ is the partial molar volume of the component.

The diffusion flux relative to the centre-of-mass velocity $\boldsymbol{v}$ will be denoted simply as

$$\boldsymbol{j}_i=\rho_i(\boldsymbol{v}_i-\boldsymbol{v}). \tag{3.25}$$

The sum of such diffusion fluxes for all k components is zero:

$$\sum_{i=1}^{k}\boldsymbol{j}_i=\sum_{i=1}^{k}\rho_i(\boldsymbol{v}_i-\boldsymbol{v})=\rho\boldsymbol{v}-\boldsymbol{v}\sum_{i=1}^{k}\rho_i=0,$$

that is,

$$\sum_{i=1}^{k} j_i = 0 . \tag{3.26}$$

Of the k diffusion fluxes, only $k-1$ are independent.

For molar diffusion fluxes referred to the centre-of-mass velocity, relations (3.19) and (3.26) lead to

$$\sum_{i=1}^{k} M_i j_{Mi} = \sum_{i=1}^{k} j_i = 0 . \tag{3.27}$$

The sum of molar diffusion fluxes referred to the mean molar velocity of all components is zero:

$$\sum_{i=1}^{k} j_{MMi} = \sum_{i=1}^{k} c_i (v_i - v_M) = c v_M - v_M \sum_{i=1}^{k} c_i = 0 . \tag{3.28}$$

Finally, when the mean volume velocity is used the molar diffusion flux of component i is defined as

$$j_{Mvi} = c_i (v_i - v_v) , \tag{3.29}$$

and these fluxes for all k components are related by

$$\sum_{i=1}^{k} \tilde{V}_{Mi} j_{Miv} = \sum_{i=1}^{k} \tilde{V}_{Mi} c_i (v_i - v_v) = v_v - v_v \sum_{i=1}^{k} \tilde{V}_{Mi} c_i = 0 . \tag{3.30}$$

3. The Balance Equations for the Amount of Substance

This chapter considers a single-phase, multicomponent fluid consisting of k components which can enter into r chemical reactions with each other. A local thermodynamic state of such a moving fluid is specified by two intensive parameters, e.g. velocity of the fluid, and the chemical composition. The fractions of components of the solution (cf. Section I.7) are used to give the chemical composition of the fluid.

For a unique description of phenomena in such a multicomponent system, balance equations must be set up for the amount of substance, momentum, energy, and entropy.

The balance equation for the amount of substance for component i is obtained from the general form of the balance equation (3.7) by setting $z = x_i$ and $J_z = j_i$, and calculating the amount of component produced inside a unit volume per unit time as a result of chemical reactions, that is, by calculating the chemical reaction rate. The equation obtained is

$$\Phi_i = M_i \sum_{j=1}^{r} \nu_{ij} J_j , \tag{3.31}$$

whereupon, bearing in mind that $\rho x_i = \rho_i$, we arrive at

$$\frac{\partial \rho_i}{\partial t} = -\nabla \cdot (\rho_i \boldsymbol{v}) - \nabla \cdot \boldsymbol{j}_i + M_i \sum_{j=1}^{r} \nu_{ij} J_j . \tag{3.32}$$

Expression (3.31) contains the stoichiometric coefficients ν_{ij} for component i in reaction j which appear in the symbolic notation of the chemical reaction:

$$\sum_{i=1}^{k} \nu_{ij} A_{ij} = 0 \quad (j = 1, 2, 3, \ldots, r), \tag{3.33}$$

where A_{ij} is the chemical symbol of component i in reaction j, denoting one kilomole of its substance in the stoichiometric equation. The stoichiometric coefficient ν_{ij} of component i in reaction j is equal to the number of kilomoles of component per kilomoles of one of its substrates. The coefficient is taken to be positive for products, and negative for substrates.

If reaction j is the reaction of hydrogen combustion

$$H_2 + \tfrac{1}{2}O_2 \rightarrow H_2O ,$$

then

$$A_{1j} = H_2 , \quad A_{2j} = O_2 , \quad A_{3j} = H_2O$$

and

$$\nu_{1j} = -1 , \quad \nu_{2j} = -\tfrac{1}{2} , \quad \nu_{3j} = 1 .$$

In equation (3.32), J_j is the chemical reaction rate of reaction j in kilomoles per unit time and volume, and hence $M_i \nu_{ij} J_j$ is the amount of component i, in units of mass, produced in reaction j per unit time and volume.

The conservation equation for the amount of substance in every individual chemical reaction can be written as

$$\sum_{i=1}^{k} \nu_{ij} M_i = 0 \quad (j = 1, 2, 3, \ldots, r), \tag{3.34}$$

and thus also as

$$J_j \sum_{i=1}^{k} \nu_{ij} M_i = 0 \quad (j = 1, 2, 3, \ldots, r).$$

If the conservation equations (3.32) for all k components are summed and if relations (3.34) and (3.26) are taken into account, the result is equation (3.11), that is,

$$\frac{\partial \rho}{\partial t} = -\nabla \cdot (\rho \boldsymbol{v}) .$$

Going over from a motionless point of observation to one moving at the centre-of-mass velocity of the fluid requires application of relations (3.2) and (3.12) for $\varphi = \rho_i$, with equation (3.32) taken into account:

$$\frac{d\rho_i}{dt}=\frac{\partial\rho_i}{\partial t}+\boldsymbol{v}\cdot\nabla\rho_i=-\nabla\cdot(\rho_i\boldsymbol{v})-\nabla\cdot\boldsymbol{j}_i+M_i\sum_{j=1}^{r}\nu_{ij}J_j+\boldsymbol{v}\cdot\nabla\rho_i,$$

$$\frac{d\rho_i}{dt}=-\rho_i(\nabla\cdot\boldsymbol{v})-\nabla\cdot\boldsymbol{j}_i+M_i\sum_{j=1}^{r}\nu_{ij}J_j. \tag{3.35}$$

For an observer moving at the centre-of-mass velocity of the fluid the conservation equation for the amount of substance of all components is obtained from the substantial derivative of the density [equation (3.2) for $\varphi=\rho$], when equation (3.11) is taken into regard:

$$\frac{d\rho}{dt}=\frac{\partial\rho}{\partial t}+\boldsymbol{v}\cdot\nabla\rho=-\nabla\cdot(\rho\boldsymbol{v})+\boldsymbol{v}\cdot\nabla\rho,$$

that is,

$$\frac{d\rho}{dt}=-\rho(\nabla\cdot\boldsymbol{v}). \tag{3.36}$$

Equation (3.36) is also obtained by summing equations (3.35) for all k components, and employing relations (3.26) and (3.34).

If the specific volume $v=1/\rho$ replaces the density, then

$$\frac{d\rho}{dt}=\frac{d}{dt}\left(\frac{1}{v}\right)=-\frac{1}{v^2}\frac{dv}{dt}=-\frac{1}{v}(\nabla\cdot\boldsymbol{v})$$

and the conservation equation (3.36) for the amount of substance goes over into the equation

$$\frac{dv}{dt}=v(\nabla\cdot\boldsymbol{v}). \tag{3.37}$$

The balance equation for the amount of substance can also be written in terms of fractions. If mass fractions are used, equation (3.32) leads to the relation

$$\frac{\partial\rho_i}{\partial t}=\frac{\partial(\rho x_i)}{\partial t}=-\nabla\cdot(\rho x_i\boldsymbol{v})-\nabla\cdot\boldsymbol{j}_i+M_i\sum_{j=1}^{r}\nu_{ij}J_j,$$

and, upon application of transformation (3.13), also to

$$\rho\frac{dx_i}{dt}=-\nabla\cdot\boldsymbol{j}_i+M_i\sum_{j=1}^{r}\nu_{ij}J_j. \tag{3.38}$$

Finally, one should consider the case of stationary states of a system in which, by relations (3.4) and (3.9),

$$\frac{dm}{dt}=\int_V\frac{\partial\rho}{\partial t}dV=0,\quad \frac{dm_i}{dt}=\int_V\frac{\partial\rho_i}{\partial t}dV=0,$$

and equation (3.11) gives

$$\nabla \cdot (\rho \boldsymbol{v}) = 0 \tag{3.39}$$

or equation (3.32) yields

$$\nabla \cdot (\rho_i \boldsymbol{v} + \boldsymbol{j}_i) = M_i \sum_{j=1}^{r} \nu_{ij} J_j . \tag{3.40}$$

There is a further simplification if convection is absent, i.e. if

$$\nabla \cdot (\rho_i \boldsymbol{v}) = 0 , \tag{3.41}$$

and if no chemical reactions occur, i.e. if

$$\sum_{j=1}^{r} \nu_{ij} J_j = 0 . \tag{3.42}$$

In a stationary state without convection and without chemical reactions (such a state can be attained in a closed vessel) the divergence of the diffusion flux is zero

$$\nabla \cdot \boldsymbol{j}_i = 0 \quad (i = 1, 2, 3, \ldots, k) . \tag{3.43}$$

The diffusion fluxes in the various directions must balance each other out.

Example 3.1. Set up a balance equation for the amount of substance, using kilomoles as the units of the amount of substance.

Solution. If the amount of substance is expressed in kilomoles, the relevant equation of conservation for a motionless observer is obtained from equation (3.32) recast into the form

$$\frac{\partial \rho_i}{\partial t} = -\nabla \cdot (\rho_i \boldsymbol{v}_i) + M_i \sum_{j=1}^{r} \nu_{ij} J_j . \tag{1}$$

On dividing this equation by

$$M_i = \frac{\rho_i}{c_i} , \tag{2}$$

we obtain the concentration instead of the density:

$$\frac{\partial c_i}{\partial t} = -\nabla \cdot (c_i \boldsymbol{v}_i) + \sum_{j=1}^{r} \nu_{ij} J_j . \tag{3}$$

By equation (3.20), when $\boldsymbol{v}_a = \boldsymbol{v}_M$, this relation yields

$$\frac{\partial c_i}{\partial t} = -\nabla \cdot (c_i \boldsymbol{v}_M) - \nabla \cdot \boldsymbol{j}_{MMi} + \sum_{j=1}^{r} \nu_{ij} J_j . \tag{4}$$

Summing k such equations for all components, and taking account of relation (3.28) and

$$\sum_{i=1}^{k} c_i = c \tag{5}$$

gives the result

$$\frac{\partial c}{\partial t} = -\nabla \cdot (c \boldsymbol{v}_M) + \sum_{i=1}^{k} \sum_{j=1}^{r} \nu_{ij} J_j . \tag{6}$$

Use of the molar fraction z_i leads to

$$c\frac{\partial z_i}{\partial t}=c\frac{\partial}{\partial t}\left(\frac{c_i}{c}\right)=\frac{\partial c_i}{\partial t}-\frac{c_i}{c}\frac{\partial c}{\partial t}=\frac{\partial c_i}{\partial t}-z_i\frac{\partial c}{\partial t} \tag{7}$$

and, on allowing for relations (4), (6), and (3.12), we obtain

$$\begin{aligned} c\frac{\partial z_i}{\partial t} &= -\nabla\cdot(cz_i\boldsymbol{v}_\mathrm{M})-\nabla\cdot\boldsymbol{j}_{\mathrm{MM}i}+z_i\nabla\cdot(c\boldsymbol{v}_\mathrm{M})+\sum_{j=1}^{r}(\nu_{ij}-z_i\sum_{i=1}^{k}\nu_{ij})J_j \\ &= -c\boldsymbol{v}_\mathrm{M}\cdot\nabla z_i-\nabla\cdot\boldsymbol{j}_{\mathrm{MM}i}+\sum_{j=1}^{r}(\nu_{ij}-z_i\sum_{i=1}^{k}\nu_{ij})J_j\,. \end{aligned} \tag{8}$$

For an observer moving at a mean molar velocity $\boldsymbol{v}_\mathrm{M}$, the equations above can be written as

$$\frac{\mathrm{d}_\mathrm{M}\varphi}{\mathrm{d}t}=\frac{\partial\varphi}{\partial t}+\boldsymbol{v}_\mathrm{M}\cdot\nabla\varphi\,, \tag{9}$$

by using the substantial derivative defined in terms of the mean molar velocity, and then equation (6) yields

$$\frac{\mathrm{d}_\mathrm{M}c}{\mathrm{d}t}=\frac{\partial c}{\partial t}+\boldsymbol{v}_\mathrm{M}\cdot\nabla c=-c(\nabla\cdot\boldsymbol{v}_\mathrm{M})+\sum_{i=1}^{k}\sum_{j=1}^{r}\nu_{ij}J_j\,. \tag{10}$$

When the defining equation (9) of the substantial derivative, multiplied by the concentration c is combined with the balance equation (6) for the amount of substance, multipied by φ, the result is an equation analogous to equation (3.13):

$$c\frac{\mathrm{d}_\mathrm{M}\varphi}{\mathrm{d}t}=\frac{\partial(c\varphi)}{\partial t}+\nabla\cdot(\varphi c\boldsymbol{v}_\mathrm{M})-\varphi\sum_{i=1}^{k}\sum_{j=1}^{r}\nu_{ij}J_j\,. \tag{11}$$

Insertion of relation (4) into equation (9) when $\varphi=c_i$ produces an equation similar to equation (3.35):

$$\frac{\mathrm{d}_\mathrm{M}c_i}{\mathrm{d}t}=-c_i(\nabla\cdot\boldsymbol{v}_\mathrm{M})-\nabla\cdot\boldsymbol{j}_{\mathrm{MM}i}+\sum_{j=1}^{r}\nu_{ij}J_j\,. \tag{12}$$

On the other hand, rewriting equation (4) as

$$\frac{\partial(cz_i)}{\partial t}=-\nabla\cdot(cz_i\boldsymbol{v}_\mathrm{M})-\nabla\cdot\boldsymbol{j}_{\mathrm{MM}i}+\sum_{j=1}^{r}\nu_{ij}J_j\,,$$

and applying transformation (11), we obtain the balance equation for the amount of substance in the form

$$c\frac{\mathrm{d}_\mathrm{M}z_i}{\mathrm{d}t}=-\nabla j_{\mathrm{MM}_i}+\sum_{j=1}^{r}(\nu_{ij}-z_i\sum_{i=1}^{k}\nu_{ij})J_j\,. \tag{13}$$

4. The Momentum Balance Equation

The momentum of fluid contained in a space of volume V is $\int_V \rho\boldsymbol{v}\,\mathrm{d}V$. The rate of change of the momentum of substance contained in a space with motionless boundaries is

$$\frac{\mathrm{d}}{\mathrm{d}t}\int_V\rho\boldsymbol{v}\,\mathrm{d}V=\int_V\frac{\partial}{\partial t}(\rho\boldsymbol{v})\,\mathrm{d}V\,.$$

The momentum flux of substance flowing across the boundary surface of the given space of fluid is equal to the product of the mass flux $\rho \boldsymbol{v}$ and the centre-of-mass velocity $\boldsymbol{v}$, and hence amounts to $\rho \boldsymbol{v}\boldsymbol{v}$. A momentum flux $\int_A \rho \boldsymbol{v}\boldsymbol{v} \cdot \mathrm{d}\boldsymbol{A}$ flows across the entire surface of area A surrounding the space under consideration.

By Newton's second law of motion, the change in the momentum of a body is equal to the resultant of all forces acting on that body. The forces acting on the volume of fluid in question can be divided into mass and surface forces.

If $\boldsymbol{F}_i$ is the force exerted per unit mass of component i, then a force $\rho_i \boldsymbol{F}_i$ is exerted on every component in a unit volume, so that a force

$$\rho \boldsymbol{F} = \sum_{i=1}^{k} \rho_i \boldsymbol{F}_i \tag{3.44}$$

acts on the whole fluid contained in a unit volume, and a force $\int_V \rho \boldsymbol{F} \, \mathrm{d}V$ acts on the fluid contained in the entire volume V. Therefore, a force

$$\boldsymbol{F} = \frac{1}{\rho} \sum_{i=1}^{k} \rho_i \boldsymbol{F}_i = \sum_{i=1}^{k} x_i \boldsymbol{F}_i \tag{3.45}$$

is exerted per unit mass of fluid.

In the special case, the following may be mass forces:
gravitational force

$$\boldsymbol{F} = \boldsymbol{g} \,, \tag{3.46}$$

where $\boldsymbol{g}$ denotes the acceleration due to gravity;
the force due to rotational motion of a system [cf. equation (2.36)]

$$\boldsymbol{F} = \sum_{i=1}^{k} x_i [\omega^2 \boldsymbol{r} + 2(\boldsymbol{v}_i \times \boldsymbol{\omega})] \,; \tag{3.47}$$

the Lorentz force (2.35)

$$\boldsymbol{F} = \sum_{i=1}^{k} x_i e_i \left[\boldsymbol{E} + \frac{1}{c} (\boldsymbol{v}_i \times \boldsymbol{B}) \right] . \tag{3.48}$$

The stress tensor $\boldsymbol{\sigma}$ gives rise to a surface force. The sign of the stress tensor, and hence the direction of its components, is opposite to that of the pressure tensor which is also often used in nonequilibrium thermodynamics [22, 25].

The labels used for the components of the stress tensor are shown in Fig. 3.2 in which the stresses acting on one surface of an element of fluid are indicated. The first index stands for the plane in which the stress acts, by giving the axis normal to that plane. The second index gives the stress direction parallel to the coordinate axis bearing the same index. Stresses with both indices the same, $\sigma_{\alpha\alpha}$, $\sigma_{\beta\beta}$, $\sigma_{\gamma\gamma}$, are normal stresses which cause a change in the linear dimensions. Stresses with two different indices, e.g. $\sigma_{\alpha\beta}$, etc. are tangential stresses which deform the element of fluid and cause a change in the angles between its surfaces.

Tangential stresses setting up a torque about the x_γ-axis are shown in Fig. 3.3. If the element of fluid is not to rotate, there must be equality of torques about the axis x_γ, i.e.

$$\sigma_{\alpha\beta}\,\mathrm{d}x_\beta\,\mathrm{d}x_\gamma\,\frac{\mathrm{d}x_\alpha}{2}=\sigma_{\beta\alpha}\,\mathrm{d}x_\alpha\,\mathrm{d}x_\gamma\,\frac{\mathrm{d}x_\beta}{2}\,.$$

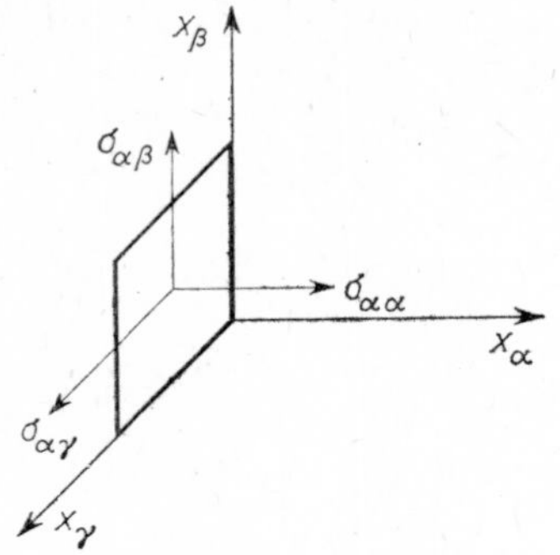

Fig. 3.2 Labelling of stress tensor components in rectangular coordinate system

In view of this

$$\sigma_{\alpha\beta}=\sigma_{\beta\alpha}\,,\qquad \sigma_{\beta\gamma}=\sigma_{\gamma\beta}\,,\qquad \sigma_{\alpha\gamma}=\sigma_{\gamma\alpha}\,, \tag{3.49}$$

and for an isotropic medium the stress tensor can be written as a symmetric matrix:

$$\boldsymbol{\sigma}=\begin{bmatrix}\sigma_{\alpha\alpha} & \sigma_{\alpha\beta} & \sigma_{\alpha\gamma}\\ \sigma_{\alpha\beta} & \sigma_{\beta\beta} & \sigma_{\beta\gamma}\\ \sigma_{\alpha\gamma} & \sigma_{\beta\gamma} & \sigma_{\gamma\gamma}\end{bmatrix}. \tag{3.50}$$

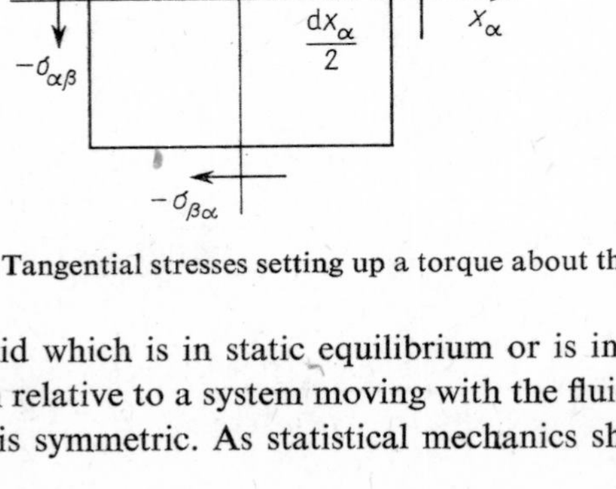

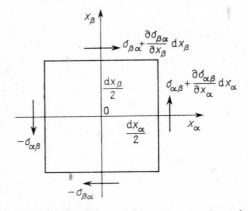

Fig. 3.3 Tangential stresses setting up a torque about the x_γ-axis

For an element of fluid which is in static equilibrium or is in uniform motion, i.e. is in static equilibrium relative to a system moving with the fluid, the torque vanishes and the stress tensor is symmetric. As statistical mechanics shows, this assumption also holds for fluids consisting of spherical molecules and having a not very high density. Except when relaxation phenomena in viscous fluids are being considered, this assumption is as a rule employed in the classical thermodynamics of non-equilibrium processes.

In a motionless fluid the tangential stresses are zero and the normal stresses are identical, regardless of the position of the surface:

$$\sigma_{\alpha\alpha}=\sigma_{\beta\beta}=\sigma_{\gamma\gamma}=-p, \tag{3.51}$$

$$-\boldsymbol{\sigma}=\begin{bmatrix} p & 0 & 0 \\ 0 & p & 0 \\ 0 & 0 & p \end{bmatrix}, \tag{3.52}$$

where p is the hydrostatic pressure of the motionless fluid.

If the concept of hydrostatic pressure is used, the stress tensor of an incompressible fluid or compressible fluid in local thermodynamic equilibrium can be decomposed into two parts

$$\boldsymbol{\sigma}=-p\boldsymbol{\delta}+\boldsymbol{\tau}, \tag{3.53}$$

where $\boldsymbol{\delta}$ is a unit tensor.

The viscosity part of the stress tensor for an isotropic fluid is of the form

$$\boldsymbol{\tau}=\begin{bmatrix} \sigma_{\alpha\alpha}+p & \sigma_{\alpha\beta} & \sigma_{\alpha\gamma} \\ \sigma_{\alpha\beta} & \sigma_{\beta\beta}+p & \sigma_{\beta\gamma} \\ \sigma_{\alpha\gamma} & \sigma_{\beta\gamma} & \sigma_{\gamma\gamma}+p \end{bmatrix}. \tag{3.54}$$

With the assumption of local thermodynamic equilibrium the pressure in equation (3.53) coincides with the thermodynamic pressure defined as

$$p=-\left(\frac{\partial u}{\partial v}\right)_s. \tag{3.55}$$

No viscosity stresses occur ($\boldsymbol{\tau}=0$) in a fluid which is in mechanical equilibrium. A scalar part

$$\tau=\tfrac{1}{3}\boldsymbol{\tau}:\boldsymbol{\delta}=\tfrac{1}{3}\tau_{\alpha\alpha} \tag{3.56}$$

can be separated from the viscosity stress tensor $\boldsymbol{\tau}$ so that the remaining part have a zero trace, that is, so that it be a deviator, and hence

$$\boldsymbol{\tau}=\tau\boldsymbol{\delta}+\overset{0}{\boldsymbol{\tau}}. \tag{3.57}$$

For a fluid which is not in a state of mechanical equilibrium use is made of the concept of hydrodynamic pressure equal to minus the arithmetic mean of the diagonal components of the stress tensor taken, that is,

$$P=-\tfrac{1}{3}\boldsymbol{\sigma}:\boldsymbol{\delta}=p-\tfrac{1}{3}\boldsymbol{\tau}:\boldsymbol{\delta}=p-\tau=-\tfrac{1}{3}\sigma_{\alpha\alpha}=p-\tfrac{1}{3}\tau_{\alpha\alpha}. \tag{3.58}$$

Since the tensor $\boldsymbol{\sigma}$ and the dyad $\boldsymbol{vv}$ are symmetric tensors, the Gauss–Ostrogradsky theorem can be applied to each component of these tensors, and it is then possible to go over from surface integrals to volume integrals. For example, for the component $\sigma_{\alpha\beta}$ of the stress tensor

$$\int_A \sigma_{\alpha\beta}\,\mathrm{d}A_\beta = \int_V \frac{\partial}{\partial x_\beta}(\sigma_{\beta\alpha})\,\mathrm{d}V\,.$$

Hence the surface forces acting on fluid of volume V under consideration are equal to

$$\int_A \boldsymbol{\sigma}\cdot\mathrm{d}\boldsymbol{A} = \int_V (\nabla\cdot\boldsymbol{\sigma})\,\mathrm{d}V\,, \tag{3.59}$$

whereas the momentum transported along with the substance is

$$-\int_A \rho\boldsymbol{vv}\cdot\mathrm{d}\boldsymbol{A} = -\int_V [\nabla\cdot(\rho\boldsymbol{vv})]\,\mathrm{d}V. \tag{3.60}$$

Summation of all changes in momentum for fluid of volume V yields

$$\int_V \frac{\partial}{\partial t}(\rho\boldsymbol{v})\,\mathrm{d}V = -\int_V [\nabla\cdot(\rho\boldsymbol{vv})]\,\mathrm{d}V + \int_V (\nabla\cdot\boldsymbol{\sigma})\,\mathrm{d}V + \int_V \rho\boldsymbol{F}\,\mathrm{d}V\,.$$

The region under consideration has arbitrary boundaries, and thus we can go over from the integral form of the equation to the differential form

$$\frac{\partial}{\partial t}(\rho\boldsymbol{v}) = -\nabla\cdot(\rho\boldsymbol{vv}) + \nabla\cdot\boldsymbol{\sigma} + \rho\boldsymbol{F} \tag{3.61}$$

or, on taking relation (3.53) into account, to the form

$$\frac{\partial}{\partial t}(\rho\boldsymbol{v}) = -\nabla\cdot(\rho\boldsymbol{vv}) - \nabla p + \nabla\cdot\boldsymbol{\tau} + \rho\boldsymbol{F}\,, \tag{3.62}$$

since

$$\nabla\cdot(p\boldsymbol{\delta}) = \nabla p\,. \tag{3.63}$$

The term on the left-hand side of equation (3.62) expresses the time variation of the momentum density, whereas the terms on the right-hand side, respectively, represent: the change in the momentum density owing to the existence of a convection momentum flux $\rho\boldsymbol{vv}$, the pressure forces per unit volume $-\nabla p$, the change in momentum owing to the action of viscous forces $\nabla\cdot\boldsymbol{\tau}$, and the mass forces per unit volume $\rho\boldsymbol{F}$.

The momentum balance equation, referred to a coordinate system moving along with the fluid, is obtained from equation (3.61) on application of relation (3.13) for $\varphi = \boldsymbol{v}$, that is

$$\rho\frac{\mathrm{d}\boldsymbol{v}}{\mathrm{d}t} = \frac{\partial(\rho\boldsymbol{v})}{\partial t} + \nabla\cdot(\rho\boldsymbol{vv})\,,$$

whence

$$\rho \frac{d\boldsymbol{v}}{dt} = \nabla\cdot\boldsymbol{\sigma} + \rho\boldsymbol{F} = -\nabla p + \nabla\cdot\boldsymbol{\tau} + \rho\boldsymbol{F}. \tag{3.64}$$

The left-hand side of equation (3.64) now contains the density, or the mass of a unit volume of fluid, multiplied by the centre-of-mass acceleration $d\boldsymbol{v}/dt$, equal to the sum of the local acceleration at the given point and the acceleration due to the motion of the centre of mass in the given element of fluid. The divergence of the convection momentum flux, $\nabla\cdot(\rho\boldsymbol{v}\boldsymbol{v})$, has vanished on the right-hand side under the transition to a coordinate system moving along with the fluid.

A state of mechanical equilibrium is characterized by the condition that the acceleration vanish:

$$\frac{d\boldsymbol{v}}{dt} = 0. \tag{3.65}$$

In a number of important cases the state of mechanical equilibrium described by the condition above is established in much less time than that taken by a thermodynamic process. Such a state as a rule already occurs in the initial state when diffusion or thermal diffusion in a closed vessel is considered. In the case of diffusion in a closed vessel, the acceleration $d\boldsymbol{v}/dt$ may differ somewhat from zero since the molecular masses of the components differ. This acceleration is very small and the pressure gradient corresponding to it is negligible. Under such conditions the pressure gradient also vanishes, and with it so does the viscous part of the stress tensor ($\boldsymbol{\tau}=0$); thus, the momentum balance equation (3.64) is limited to the momentum conservation equation,

$$\nabla p = \sum_{i=1}^{k} \rho_i \boldsymbol{F}_i = \rho\boldsymbol{F}, \tag{3.66}$$

i.e. the pressure gradient is equal to the sum of the mass forces acting on the substance in a unit volume.

The general form of the viscous stress tensor $\boldsymbol{\tau}$ can be defined on the basis of the following reasoning. Internal friction, associated with viscosity, arises in the fluid only if its molecules move with different velocities. For this reason the tensor component must depend on the derivatives of the velocity with respect to the coordinates, and in the first approximation must depend linearly on the gradients $\partial v_\alpha/\partial x_\beta$ and on the divergence $\partial v_\gamma/\partial x_\gamma$ of the velocity.

The expression for the component $\tau_{\alpha\beta}$ cannot contain terms independent of the derivatives of the velocity since the viscous stress tensor must vanish for $\boldsymbol{v}=\text{const}$. This tensor becomes equal to zero if the fluid performs rotational motion as a whole with angular velocity $\boldsymbol{\omega}$ and linear velocity $\boldsymbol{v}$ on the radius $\boldsymbol{r}$, the linear velocity being equal to the vector product $\boldsymbol{v}=\boldsymbol{\omega}\times\boldsymbol{r}$. The sum $\dfrac{\partial v_\alpha}{\partial x_\beta}+\dfrac{\partial v_\beta}{\partial x_\alpha}$, which are symmetric combinations of gradients, become equal to zero in this case. The most general form in which the viscous stress tensor can be written so as to satisfy the conditions above is

$$\tau_{\alpha\beta}=a\left(\frac{\partial v_\alpha}{\partial x_\beta}+\frac{\partial v_\beta}{\partial x_\alpha}\right)+b\,\frac{\partial v_\gamma}{\partial x_\gamma}\,\delta_{\alpha\beta},$$

where for an isotropic fluid the positive quantities a and b are scalar properties of the fluid, whereas the symmetric part of the dyadic $\nabla \boldsymbol{v}$,

$$(\nabla \boldsymbol{v})^{\mathrm{s}}_{\alpha\beta}=\frac{1}{2}\left[(\nabla \boldsymbol{v})_{\alpha\beta}+(\nabla \boldsymbol{v})_{\beta\alpha}\right]=\frac{1}{2}\left(\frac{\partial v_\alpha}{\partial x_\beta}+\frac{\partial v_\beta}{\partial x_\alpha}\right), \tag{3.67}$$

is the deformation rate tensor. The viscous stress tensor $\boldsymbol{\tau}$ thus is proportional to the deformation rate tensor.

For Newtonian fluids the viscous stress tensor takes the form

$$\boldsymbol{\tau}=2\eta\,(\nabla \boldsymbol{v})^{\mathrm{s}}-(\tfrac{2}{3}\eta-\eta_v)(\nabla\cdot\boldsymbol{v})\,\boldsymbol{\delta}, \tag{3.68}$$

and the total stress tensor

$$\boldsymbol{\sigma}=2\eta\,(\nabla \boldsymbol{v})^{\mathrm{s}}-\left[p+(\tfrac{2}{3}\eta-\eta_v)(\nabla\cdot\boldsymbol{v})\right]\boldsymbol{\delta}, \tag{3.69}$$

where η is the coefficient of shear (or ordinary) viscosity, and η_v is the coefficient of bulk (or volume) viscosity, or the second coefficient of viscosity.

The tensor $\nabla \boldsymbol{v}$ can be decomposed into two parts by separating a scalar part so that the second part has trace zero:

$$\nabla \boldsymbol{v}=\tfrac{1}{3}(\nabla\cdot\boldsymbol{v})\,\boldsymbol{\delta}+(\overset{0}{\nabla \boldsymbol{v}})^{\mathrm{s}}+(\nabla \boldsymbol{v})^{\mathrm{a}}. \tag{3.70}$$

The symmetric part of the deviator $(\overset{0}{\nabla \boldsymbol{v}})^{\mathrm{s}}$ is of the form

$$(\overset{0}{\nabla \boldsymbol{v}})^{\mathrm{s}}_{\alpha\beta}=\frac{1}{2}\left[\left(\frac{\partial v_\beta}{\partial x_\alpha}+\frac{\partial v_\alpha}{\partial x_\beta}\right)-\frac{2}{3}\,\frac{\partial v_\gamma}{\partial x_\gamma}\,\delta_{\alpha\beta}\right], \tag{3.71}$$

whereas the antisymmetric part is of the form

$$(\nabla \boldsymbol{v})^{\mathrm{a}}_{\alpha\beta}=\frac{1}{2}\left(\frac{\partial v_\beta}{\partial x_\alpha}-\frac{\partial v_\alpha}{\partial x_\beta}\right). \tag{3.72}$$

In view of the above, equation (3.68) can be rewritten as

$$\boldsymbol{\tau}=2\eta\,(\overset{0}{\nabla \boldsymbol{v}})^{\mathrm{s}}+\eta_v(\nabla\cdot\boldsymbol{v})\,\boldsymbol{\delta}. \tag{3.73}$$

The coefficient of bulk viscosity is zero for dilute gases, and is small for condensed gases and liquids. Accordingly, it is often possible to assume for fluids (in conformity with Stoke's hypothesis) that $\eta_v=0$ and that

$$\boldsymbol{\tau}=2\eta\,(\nabla \boldsymbol{v})^{\mathrm{s}}-\tfrac{2}{3}\eta\,(\nabla\cdot\boldsymbol{v})\,\boldsymbol{\delta}=2\eta\,(\overset{0}{\nabla \boldsymbol{v}})^{\mathrm{s}}, \tag{3.74}$$

and by equation (3.57) we also have $\tau=0$.

Example 3.2. Derive the momentum balance equation, set up by a motionless observer, for Newtonian fluids, assuming that the viscosity coefficients η and η_v are constant, and then simplify it for fluids which satisfy Stokes's hypothesis and for incompressible fluids.

Solution. The momentum balance equation was derived for the general case as equation (3.64),

$$\rho \frac{d\boldsymbol{v}}{dt} = -\nabla p + \nabla \cdot \boldsymbol{\tau} + \rho \boldsymbol{F}. \tag{1}$$

The viscous stress tensor for Newtonian fluids is of a form defined by equation (3.68), that is its components can be written as

$$\tau_{\alpha\beta} = \eta \left(\frac{\partial v_\alpha}{\partial x_\beta} + \frac{\partial v_\beta}{\partial x_\alpha} \right) - \left(\frac{2}{3}\eta - \eta_v \right) \frac{\partial v_\gamma}{\partial x_\gamma} \delta_{\alpha\beta}. \tag{2}$$

For constant viscosity coefficients η and η_v, the component α of the divergence of the viscous stress tensor $\boldsymbol{\tau}$ is of the form

$$(\nabla \cdot \boldsymbol{\tau})_\alpha = \frac{\partial \tau_{\beta\alpha}}{\partial x_\beta} = \eta \left[\frac{\partial^2 v_\alpha}{\partial x_\beta^2} + \frac{\partial}{\partial x_\alpha} \left(\frac{\partial v_\beta}{\partial x_\beta} \right) \right] - \left(\frac{2}{3}\eta - \eta_v \right) \frac{\partial}{\partial x_\alpha} \left(\frac{\partial v_\gamma}{\partial x_\gamma} \right), \tag{3}$$

or

$$\nabla \cdot \boldsymbol{\tau} = \eta \nabla^2 \boldsymbol{v} + (\tfrac{1}{3}\eta + \eta_v) \nabla (\nabla \cdot \boldsymbol{v}). \tag{4}$$

The momentum balance equation now assumes the form called the Navier–Stokes equation (first given by Navier in 1822)

$$\rho \frac{d\boldsymbol{v}}{dt} = \rho \boldsymbol{F} - \nabla p + \eta \nabla^2 \boldsymbol{v} + (\tfrac{1}{3}\eta + \eta_v) \nabla (\nabla \cdot \boldsymbol{v}). \tag{5}$$

For fluids which satisfy Stokes's hypothesis $\eta_v = 0$ and for incompressible fluids the density is constant. The conservation equation (3.36) for the amount of substance implies that for $\rho = \text{const}$ the divergence of the velocity is zero,

$$\nabla \cdot \boldsymbol{v} = 0. \tag{6}$$

In this case the Navier–Stokes equation for natural convection due to a gravitational field of acceleration g is limited to the form

$$\rho \frac{d\boldsymbol{v}}{dt} = \rho \boldsymbol{g} - \nabla p + \eta \nabla^2 \boldsymbol{v}. \tag{7}$$

Finally, for nonviscous fluids, Euler's equation (first derived in 1755) is obtained with $\eta = 0$:

$$\rho \frac{d\boldsymbol{v}}{dt} = \rho \boldsymbol{g} - \nabla p. \tag{8}$$

5. The Energy Balance Equations

The total energy of a fluid is subject to a law of conservation and equation (3.8) for $z = e$ can be applied to it in the form

$$\frac{\partial (\rho e)}{\partial t} = -\nabla \cdot (\rho e \boldsymbol{v}) - \nabla \cdot \boldsymbol{J}_{\text{et}}. \tag{3.75}$$

The total specific energy of a substance

$$e = u + \tfrac{1}{2} v^2 + \psi \tag{3.76}$$

comprises the specific internal energy u, the specific kinetic energy $\frac{1}{2}v^2$, and the specific potential energy ψ. The specific kinetic energy was determined on the basis of the centre-of-mass velocity of every element of volume, v^2 being the shorthand for $\boldsymbol{v}\cdot\boldsymbol{v}$.

The potential energy of all components in a unit volume is obtained by summing the potential energies of the individual components:

$$\rho\psi = \sum_{i=1}^{k} \rho_i \psi_i . \tag{3.77}$$

The specific potential energy of all components thus is equal to

$$\psi = \sum_{i=1}^{k} x_i \psi_i . \tag{3.78}$$

The mass forces acting on a unit mass of component are related to the specific potential energy of the component by

$$\boldsymbol{F}_i = -\nabla\psi_i . \tag{3.79}$$

In conformity with the general form of the conservation equation (3.8) for an extensive quantity, the time variation of the total energy per unit volume, $\partial(\rho e)/\partial t$, is due to a convection flux $\rho e\boldsymbol{v}$ and a conduction flux

$$\boldsymbol{J}_{\mathrm{et}} = \boldsymbol{J}_u + \sum_{i=1}^{k} \psi_i \boldsymbol{j}_i - \boldsymbol{v}\cdot\boldsymbol{\sigma} . \tag{3.80}$$

The conduction flux $\boldsymbol{J}_{\mathrm{et}}$ of the total energy consists of the conduction flux $\boldsymbol{J}_u$ of the internal energy, the potential energy flux $\sum_{i=1}^{k} \psi_i \boldsymbol{j}_i$ due to diffusion of components, and the term $-\boldsymbol{v}\cdot\boldsymbol{\sigma}$ resulting from the work of surface forces per unit surface area.

The work of the mass forces, per unit time, is

$$\int_V \sum_{i=1}^{k} \boldsymbol{v}_i \cdot \rho \boldsymbol{F}_i \, \mathrm{d}V ,$$

and per unit volume and time,

$$\sum_{i=1}^{k} \boldsymbol{v}_i \cdot \rho \boldsymbol{F}_i = \sum_{i=1}^{k} \boldsymbol{j}_i \cdot \boldsymbol{F}_i + \rho \boldsymbol{v}\cdot\boldsymbol{F} . \tag{3.81}$$

The work done by surface forces, per unit time, is equal to

$$-\int_A \boldsymbol{v}\cdot\boldsymbol{\sigma}\cdot \mathrm{d}\boldsymbol{A} = -\int_V [\nabla\cdot(\boldsymbol{v}\cdot\boldsymbol{\sigma})]\,\mathrm{d}V , \tag{3.82}$$

and, per unit volume and time, is equal to $-\nabla\cdot(\boldsymbol{v}\cdot\boldsymbol{\sigma})$. The divergence of the total energy flux is thus defined by the expression

$$\nabla\cdot\boldsymbol{J}_{\mathrm{et}} = \nabla\cdot(\boldsymbol{J}_u + \sum_{i=1}^{k} \psi_i \boldsymbol{j}_i - \boldsymbol{v}\cdot\boldsymbol{\sigma}) \tag{3.83}$$

which is consistent with equation (3.80).

The balance equation for kinetic energy is obtained by scalar multiplication of the momentum balance equation (3.64) by the centre-of-mass velocity:

$$\rho \frac{\mathrm{d}(\frac{1}{2}v^2)}{\mathrm{d}t} = \rho \boldsymbol{v}\cdot\boldsymbol{F} + \boldsymbol{v}\cdot(\nabla\cdot\boldsymbol{\sigma}) = \rho\boldsymbol{v}\cdot\boldsymbol{F} - \boldsymbol{v}\cdot\nabla p + \boldsymbol{v}\cdot(\nabla\cdot\boldsymbol{\tau}). \tag{3.84}$$

The symmetry of the stress tensor $\boldsymbol{\sigma}$ can be invoked to carry out the transformations

$$v_\alpha \frac{\partial\sigma_{\beta\alpha}}{\partial x_\beta} = v_\alpha \frac{\partial\sigma_{\alpha\beta}}{\partial x_\beta} = \frac{\partial}{\partial x_\beta}(v_\alpha\sigma_{\alpha\beta}) - \sigma_{\alpha\beta}\frac{\partial v_\alpha}{\partial x_\beta},$$

or

$$\boldsymbol{v}\cdot(\nabla\cdot\boldsymbol{\sigma}) = \nabla\cdot(\boldsymbol{v}\cdot\boldsymbol{\sigma}) - \boldsymbol{\sigma} : (\nabla\boldsymbol{v}) \tag{3.85}$$

and similarly for the viscous stress tensor $\boldsymbol{\tau}$ we obtain

$$\boldsymbol{v}\cdot(\nabla\cdot\boldsymbol{\tau}) = \nabla\cdot(\boldsymbol{v}\cdot\boldsymbol{\tau}) - \boldsymbol{\tau} : (\nabla\boldsymbol{v}). \tag{3.86}$$

By relation (3.12)

$$\boldsymbol{v}\cdot\nabla p = \nabla\cdot(p\boldsymbol{v}) - p(\nabla\cdot\boldsymbol{v}). \tag{3.87}$$

In view of the above, the kinetic energy balance equation for a fluid per unit volume and unit time, for an observer moving together with the fluid, assumes the form

$$\rho \frac{\mathrm{d}(\frac{1}{2}v^2)}{\mathrm{d}t} = \rho\boldsymbol{v}\cdot\boldsymbol{F} + \nabla\cdot(\boldsymbol{v}\cdot\boldsymbol{\sigma}) - \boldsymbol{\sigma} : (\nabla\boldsymbol{v}) \tag{3.88}$$

or

$$\rho \frac{\mathrm{d}(\frac{1}{2}v^2)}{\mathrm{d}t} = \rho\boldsymbol{v}\cdot\boldsymbol{F} - \nabla\cdot(p\boldsymbol{v}) + p(\nabla\cdot\boldsymbol{v}) + \nabla\cdot(\boldsymbol{v}\cdot\boldsymbol{\tau}) - \boldsymbol{\tau} : (\nabla\boldsymbol{v}), \tag{3.89}$$

where relation (3.53) has been taken into account under transition to the second form of the equation.

The reasoning above can be carried over to a motionless reference frame by using relation (3.13) for $\varphi = \frac{1}{2}v^2$:

$$\rho \frac{\mathrm{d}(\frac{1}{2}v^2)}{\mathrm{d}t} = \frac{\partial(\frac{1}{2}\rho v^2)}{\partial t} + \nabla\cdot(\tfrac{1}{2}\rho v^2\boldsymbol{v}),$$

and then

$$\begin{aligned}\frac{\partial(\frac{1}{2}\rho v^2)}{\partial t} &= -\nabla\cdot(\tfrac{1}{2}\rho v^2\boldsymbol{v}) + \nabla\cdot(\boldsymbol{v}\cdot\boldsymbol{\sigma}) - \boldsymbol{\sigma} : (\nabla\boldsymbol{v}) + \rho\boldsymbol{v}\cdot\boldsymbol{F}\\ &= -\nabla\cdot(\tfrac{1}{2}\rho v^2\boldsymbol{v}) - \nabla\cdot(p\boldsymbol{v}) + \nabla\cdot(\boldsymbol{v}\cdot\boldsymbol{\tau}) + p(\nabla\cdot\boldsymbol{v}) - \boldsymbol{\tau} : (\nabla\boldsymbol{v}) + \rho\boldsymbol{v}\cdot\boldsymbol{F}.\end{aligned} \tag{3.90}$$

The time variation $\partial(\frac{1}{2}\rho v^2)/\partial t$ of the kinetic energy per unit volume consists of: $-\nabla\cdot(\frac{1}{2}\rho v^2\boldsymbol{v})$, the convection transport of kinetic energy along with the fluid across the system boundary; $-\nabla\cdot(p\boldsymbol{v})$, the work of the pressure acting on the surface surrounding the volume under consideration, per unit volume and time; $\nabla\cdot(\boldsymbol{v}\cdot\boldsymbol{\tau})$, the work of the viscous forces, per unit volume and time; $\rho\boldsymbol{v}\cdot\boldsymbol{F}$, the work of the mass forces, per unit volume and time. Moreover, part of the kinetic energy, $p(\nabla\cdot\boldsymbol{v})$, is transformed reversibly into internal energy, and part, $-\boldsymbol{\tau} : (\nabla\boldsymbol{v})$, is transformed irreversibly, that is to say, is dissipated.

The part of the kinetic energy which is dissipated, i.e. transformed irreversibly into internal energy, will now be calculated for Newtonian fluids, for which by equation (3.73)

$$\boldsymbol{\tau} : (\nabla \boldsymbol{v}) = 2\eta (\nabla \overset{0}{\boldsymbol{v}})^{\mathrm{s}} : (\nabla \boldsymbol{v}) + \eta_v (\nabla \cdot \boldsymbol{v}) \boldsymbol{\delta} : (\nabla \boldsymbol{v}) . \tag{3.91}$$

A scalar part equal to

$$\tfrac{1}{3} (\nabla \boldsymbol{v}) : \boldsymbol{\delta} = \tfrac{1}{3} (\nabla \cdot \boldsymbol{v}) \tag{3.92}$$

can be separated from the tensor constituted by the velocity gradient $\nabla \boldsymbol{v}$ so that the remaining part

$$(\nabla \overset{0}{\boldsymbol{v}}) = \nabla \boldsymbol{v} - \tfrac{1}{3} (\nabla \cdot \boldsymbol{v}) \boldsymbol{\delta} \tag{3.93}$$

is a deviator. This deviator is decomposable into a symmetric part

$$(\nabla \overset{0}{\boldsymbol{v}})^{\mathrm{s}}_{\alpha\beta} = \frac{1}{2} \left(\frac{\partial v_\beta}{\partial x_\alpha} + \frac{\partial v_\alpha}{\partial x_\beta} \right) - \frac{1}{3} \frac{\partial v_\gamma}{\partial x_\gamma} \delta_{\alpha\beta} \tag{3.94}$$

and antisymmetric part

$$(\nabla \boldsymbol{v})^{\mathrm{a}}_{\alpha\beta} = \frac{1}{2} \left(\frac{\partial v_\beta}{\partial x_\alpha} - \frac{\partial v_\alpha}{\partial x_\beta} \right) . \tag{3.95}$$

Accordingly, the velocity gradient $\nabla \boldsymbol{v}$ is expressible as the sum of three components

$$\nabla \boldsymbol{v} = (\nabla \overset{0}{\boldsymbol{v}})^{\mathrm{s}} + (\nabla \boldsymbol{v})^{\mathrm{a}} + \tfrac{1}{3} (\nabla \cdot \boldsymbol{v}) \boldsymbol{\delta} . \tag{3.96}$$

The double dot product of a symmetric and an antisymmetric tensor is zero,

$$(\nabla \overset{0}{\boldsymbol{v}})^{\mathrm{s}} : (\nabla \boldsymbol{v})^{\mathrm{a}} = 0 , \tag{3.97}$$

and similarly so is the double dot product of the symmetric part of a deviator and a unit tensor,

$$(\nabla \overset{0}{\boldsymbol{v}})^{\mathrm{s}} : \boldsymbol{\delta} = 0 . \tag{3.98}$$

It should also be borne in mind that

$$(\nabla \boldsymbol{v}) : \boldsymbol{\delta} = \nabla \cdot \boldsymbol{v} , \tag{3.99}$$

and hence finally

$$\boldsymbol{\tau} : (\nabla \boldsymbol{v}) = 2\eta (\nabla \overset{0}{\boldsymbol{v}})^{\mathrm{s}} : (\nabla \overset{0}{\boldsymbol{v}})^{\mathrm{s}} + \eta_v (\nabla \cdot \boldsymbol{v})^2 \tag{3.100}$$

or, in indicial notation,

$$\begin{aligned}
[\boldsymbol{\tau} : (\nabla \boldsymbol{v})]_{\alpha,\beta} &= \tau_{\alpha\beta} \frac{\partial v_\alpha}{\partial x_\beta} \\
&= \eta \left(\frac{\partial v_\alpha}{\partial x_\beta} + \frac{\partial v_\beta}{\partial x_\alpha} - \frac{2}{3} \frac{\partial v_\gamma}{\partial x_\gamma} \delta_{\alpha\beta} \right) \frac{\partial v_\alpha}{\partial x_\beta} + \eta_v \frac{\partial v_\gamma}{\partial x_\gamma} \delta_{\alpha\beta} \frac{\partial v_\alpha}{\partial x_\beta} \\
&= \frac{1}{2} \eta \left(\frac{\partial v_\alpha}{\partial x_\beta} + \frac{\partial v_\beta}{\partial x_\alpha} - \frac{2}{3} \frac{\partial v_\gamma}{\partial x_\gamma} \delta_{\alpha\beta} \right)^2 + \eta_v \left(\frac{\partial v_\gamma}{\partial x_\gamma} \right)^2 .
\end{aligned} \tag{3.101}$$

The right-hand side of this equation is the sum of two squares, and hence the term $-\boldsymbol{\tau}:(\nabla\boldsymbol{v})$ in the kinetic energy balance equation cannot be positive. Dissipation of kinetic energy can thus proceed only in one direction; therefore it is irreversible.

The coefficient of bulk viscosity η_v can be neglected, as before, and then

$$\boldsymbol{\tau}:(\nabla\boldsymbol{v})=2\eta(\nabla\overset{0}{\boldsymbol{v}})^s:(\nabla\overset{0}{\boldsymbol{v}})^s=\eta\Phi_v, \tag{3.102}$$

where Φ_v is the Rayleigh dissipation function, defined as

$$\Phi_v=2(\nabla\overset{0}{\boldsymbol{v}})^s:(\nabla\overset{0}{\boldsymbol{v}})^s=\frac{1}{2}\left(\frac{\partial v_\alpha}{\partial x_\beta}+\frac{\partial v_\beta}{\partial x_\alpha}-\frac{2}{3}\frac{\partial v_\gamma}{\partial x_\gamma}\delta_{\alpha\beta}\right)^2. \tag{3.103}$$

Next, we set up the balance equation for the potential energy. The specific potential energy ψ_i of component i is associated with the mass force $\boldsymbol{F}_i$ by relation (3.79), that is, by

$$\boldsymbol{F}_i=-\nabla\psi_i.$$

The potential energy of a multicomponent fluid is obtained by summing the potential energies of all the components. Per unit volume, by equation (3.77) we have

$$\rho\psi=\sum_{i=1}^{k}\rho_i\psi_i.$$

Further considerations will be confined to the case of mass forces which are conservative, that is, are such that they have time-independent potentials:

$$\frac{\partial\psi_i}{\partial t}=0,\qquad \frac{\partial\boldsymbol{F}_i}{\partial t}=0. \tag{3.104}$$

If equations (3.104), (3.32), and (3.12) are taken into account, the time variation of the potential energy of component i, per unit volume, is

$$\frac{\partial(\rho_i\psi_i)}{\partial t}=\psi_i\frac{\partial\rho_i}{\partial t}=-\psi_i\nabla\cdot(\rho_i\boldsymbol{v}_i)+\psi_iM_i\sum_{j=1}^{r}\nu_{ij}J_j$$

$$=\rho_i\boldsymbol{v}_i\cdot\nabla\psi_i-\nabla\cdot(\rho_i\psi_i\boldsymbol{v}_i)+\psi_iM_i\sum_{j=1}^{r}\nu_{ij}J_j. \tag{3.105}$$

Introduction of the diffusion flux (3.25) relative to the centre-of-mass velocity gives

$$\frac{\partial(\rho_i\psi_i)}{\partial t}=-\nabla\cdot(\rho_i\psi_i\boldsymbol{v}+\psi_i\boldsymbol{j}_i)-\rho_i\boldsymbol{F}_i\cdot\boldsymbol{v}-\boldsymbol{j}_i\cdot\boldsymbol{F}_i+\psi_iM_i\sum_{j=1}^{r}\nu_{ij}J_j. \tag{3.106}$$

Summing the equations above for all components yields the time variation of the potential energy of a unit volume of the entire fluid

$$\frac{\partial(\rho\psi)}{\partial t}=-\nabla\cdot\left(\rho\psi\boldsymbol{v}+\sum_{i=1}^{k}\psi_i\boldsymbol{j}_i\right)-\rho\boldsymbol{F}\cdot\boldsymbol{v}-\sum_{i=1}^{k}\boldsymbol{j}_i\cdot\boldsymbol{F}_i$$

$$+\sum_{i=1}^{k}\psi_iM_i\sum_{j=1}^{r}\nu_{ij}J_j. \tag{3.107}$$

The last term of this equation is zero if the potential energy is conserved during the chemical reaction:

$$\sum_{i=1}^{k} \psi_i M_i \nu_{ij} = 0 . \tag{3.108}$$

This case occurs if there is conservation of the molecular property which is decisive about the action of a force field on the molecules, e.g. mass in the case of a gravitational field.

The total change in kinetic and potential energy per unit volume is obtained by adding equations (3.90) and (3.107) and taking account of condition (3.108):

$$\frac{\partial \rho (\frac{1}{2}v^2 + \psi)}{\partial t} = -\nabla \cdot [\rho (\tfrac{1}{2}v^2 + \psi) \boldsymbol{v} - \boldsymbol{v} \cdot \boldsymbol{\sigma} + \sum_{i=1}^{k} \psi_i \boldsymbol{j}_i] - \boldsymbol{\sigma} : (\nabla \boldsymbol{v}) - \sum_{i=1}^{k} \boldsymbol{j}_i \cdot \boldsymbol{F}_i . \tag{3.109}$$

Subtraction of this equation from the total energy conservation equation (3.75) yields the internal energy balance equation for a motionless observer:

$$\frac{\partial (\rho u)}{\partial t} = -\nabla \cdot (\rho u \boldsymbol{v}) - \nabla \cdot \boldsymbol{J}_u + \boldsymbol{\sigma} : (\nabla \boldsymbol{v}) + \sum_{i=1}^{k} \boldsymbol{j}_i \cdot \boldsymbol{F}_i$$
$$= -\nabla \cdot (\rho u \boldsymbol{v}) - \nabla \cdot \boldsymbol{J}_u - p (\nabla \cdot \boldsymbol{v}) + \boldsymbol{\tau} : (\nabla \boldsymbol{v}) + \sum_{i=1}^{k} \boldsymbol{j}_i \cdot \boldsymbol{F}_i . \tag{3.110}$$

The left-hand side of this equation represents the rate of change in the internal energy of a unit volume. The terms on the right-hand side of the equation are: $-\nabla \cdot (\rho u \boldsymbol{v})$, the divergence of the convection internal energy flux: $-\nabla \cdot \boldsymbol{J}_u$, the divergence of the conduction internal energy flux; $-p(\nabla \cdot \boldsymbol{v})$, the reversible increment of internal energy per unit volume and time owing to the work of change in volume; $\boldsymbol{\tau} : (\nabla \boldsymbol{v})$, the irreversible increment of internal energy per unit volume and time owing to work dissipation because of viscosity; and $\sum_{i=1}^{k} \boldsymbol{j}_i \cdot \boldsymbol{F}_i$, the internal energy change associated with the transport of potential energy by diffusion fluxes.

Application of relation (3.13) to the specific internal energy makes it possible to go over from the position of a motionless observer to that of an observer who is moving along with the fluid with its centre-of-mass velocity:

$$\rho \frac{\mathrm{d}u}{\mathrm{d}t} = -\nabla \cdot \boldsymbol{J}_u + \boldsymbol{\sigma} : (\nabla \boldsymbol{v}) + \sum_{i=1}^{k} \boldsymbol{j}_i \cdot \boldsymbol{F}_i$$
$$= -\nabla \cdot \boldsymbol{J}_u - p (\nabla \cdot \boldsymbol{v}) + \boldsymbol{\tau} : (\nabla \boldsymbol{v}) + \sum_{i=1}^{k} \boldsymbol{j}_i \cdot \boldsymbol{F}_i . \tag{3.111}$$

The considerations above concerning the internal energy balance equation for all components of the fluid were based directly on the momentum balance equation. Invoking the postulate of local thermodynamic equilibrium will enable us to introduce thermodynamic relationships linking intensive quantities in the state of equilibrium and to derive the internal energy balance equation on the basis of partial quantities defined in the same way as for a medium in equilibrium (cf. Section 1.7).

Specific quantities z, referring to the entire fluid, are related to the partial quantities $\tilde{z}_i$ for individual components by the weighted-sum law (1.40); hence, per unit volume we get relationship (1.41),

$$\rho z = \sum_{i=1}^{k} \rho_i \tilde{z}_i . \tag{3.112}$$

In accordance with the above:
the internal energy per unit volume is

$$\rho u = \sum_{i=1}^{k} \rho_i \tilde{u}_i , \tag{3.113}$$

the enthalpy per unit volume is

$$\rho h = \sum_{i=1}^{k} \rho_i \tilde{h}_i , \tag{3.114}$$

the entropy per unit volume is

$$\rho s = \sum_{i=1}^{k} \rho_i \tilde{s}_i . \tag{3.115}$$

By the equation defining enthalpy and by the definition of partial quantities determined at the temperature and pressure of the multicomponent fluid, the partial enthalpy is related to the partial internal energy and partial specific volume by

$$\tilde{h}_i = \tilde{u}_i + p\tilde{v}_i , \tag{3.116}$$

and hence the chemical potential coincides with the partial free enthalpy

$$\mu_i = \tilde{g}_i = \tilde{h}_i - T\tilde{s}_i = \tilde{u}_i + p\tilde{v}_i - T\tilde{s}_i . \tag{3.117}$$

The rate of change of the total energy of all components of the fluid contained in a unit volume is

$$\frac{\partial}{\partial t}\Big[\sum_{i=1}^{k} \rho_i (\tilde{u}_i + \tfrac{1}{2} v_i^2 + \psi_i) \Big] .$$

By relation (3.25), which gives

$$\boldsymbol{v}_i = \boldsymbol{v} + \frac{\boldsymbol{j}_i}{\rho_i} ,$$

that is

$$v_i^2 = v^2 + \frac{2}{\rho_i} \boldsymbol{j}_i \cdot \boldsymbol{v} + \frac{1}{\rho_i^2} j_i^2, \tag{3.118}$$

if account is taken of equation (3.26)

$$\sum_{i=1}^{k} \boldsymbol{j}_i = 0$$

one obtains

$$\sum_{i=1}^{k} \tfrac{1}{2}\rho_i v_i^2 = \tfrac{1}{2}\left(\rho v^2 + \sum_{i=1}^{k} \frac{j_i^2}{\rho_i}\right).$$

Diffusion is a slow phenomenon, and thus it may be assumed that $\sum_{i=1}^{k} j_i^2/\rho_i \ll \rho v^2$, and then the rate of change of the total energy of all components per unit volume is

$$\frac{\partial}{\partial t}\left[\sum_{i=1}^{k} \rho_i(\tilde{u}_i + \tfrac{1}{2}v_i^2 + \psi_i)\right] = \frac{\partial}{\partial t}\left[\rho(u + \tfrac{1}{2}v^2 + \psi)\right], \tag{3.119}$$

that is to say, takes on the same form as before. As has now been established, this form is the result of these quantities being summed for the individual components only if the phenomenon is slow enough, i.e. if the condition of local thermodynamic equilibrium is also satisfied.

When relations (3.25) and (3.118) are taken into account the convection flux of the total energy is

$$\sum_{i=1}^{k} \rho_i(\tilde{u}_i + \tfrac{1}{2}v_i^2 + \psi_i)\boldsymbol{v}_i = \sum_{i=1}^{k} (\boldsymbol{j}_i + \rho_i \boldsymbol{v})\left(\tilde{u}_i + \tfrac{1}{2}v^2 + \frac{1}{\rho_i}\boldsymbol{j}_i \cdot \boldsymbol{v} + \frac{j_i^2}{2\rho_i^2} + \psi_i\right).$$

If in the foregoing expression we discard terms containing j_i^2 and j_i^3 as being negligibly small, and then use equations (3.26) and (3.113), what is left is

$$\sum_{i=1}^{k} \rho_i(\tilde{u}_i + \tfrac{1}{2}v_i^2 + \psi_i)\boldsymbol{v}_i = \sum_{i=1}^{k} \tilde{u}_i \boldsymbol{j}_i + \rho(u + \tfrac{1}{2}v^2 + \rho\psi)\boldsymbol{v}$$

$$+ \sum_{i=1}^{k} \psi_i \boldsymbol{j}_i = \rho e \boldsymbol{v} + \sum_{i=1}^{k} \tilde{u}_i \boldsymbol{j}_i + \sum_{i=1}^{k} \psi_i \boldsymbol{j}_i. \tag{3.120}$$

The expression derived above contains energy changes due to diffusion flows, in addition to the convection flux of total energy. In this case the conduction energy flux is confined to 'pure' heat conduction without a flow of internal energy along with the substance, and its flux will be denoted by $\boldsymbol{J}_q'$. Since this quantity occurs in the derivation of the energy balance equation, it will henceforth be called the heat flux in the energy balance equation.

The total energy balance equation now assumes the form:

$$\frac{\partial(\rho e)}{\partial t} = -\nabla \cdot (\rho e \boldsymbol{v}) - \nabla \cdot \left(\boldsymbol{J}_q' + \sum_{i=1}^{k} \tilde{u}_i \boldsymbol{j}_i + \sum_{i=1}^{k} \psi_i \boldsymbol{j}_i - \boldsymbol{v} \cdot \boldsymbol{\sigma}\right). \tag{3.121}$$

Comparison of this equation with equations (3.75) and (3.80), rewritten as

$$\frac{\partial(\rho e)}{\partial t}=-\nabla\cdot(\rho e\boldsymbol{v})-\nabla\cdot\left(\boldsymbol{J}_u+\sum_{i=1}^{k}\psi_i\boldsymbol{j}_i-\boldsymbol{v}\cdot\boldsymbol{\sigma}\right), \tag{3.122}$$

shows that

$$\boldsymbol{J}_u=\boldsymbol{J}_q'+\sum_{i=1}^{k}\tilde{u}_i\boldsymbol{j}_i. \tag{3.123}$$

The second term on the right-hand side of this expression is the net flux of the internal energy transported along with the substance of components undergoing diffusion.

Example 3.3. Rearrange the internal energy balance equation for a multicomponent fluid by introducing specific heat at constant volume c_v, and then give different versions of this equation for particular cases.

Solution. Equation (3.111), given in the form

$$\rho\frac{\mathrm{d}u}{\mathrm{d}t}=-\nabla\cdot\boldsymbol{J}_u-p(\nabla\cdot\boldsymbol{v})+\boldsymbol{\tau}:(\nabla\boldsymbol{v})+\sum_{i=1}^{k}\boldsymbol{j}_i\cdot\boldsymbol{F}_i \tag{1}$$

will serve as the starting point for the manipulations here.

The total differential of the specific internal energy is expressible in terms of the internal energy of the components as

$$\frac{\mathrm{d}u}{\mathrm{d}t}=\frac{\mathrm{d}}{\mathrm{d}t}\left(\sum_{i=1}^{k}x_i\tilde{u}_i\right)=\sum_{i=1}^{k}\tilde{u}_i\frac{\mathrm{d}x_i}{\mathrm{d}t}+\sum_{i=1}^{k}x_i\frac{\mathrm{d}\tilde{u}_i}{\mathrm{d}t}. \tag{2}$$

By the balance equation (3.38) for the amount of substance it may be said that

$$\rho\sum_{i=1}^{k}\tilde{u}_i\frac{\mathrm{d}x_i}{\mathrm{d}t}=\sum_{i=1}^{k}\tilde{u}_i\left(M_i\sum_{j=1}^{r}\nu_{ij}J_j-\nabla\cdot\boldsymbol{j}_i\right). \tag{3}$$

Since the specific heat c_v of a multicomponent fluid at constant volume depends on the specific heats $\tilde{c}_{vi}$ of the components and on the partial internal energies $\tilde{u}_i$ of the components,

$$c_v=\sum_{i=1}^{k}x_i\tilde{c}_{vi}=\sum_{i=1}^{k}x_i\left(\frac{\partial\tilde{u}_i}{\partial T}\right)_v, \tag{4}$$

and since

$$\sum_{i=1}^{k}x_i\left(\frac{\partial\tilde{u}_i}{\partial v}\right)_T=T\left(\frac{\partial p}{\partial T}\right)_v-p, \tag{5}$$

the rearrangement

$$\begin{aligned}\sum_{i=1}^{k}x_i\frac{\mathrm{d}\tilde{u}_i}{\mathrm{d}t}&=\sum_{i=1}^{k}x_i\left(\frac{\partial\tilde{u}_i}{\partial T}\right)_v\frac{\mathrm{d}T}{\mathrm{d}t}+\sum_{i=1}^{k}x_i\left(\frac{\partial\tilde{u}_i}{\partial v}\right)\frac{\mathrm{d}v}{\mathrm{d}t}\\&=c_v\frac{\mathrm{d}T}{\mathrm{d}t}+\left[T\left(\frac{\partial p}{\partial T}\right)_v-p\right]\frac{\mathrm{d}v}{\mathrm{d}t}\end{aligned} \tag{6}$$

can be carried out. Finally, once these transformations have been taken into account, the result for multicomponent fluids is

$$\begin{aligned}\rho c_v\frac{\mathrm{d}T}{\mathrm{d}t}=&-\nabla\cdot\boldsymbol{J}_u+\sum_{i=1}^{k}\tilde{u}_i(\nabla\cdot\boldsymbol{j}_i)-\rho T\left(\frac{\partial p}{\partial T}\right)_v\frac{\mathrm{d}v}{\mathrm{d}t}-\sum_{j=1}^{r}\sum_{i=1}^{k}\tilde{u}_i\nu_{ij}M_iJ_j\\&+\boldsymbol{\tau}:(\nabla\boldsymbol{v})+\sum_{i=1}^{k}\boldsymbol{j}_i\cdot\boldsymbol{F}_i,\end{aligned} \tag{7}$$

whereas for a one-component fluid it is

$$\rho c_v \frac{\mathrm{d}T}{\mathrm{d}t} = -\nabla \cdot \boldsymbol{J}_u - \rho T \left(\frac{\partial p}{\partial T} \right)_v \frac{\mathrm{d}v}{\mathrm{d}t} + \boldsymbol{\tau} : (\nabla \boldsymbol{v}). \tag{8}$$

For ideal gases, when use has been made of the equation of state

$$pv = RT, \tag{9}$$

the relation obtained is

$$T \left(\frac{\partial p}{\partial T} \right)_v - p = 0. \tag{10}$$

For Newtonian fluids (where Stokes's hypothesis that $\eta_v = 0$ is also satisfied)

$$\boldsymbol{\tau} : (\nabla \boldsymbol{v}) = \eta \Phi_v. \tag{11}$$

Hence, for an ideal multicomponent gas

$$\rho c_v \frac{\mathrm{d}T}{\mathrm{d}t} = -\nabla \cdot \boldsymbol{J}_u + \sum_{i=1}^{k} \tilde{u}_i (\nabla \cdot \boldsymbol{j}_i) - \rho p \frac{\mathrm{d}v}{\mathrm{d}t} + \rho \dot{Q}_{v,T} + \eta \Phi_v + \sum_{i=1}^{k} \boldsymbol{j}_i \cdot \boldsymbol{F}_i, \tag{12}$$

and for an ideal one-component gas

$$\rho c_v \frac{\mathrm{d}T}{\mathrm{d}t} = -\nabla \cdot \boldsymbol{J}_u - \rho p \frac{\mathrm{d}v}{\mathrm{d}t} + \eta \Phi_v. \tag{13}$$

6. The Entropy Balance Equation

The entropy balance equation for a motionless observer can be written directly from the general form of the extensive quantity balance equation (3.7) for a unit volume, on setting $z = s$, and this yields

$$\frac{\partial (\rho s)}{\partial t} = -\nabla \cdot (\rho s \boldsymbol{v}) - \nabla \cdot \boldsymbol{J}_s + \Phi_s, \tag{3.124}$$

which means that the rate of change of the entropy of a unit volume of substance is due to the convection entropy flux $\rho s \boldsymbol{v}$, the conduction entropy flux $\boldsymbol{J}_s$, and entropy source with a strength of Φ_s.

The conduction entropy flux is defined as

$$\boldsymbol{J}_s = \frac{\boldsymbol{J}_q''}{T} + \sum_{i=1}^{k} \tilde{s}_i \boldsymbol{j}_i, \tag{3.125}$$

where $\boldsymbol{J}_q''$ is the heat flux in the entropy balance equation, and $\sum_{i=1}^{k} \tilde{s}_i \boldsymbol{j}_i$ is the resultant entropy flux transported along with the substance of the diffusible components.

Application of transformations (3.13) enables us to go over from a reference frame at rest to one moving along with the fluid with its centre-of-mass velocity,

$$\rho \frac{\mathrm{d}s}{\mathrm{d}t} = -\nabla \cdot \left(\frac{\boldsymbol{J}_q''}{T} + \sum_{i=1}^{k} \tilde{s}_i \boldsymbol{j}_i \right) + \Phi_s. \tag{3.126}$$

To determine the relationship between the heat flux J_q'' in the entropy balance equation now introduced here and the conduction internal energy flux J_u and the heat flux J_q' in the energy balance equation, and to determine the form of the entropy source strength Φ_s, the entropy balance equation must be derived on the basis of the postulate of local thermodynamic equilibrium.

The point of departure for determining the expression $\rho\,\mathrm{d}s/\mathrm{d}t$ is the Gibbs relation (1.53), written in terms of specific quantities, i.e. per unit mass

$$T\,\mathrm{d}s=\mathrm{d}u+p\,\mathrm{d}v-\sum_{i=1}^{k}\mu_i\,\mathrm{d}x_i, \tag{3.127}$$

which yields

$$\rho\frac{\mathrm{d}s}{\mathrm{d}t}=\frac{\rho}{T}\frac{\mathrm{d}u}{\mathrm{d}t}+\frac{\rho p}{T}\frac{\mathrm{d}v}{\mathrm{d}t}-\frac{\rho}{T}\sum_{i=1}^{k}\mu_i\frac{\mathrm{d}x_i}{\mathrm{d}t}. \tag{3.128}$$

The individual terms in this equation are transformed:
by relation (3.111)

$$\rho\frac{\mathrm{d}u}{\mathrm{d}t}=-\nabla\cdot\boldsymbol{J}_u-p(\nabla\cdot\boldsymbol{v})+\boldsymbol{\tau}:(\nabla\boldsymbol{v})+\sum_{i=1}^{k}\boldsymbol{j}_i\cdot\boldsymbol{F}_i,$$

by relation (3.37)

$$\rho p\frac{\mathrm{d}v}{\mathrm{d}t}=p(\nabla\cdot\boldsymbol{v}),$$

and by relation (3.38)

$$\rho\sum_{i=1}^{k}\mu_i\frac{\mathrm{d}x_i}{\mathrm{d}t}=-\sum_{i=1}^{k}\mu_i(\nabla\cdot\boldsymbol{j}_i)+\sum_{i=1}^{k}\mu_i M_i\sum_{j=1}^{r}\nu_{ij}J_j$$

$$=-\sum_{i=1}^{k}\mu_i(\nabla\cdot\boldsymbol{j}_i)+\sum_{j=1}^{r}A_jJ_j,$$

where the affinity of chemical reaction j, defined as

$$A_j=\sum_{i=1}^{k}\mu_i\nu_{ij}M_i, \tag{3.129}$$

has been introduced for brevity of notation. Insertion of the expressions above into equation (3.128) gives

$$\rho\frac{\mathrm{d}s}{\mathrm{d}t}=-\frac{\nabla\cdot\boldsymbol{J}_u}{T}+\frac{1}{T}\boldsymbol{\tau}:(\nabla\boldsymbol{v})+\frac{1}{T}\sum_{i=1}^{k}\boldsymbol{j}_i\cdot\boldsymbol{F}_i$$

$$+\frac{1}{T}\sum_{i=1}^{k}\mu_i(\nabla\cdot\boldsymbol{j}_i)-\frac{1}{T}\sum_{j=1}^{r}A_jJ_j. \tag{3.130}$$

Transformation (3.12) can be used to form the relationship

$$\frac{\nabla \cdot \boldsymbol{J}_u}{T} = \nabla \cdot \left(\frac{\boldsymbol{J}_u}{T}\right) + \frac{1}{T^2} \boldsymbol{J}_u \cdot \nabla T ,$$

$$\frac{\mu_i}{T}(\nabla \cdot \boldsymbol{j}_i) = \nabla \cdot \left(\frac{\mu_i}{T} \boldsymbol{j}_i\right) - \boldsymbol{j}_i \cdot \nabla \left(\frac{\mu_i}{T}\right).$$

Finally, the entropy balance equation takes on the form

$$\rho \frac{ds}{dt} = -\nabla \cdot \left(\frac{\boldsymbol{J}_u - \sum_{i=1}^{k} \mu_i \boldsymbol{j}_i}{T} \right) - \frac{1}{T^2} \boldsymbol{J}_u \cdot \nabla T$$

$$- \frac{1}{T} \sum_{i=1}^{k} \boldsymbol{j}_i \cdot \left[T \nabla \left(\frac{\mu_i}{T}\right) - \boldsymbol{F}_i \right] + \frac{1}{T} \boldsymbol{\tau} : (\nabla \boldsymbol{v}) - \frac{1}{T} \sum_{j=1}^{r} A_j J_j . \qquad (3.131)$$

Comparison of this equation with equation (3.126) allows the conduction entropy flux to be expressed in the form

$$\boldsymbol{J}_s = \frac{\boldsymbol{J}_q''}{T} + \sum_{i=1}^{k} \tilde{s}_i \boldsymbol{j}_i = \frac{1}{T}(\boldsymbol{J}_u - \sum_{i=1}^{k} \mu_i \boldsymbol{j}_i) \qquad (3.132)$$

whence, if (3.117) is taken into account, the relation is obtained between the heat flux $\boldsymbol{J}_q''$ in the entropy balance equation, the conduction energy flux $\boldsymbol{J}_u$, and the heat flux $\boldsymbol{J}_q'$ which occurs in the energy balance equation:

$$\boldsymbol{J}_q'' = \boldsymbol{J}_u - \sum_{i=1}^{k} \tilde{h}_i \boldsymbol{j}_i = \boldsymbol{J}_q' - \sum_{i=1}^{k} p \tilde{v}_i \boldsymbol{j}_i . \qquad (3.133)$$

As is seen, the heat flux can be defined in a variety of ways if diffusion occurs in multicomponent fluids. It seems more logical to separate the enthalpy flux transported along with the substance of diffusible components from the conduction energy flux $\boldsymbol{J}_u$ than it does to separate the internal energy flux transported in the same way since enthalpy lends itself to an energetic evaluation of substance flowing across a control surface. The heat flux in the entropy balance equation in general yields relations which most conveniently admit physical interpretation and consequently will be used most often in subsequent discussions.

The concept of heat flux emerges from a macroscopic treatment of the energy balance or the entropy balance. A clearer picture is obtained by looking at the phenomenon from the molecular point of view. The internal energy of a substance contained in a given volume is in a strict relation with the molecular kinetic energy and the potential energy of the intermolecular interactions. If a molecule leaves the region under study without colliding with other molecules, the loss of kinetic energy is due to diffusion. If the kinetic energy loss is the result of molecular collisions on either side of a surface bounding the volume under study, it is classified as heat conduction.

Changes in the potential energy of intermolecular interactions are not uniquely separable into those two cases.

As follows from equations (3.131) and (3.126), the entropy source strength is defined by

$$\Phi_s = \boldsymbol{J}_u \cdot \nabla\left(\frac{1}{T}\right) - \frac{1}{T}\sum_{i=1}^{k} \boldsymbol{j}_i \cdot \left[T\nabla\left(\frac{\mu_i}{T}\right) - \boldsymbol{F}_i\right]$$

$$+\frac{1}{T}\boldsymbol{\tau} : (\nabla \boldsymbol{v}) - \frac{1}{T}\sum_{j=1}^{r} A_j J_j . \tag{3.134}$$

This expression can be decomposed into three non-negative terms which are the sums of the products of tensors of the same rank. To this end, the scalar parts should be separated from the tensors $\boldsymbol{\tau}$ and $\nabla\boldsymbol{v}$ so that the remaining part have a zero trace, that is, so that it be a deviator. Hence by equations (3.56) and (3.57)

$$\boldsymbol{\tau} = \tau\boldsymbol{\delta} + \overset{0}{\boldsymbol{\tau}},$$

where

$$\tau = \tfrac{1}{3}\mathrm{Tr}\,\boldsymbol{\tau} = \tfrac{1}{3}\boldsymbol{\tau} : \boldsymbol{\delta}$$

and

$$\nabla\boldsymbol{v} = \tfrac{1}{3}\mathrm{Tr}(\nabla\boldsymbol{v})\,\boldsymbol{\delta} + (\overset{0}{\nabla\boldsymbol{v}}) = \tfrac{1}{3}(\nabla \cdot \boldsymbol{v})\,\boldsymbol{\delta} + (\overset{0}{\nabla\boldsymbol{v}}), \tag{3.135}$$

since

$$\mathrm{Tr}(\nabla\boldsymbol{v}) = (\nabla\boldsymbol{v}) : \boldsymbol{\delta} = \nabla \cdot \boldsymbol{v} . \tag{3.136}$$

The tensor $(\overset{0}{\nabla\boldsymbol{v}})$ can be decomposed into a symmetric part $(\overset{0}{\nabla\boldsymbol{v}})^{\mathrm{s}}$ and an antisymmetric part $(\nabla\boldsymbol{v})^{\mathrm{a}}$, defined by equations (3.94) and (3.95), respectively:

$$(\overset{0}{\nabla\boldsymbol{v}}) = (\overset{0}{\nabla\boldsymbol{v}})^{\mathrm{s}} + (\nabla\boldsymbol{v})^{\mathrm{a}} .$$

Since the double dot product of a symmetric and an antisymmetric tensor is zero, we have

$$\boldsymbol{\tau} : (\nabla\boldsymbol{v}) = \overset{0}{\boldsymbol{\tau}} : (\overset{0}{\nabla\boldsymbol{v}})^{\mathrm{s}} + \tau(\nabla \cdot \boldsymbol{v}) . \tag{3.137}$$

For Newtonian fluids and with Stokes's hypothesis that the bulk viscosity is zero, by equation (3.74)

$$\boldsymbol{\tau} = 2\eta\,(\overset{0}{\nabla\boldsymbol{v}})^{\mathrm{s}} \tag{3.138}$$

and $\tau = 0$, that is, the second term of equation (3.137) vanishes.

In view of the manipulations above, the individual terms of the entropy source strength take on the form:

$$\Phi_{s0} = \frac{1}{T}\tau(\nabla \cdot \boldsymbol{v}) - \frac{1}{T}\sum_{j=1}^{r} A_j J_j \geqslant 0, \tag{3.139}$$

$$\Phi_{s1} = \boldsymbol{J}_u \cdot \nabla\left(\frac{1}{T}\right) - \frac{1}{T}\sum_{i=1}^{k} \boldsymbol{j}_i \cdot \left[T\nabla\left(\frac{\mu_i}{T}\right) - \boldsymbol{F}_i\right] \geqslant 0, \tag{3.140}$$

$$\Phi_{s2} = \frac{1}{T}\overset{0}{\boldsymbol{\tau}} : (\overset{0}{\nabla\boldsymbol{v}})^{\mathrm{s}} \geqslant 0. \tag{3.141}$$

Expression (3.140) can be transformed further: by replacing the conduction energy flux $\boldsymbol{J}_u$ by the heat flux $\boldsymbol{J}_q''$ which occurs in the entropy balance equation, in accordance with relation (3.133),

$$\boldsymbol{J}_u = \boldsymbol{J}_q'' + \sum_{i=1}^{k} \tilde{h}_i \boldsymbol{j}_i\,;$$

by introducing the total potential comprising the chemical potential and the potential energy per unit mass of component,

$$\mu_i' = \mu_i + \psi_i \qquad (\nabla\psi_i = -\boldsymbol{F}_i)\,; \tag{3.142}$$

and by performing the transformation

$$T\nabla\left(\frac{\mu_i}{T}\right) - \boldsymbol{F}_i = \nabla\mu_i - \mu_i\frac{\nabla T}{T} + \nabla\psi_i$$

$$= \nabla\mu_i + \tilde{s}_i\nabla T + \nabla\psi_i - \frac{\tilde{h}_i\nabla T}{T} = \nabla_T\mu_i' - \tilde{h}_i\frac{\nabla T}{T}\,.$$

The isothermal gradient of the total potential has been introduced in this last transformation

$$\nabla_T\mu_i' = \nabla\mu_i + \tilde{s}_i\nabla T + \nabla\psi_i\,. \tag{3.143}$$

In view of the above,

$$\Phi_{s1} = \boldsymbol{J}_q'' \cdot \nabla\left(\frac{1}{T}\right) - \frac{1}{T}\sum_{i=1}^{k} \boldsymbol{j}_i \cdot \nabla_T\mu_i' \geqslant 0\,. \tag{3.144}$$

Since only $k-1$ diffusion fluxes $\boldsymbol{j}_i$ are independent, the relation

$$\sum_{i=1}^{k} \boldsymbol{j}_i \cdot \nabla_T\mu_i' = \sum_{i=1}^{k-1} \boldsymbol{j}_i \cdot \nabla_T(\mu_i' - \mu_k') \tag{3.145}$$

is satisfied on the basis of equation (3.26), so that finally only independent diffusion fluxes remain:

$$\Phi_{s1} = \boldsymbol{J}_q'' \cdot \nabla\left(\frac{1}{T}\right) - \frac{1}{T}\sum_{i=1}^{k-1} \boldsymbol{j}_i \cdot \nabla_T(\mu_i' - \mu_k')\,. \tag{3.146}$$

If the dissipation function

$$\Psi = T\Phi_s = \Psi_0 + \Psi_1 + \Psi_2 \tag{3.147}$$

is the basis for determining the thermodynamic forces, then the following equations are used instead of equations (3.139), (3.146), and (3.141):

$$\Psi_0 = T\Phi_{s0} = \tau(\nabla\cdot\boldsymbol{v}) - \sum_{j=1}^{r} A_j J_j \geqslant 0\,, \tag{3.148}$$

$$\Psi_1 = T\Phi_{s1} = -\boldsymbol{J}_q'' \cdot \nabla \ln T - \sum_{i=1}^{k-1} \boldsymbol{j}_i \cdot \nabla_T(\mu_i' - \mu_k') \geqslant 0\,, \tag{3.149}$$

$$\Psi_2 = T\Phi_{s2} = \overset{0}{\tau} : \overset{0}{(\nabla\boldsymbol{v})^s} \geqslant 0\,. \tag{3.150}$$

Equation (3.149) can be written in still another form if equation (3.140) is operated on by the transformations

$$T\nabla\left(\frac{\mu_i}{T}\right)-F_i=\nabla\mu_i-\frac{\mu_i}{T}\nabla T+\nabla\psi_i=\nabla\mu_i'-\frac{\mu_i}{T}\nabla T, \tag{3.151}$$

$$\sum_{i=1}^{k} j_i\cdot\nabla\mu_i'=\sum_{i=1}^{k-1} j_i\cdot\nabla(\mu_i'-\mu_k'), \tag{3.152}$$

and then

$$\Psi_1=-J_s\cdot\nabla T-\sum_{i=1}^{k-1} j_i\cdot\nabla(\mu_i'-\mu_k'). \tag{3.153}$$

Example 3.4. Derive the enthalpy balance equation on the basis of the energy balance equation.

Solution. If the defining equation for specific enthalpy,

$$h=u+pv, \tag{1}$$

is differentiated with respect to time and multiplied by the density ρ, the result is

$$\rho\frac{dh}{dt}=\rho\frac{du}{dt}+\rho p\frac{dv}{dt}+\frac{dp}{dt}. \tag{2}$$

Equations (3.111), (3.133) and (3.37) can be used to rewrite the internal energy balance equation as

$$\rho\frac{du}{dt}=-\nabla\cdot(J_q''+\sum_{i=1}^{k}\tilde{h}_i j_i)-\rho p\frac{dv}{dt}+\tau:(\nabla v)+\sum_{i=1}^{k} j_i\cdot F_i, \tag{3}$$

whereby

$$\rho\frac{dh}{dt}=-\nabla\cdot J_q''-\nabla\cdot(\sum_{i=1}^{k}\tilde{h}_i j_i)+\frac{dp}{dt}+\tau:(\nabla v)+\sum_{i=1}^{k} j_i\cdot F_i. \tag{4}$$

The total differential of the specific enthalpy can be expressed in terms of the partial enthalpies of the components as

$$\frac{dh}{dt}=\frac{d}{dt}(\sum_{i=1}^{k} x_i\tilde{h}_i)=\sum_{i=1}^{k}\tilde{h}_i\frac{dx_i}{dt}+\sum_{i=1}^{k} x_i\frac{d\tilde{h}_i}{dt}. \tag{5}$$

The balance equation (3.38) for the amount of substance implies

$$\rho\sum_{i=1}^{k}\tilde{h}_i\frac{dx_i}{dt}=\sum_{i=1}^{k}\tilde{h}_i(M_i\sum_{j=1}^{r}\nu_{ij}J_j-\nabla\cdot j_i). \tag{6}$$

Since

$$c_p=\sum_{i=1}^{k} x_i\tilde{c}_{pi}=\sum_{i=1}^{k} x_i\left(\frac{\partial\tilde{h}_i}{\partial T}\right)_p, \tag{7}$$

$$\left(\frac{\partial\tilde{h}_i}{\partial p}\right)_T=\tilde{v}_i-T\left(\frac{\partial\tilde{v}_i}{\partial T}\right)_p, \tag{8}$$

it is possible to transform the expressions

$$\sum_{i=1}^{k} x_i\frac{d\tilde{h}_i}{dt}=\sum_{i=1}^{k} x_i\left(\frac{\partial\tilde{h}_i}{\partial T}\right)_p\frac{dT}{dt}+\sum_{i=1}^{k} x_i\left(\frac{\partial\tilde{h}_i}{\partial p}\right)_T\frac{dp}{dt}$$

$$=c_p\frac{dT}{dt}+\left[v-T\left(\frac{\partial v}{\partial}\right)_p\right]\frac{dp}{dt}. \tag{9}$$

By formula (3.12)

$$\nabla \cdot \left(\sum_{i=1}^{k} \tilde{h}_i \boldsymbol{j}_i\right) = \sum_{i=1}^{k} \tilde{h}_i (\nabla \cdot \boldsymbol{j}_i) + \sum_{i=1}^{k} \boldsymbol{j}_i \cdot \nabla \tilde{h}_i. \tag{10}$$

Finally, after all the transformations, for multicomponent fluids

$$\rho c_p \frac{\mathrm{d}T}{\mathrm{d}t} = -\nabla \cdot \boldsymbol{J}_q'' - \sum_{i=1}^{k} \boldsymbol{j}_i \cdot \nabla \tilde{h}_i + \left(\frac{\partial \ln v}{\partial \ln T}\right)_p \frac{\mathrm{d}p}{\mathrm{d}t} - \sum_{j=1}^{r} \sum_{i=1}^{k} \tilde{h}_i \nu_{ij} M_i J_j + \boldsymbol{\tau} : (\nabla \boldsymbol{v}) + \sum_{i=1}^{k} \boldsymbol{j}_i \cdot \boldsymbol{F}_i, \tag{11}$$

and for a one-component fluid

$$\rho c_p \frac{\mathrm{d}T}{\mathrm{d}t} = -\nabla \cdot \boldsymbol{J}_u + \left(\frac{\partial \ln v}{\partial \ln T}\right)_p \frac{\mathrm{d}p}{\mathrm{d}t} + \boldsymbol{\tau} : (\nabla \boldsymbol{v}). \tag{12}$$

For ideal gases

$$\left(\frac{\partial \ln v}{\partial \ln T}\right)_p = 1. \tag{13}$$

Ideal gases can, as a rule, be treated as Newtonian fluids which satisfy Stokes's hypothesis and for which, by relation (3.102), the stress-tensor term can be replaced by a term containing the Rayleigh dissipation function Φ_v,

$$\boldsymbol{\tau} : (\nabla \boldsymbol{v}) = \eta \Phi_v. \tag{14}$$

For an ideal multicomponent gas

$$\rho c_p \frac{\mathrm{d}T}{\mathrm{d}t} = -\nabla \cdot \boldsymbol{J}_q'' - \sum_{i=1}^{k} \boldsymbol{j}_i \cdot \nabla \tilde{h}_i + \frac{\mathrm{d}p}{\mathrm{d}t} + \rho \dot{Q}_{p,T} + \eta \Phi_v + \sum_{i=1}^{k} \boldsymbol{j}_i \cdot \boldsymbol{F}_i, \tag{15}$$

and for a one-component gas

$$\rho c_p \frac{\mathrm{d}T}{\mathrm{d}t} = -\nabla \cdot \boldsymbol{J}_u + \frac{\mathrm{d}p}{\mathrm{d}t} + \eta \Phi_v. \tag{16}$$

For solids with constant thermal conductivity λ and constant specific heat ρc_p, when Fourier's law

$$\boldsymbol{J}_u = -\lambda \nabla T \tag{17}$$

and the definition of thermal diffusivity

$$a = \frac{\lambda}{\rho c_p} \tag{18}$$

are taken into account,

$$\frac{\partial T}{\partial t} = a \nabla^2 T. \tag{19}$$

7. The Phenomenological Equations

The equations determining the entropy source strength or dissipation function are used to choose the thermodynamic forces and generalized flows. If the Onsager reciprocal relations are to be applied next to the phenomenological coefficients, the thermodynamic forces must be chosen appropriately. The entropy source strength and dissipation function are best rewritten so that they contain only independent thermodynamic forces and only independent generalized flows, and then the Onsager reciprocal relations are satisfied for phenomenological coefficients in linear equations which relate the generalized flows and thermodynamic forces.

Phenomenological coefficients are functions of the local state of the substance and hence are also functions of the temperature. A choice of phenomenological coefficients may be considered justifiable if the coefficients depend little on the temperature, or best of all if they can be treated as constants. In general, the coefficients of phenomenological equations with thermodynamic forces determined on the basis of the dissipation function are less temperature-dependent than are the coefficients of equations determined on the basis of the entropy source strength. Accordingly, phenomenological equations are set up below for multicomponent fluids with thermodynamic forces determined by means of the dissipation function. Phenomenological equations with thermodynamic forces which occur in the equation for the entropy source strength are given in Example 3.5.

In the case of reasoning starting from the dissipation function, it is seen from equation (3.148) that

$$\Psi_0 = T\Phi_{s0} = \tau(\nabla\cdot\boldsymbol{v}) - \sum_{j=1}^{r} A_j J_j .$$

For phenomena of a scalar nature the following may be used as thermodynamic forces: $\nabla\cdot\boldsymbol{v}$ the divergence of the velocity and $-A_j$, the chemical affinity of reaction j taken with a minus sign, as well as the generalized flows in the form of τ equal to one-third the trace of the viscous part of the stress tensor and J_j, the chemical reaction rate.

The phenomenological equations for the foregoing generalized flows are of the form:

$$\tau = L_{vv}(\nabla\cdot\boldsymbol{v}) - \sum_{m=1}^{r} L_{vm} A_m , \tag{3.154}$$

$$J_j = L_{jv}(\nabla\cdot\boldsymbol{v}) - \sum_{m=1}^{r} L_{jm} A_m \quad (j=1, 2, 3, \ldots, r). \tag{3.155}$$

Equation (3.154) implies that the phenomenon of viscosity is affected not only by the velocity field (first term) but also by chemical reactions (the other terms). As is seen from equation (3.155), however, the chemical reaction rate also depends on the divergence of the velocity field (first term) and not only on the chemical affinities of the reactions (the other terms). The phenomenological coefficients L_{jv} and L_{vm} pertaining to cross-phenomena are not large, as experiments show. The cross-effects between chemical reactions and the flows of a viscous fluid are weak enough so as not to have been observed.

For Newtonian fluids without chemical reactions, by equations (3.56) and (3.73)

$$\tau = \tfrac{1}{3}\boldsymbol{\tau} : \boldsymbol{\delta} = \eta_v(\nabla\cdot\boldsymbol{v}) . \tag{3.156}$$

Hence, in accordance with equation (3.154) the phenomenological coefficient L_{vv} coincides with the bulk viscosity η_v,

$$L_{vv}=\eta_v \geqslant 0, \tag{3.157}$$

which is equal to zero with Stokes's hypothesis, and positive otherwise.

For the phenomenological coefficients occurring in equations (3.154) and (3.155) the Onsager reciprocal relations take on the form:

$$L_{vj}=-L_{jv} \qquad (j=1,2,3,\ldots,r), \tag{3.158}$$

$$L_{jm}=L_{mj} \qquad (j,m=1,2,3,\ldots,r). \tag{3.159}$$

By the condition that the dissipation function be non-negative, the phenomenological coefficients must satisfy the inequalities

$$\left.\begin{array}{l} L_{vv}>0, \quad L_{jj}>0 \qquad (3.160) \\ L_{vv}L_{jj}-L_{vj}^2>0 \qquad (3.161) \\ L_{mm}L_{jj}-L_{mj}^2>0 \qquad (3.162) \end{array}\right\} \quad (m,j=1,2,3,\ldots,r).$$

Suppose that use is made of that part of the dissipation thermodynamic forces and fluxes (3.149) which is the sum of the products of vectorial thermodynamic forces and fluxes in the form

$$\Psi_1=T\Phi_{s1}=-J_q''\cdot\nabla\ln T-\sum_{i=1}^{k-1} j_i\cdot\nabla_T(\mu_i'-\mu_k'), \tag{3.163}$$

Then, for phenomena of a vectorial character the thermodynamic forces obtained are $-\nabla \ln T$, the gradient of the temperature logarithm taken with the opposite sign, and $-\nabla_T(\mu_i'-\mu_k')$, the isothermal gradients of the total potential, also taken with the opposite sign. The generalized flows obtained are J_q'', the heat flux in the entropy balance equation, and j_i, the diffusion fluxes of the individual components. Since all the thermodynamic forces are now negative, it is convenient to write the phenomenological equations in a form for negative fluxes.

The heat flux in the entropy balance equation is

$$-J_q''=L_{qq}\nabla\ln T+\sum_{i=1}^{k-1} L_{qi}\nabla_T(\mu_i'-\mu_k'). \tag{3.164}$$

The diffusion flux of component n is

$$-j_n=L_{nq}\nabla\ln T+\sum_{i=1}^{k-1} L_{ni}\nabla_T(\mu_i'-\mu_k') \qquad (n=1,2,3,\ldots,k-1). \tag{3.165}$$

As follows from equation (3.164), conduction of heat is due not only to a temperature gradient (first term), but also to gradients of the total potentials of the individual components (all the other terms). The phenomenological coefficient L_{qq} occurs in the term corresponding to ordinary heat conduction under the influence of the temperature gradient, whereas the phenomenological coefficients L_{qi} appear in the terms corresponding to the Dufour effect.

It is seen from equation (3.165) that in addition to ordinary diffusion caused by the gradients of the total potentials, described by the terms with the phenomenological coefficients L_{ni}, thermal diffusion—also called the *Soret effect*—occurs in a multicomponent fluid under the influence of the temperature gradient. The phenomenological coefficient L_{nq} appears in the first, thermal-diffusion, term of equation (3.165). Thermal diffusion and the Dufour effect are cross-effects.

The Onsager reciprocal relations for the phenomenological coefficients in equations (3.164) and (3.165) are of the form:

$$L_{qi} = L_{iq} \qquad (i = 1, 2, 3, \ldots, k-1), \tag{3.166}$$

$$L_{ni} = L_{in} \qquad (i, n = 1, 2, 3, \ldots, k-1). \tag{3.167}$$

By the condition requiring the dissipation function to be non-negative, the phenomenological coefficients must satisfy the inequalities:

$$\left.\begin{array}{ll} L_{qq} > 0, \quad L_{ii} > 0 & (3.168) \\ L_{qq} L_{ii} - L_{qi}^2 > 0 & (3.169) \\ L_{nn} L_{ii} - L_{ni}^2 > 0 & (3.170) \end{array}\right\} \quad (i, n = 1, 2, 3, \ldots, k-1).$$

If that part of the dissipation function (3.150) which is the product of tensors of rank two,

$$\Psi_2 = \Phi_{s2} T = \overset{0}{\boldsymbol{\tau}} : (\overset{0}{\nabla \boldsymbol{v}})^{\mathrm{s}}, \tag{3.171}$$

is taken as the starting point, then the symmetric part of the deviator formed from the velocity gradient $(\overset{0}{\nabla \boldsymbol{v}})^{\mathrm{s}}$ can be taken as the thermodynamic forces, whereas the deviator $\overset{0}{\tau}$ formed from the viscous part of the stress tensor is the generalized flow.

The phenomenological equation will be of the form

$$\overset{0}{\boldsymbol{\tau}} = L (\overset{0}{\nabla \boldsymbol{v}})^{\mathrm{s}}. \tag{3.172}$$

Provided that Stokes's hypothesis is satisfied, the bulk viscosity for Newtonian fluids is zero by equation (3.138),

$$\overset{0}{\boldsymbol{\tau}} = \boldsymbol{\tau} = 2\eta (\overset{0}{\nabla \boldsymbol{v}})^{\mathrm{s}}, \tag{3.173}$$

whereby the phenomenological coefficient

$$L = 2\eta \geq 0 \tag{3.174}$$

is double the shear viscosity η. The linear phenomenological equation (3.172) thus is rigorously valid under the assumptions made. The positive value of the term Ψ_2 in the dissipation function implies that the shear viscosity η is positive.

Finally, it would be useful to summarize the assumptions made in the foregoing considerations as to the limits of applicability of the linear phenomenological equations above:

- the postulate of local thermodynamic equilibrium is valid,
- the fluid is sufficiently close to equilibrium for linear relationships to hold in the phenomenological equations,
- the fluid is isotropic,
- the mass forces do not vary with time,
- there is no radiation,
- electric and magnetic polarization of the substance are absent,
- the Coriolis force is negligible compared to the centrifugal force, and
- the Lorentz force may be neglected.

Example 3.5. Set up the phenomenological equations for heat conduction and diffusion in a multicompoment fluid by using thermodynamic forces defined on the basis of the entropy source strength.

Solution. By equation (3.146) the entropy source strength for phenomena of a vectorial character is of the form

$$\Phi_{s1}=\boldsymbol{J}_q''\cdot\nabla\left(\frac{1}{T}\right)-\frac{1}{T}\sum_{i=1}^{k-1}\boldsymbol{j}_i\cdot\nabla_T(\mu_i'-\mu_k'). \tag{1}$$

This equation can be recast into a simpler form by replacing the chemical potential by the Planck function

$$y_i=-\frac{\mu_i}{T}, \tag{2}$$

and the total potential by the generalized Planck function defined as

$$y_i'=-\frac{\mu_i'}{T}=-\frac{\mu_i+\psi_i}{T}=y_i-\frac{\psi_i}{T}, \tag{3}$$

which yields

$$\Phi_{s1}=\boldsymbol{J}_q''\cdot\nabla\left(\frac{1}{T}\right)+\sum_{i=1}^{k-1}\boldsymbol{j}_i\cdot\nabla_T(y_i'-y_k'). \tag{4}$$

The inverse temperature gradient $\nabla(1/T)$ and the isothermal gradients $\nabla_T(y_i'-y_k')$ of the generalized Planck functions now are the thermodynamic forces, whereas the heat flux $\boldsymbol{J}_q''$ in the entropy balance equation and the diffusion fluxes $\boldsymbol{j}_i$ of the individual components are the generalized flows.

With these thermodynamic forces, the phenomenological equations for the fluxes are of the following forms:
the heat flux in the entropy balance equation

$$\boldsymbol{J}_q''=l_{qq}\nabla\left(\frac{1}{T}\right)+\sum_{i=1}^{k-1}l_{qi}\nabla_T(y_i'-y_k'), \tag{5}$$

the diffusion flux of component n

$$\boldsymbol{j}_n=l_{nq}\nabla\left(\frac{1}{T}\right)+\sum_{i=1}^{k-1}l_{ni}\nabla_T(y_i'-y_k')\qquad(n=1,2,3,\dots,k-1). \tag{6}$$

The Onsager reciprocal relations for the phenomenological equations in the foregoing equations become:

$$l_{qi}=l_{iq} \qquad (i=1, 2, 3, \ldots, k-1), \tag{7}$$

$$l_{ni}=l_{in} \qquad (i, n=1, 2, 3, \ldots, k-1). \tag{8}$$

In view of the non-negative value of the entropy source strength, the phenomenological coefficients must satisfy the inequalities:

$$\left.\begin{array}{l} l_{qq}>0, \quad l_{ii}>0 \\ l_{qq}l_{ii}-l_{qi}^2>0 \\ l_{nn}l_{ii}-l_{ni}^2>0 \end{array}\right\} \quad (i, n=1, 2, 3, \ldots, k-1). \qquad \begin{array}{r}(9)\\(10)\\(11)\end{array}$$

Example 3.6. Set up the phenomenological equations for the entropy flux and diffusion flux in a multicomponent fluid.

Solution. The dissipation function for vectorial phenomena which, by equation (3.153), is written as

$$\Psi_1=-\boldsymbol{J}_s\cdot\nabla T-\sum_{i=1}^{k-1}\boldsymbol{j}_i\cdot\nabla(\mu_i'-\mu_k') \tag{1}$$

is the starting point for our considerations.

The entropy flux $\boldsymbol{J}_s$ and the diffusion fluxes $\boldsymbol{j}_i$ of the individual components have been taken here for the generalized flows. Hence, $-\nabla T$ and $-\nabla(\mu_i'-\mu_k')$, respectively the temperature gradient and the gradients of the total potentials taken with the opposite sign, are the thermodynamic forces.

Since all thermodynamic forces are negative, the phenomenological equations can be written as:

entropy flux

$$-\boldsymbol{J}_s=\mathscr{L}_{ss}\nabla T+\sum_{i=1}^{k-1}\mathscr{L}_{si}\nabla(\mu_i'-\mu_k'), \tag{2}$$

the diffusion flux of component n

$$-\boldsymbol{j}_n=\mathscr{L}_{ns}\nabla T+\sum_{i=1}^{k-1}\mathscr{L}_{ni}\nabla(\mu_i'-\mu_k') \qquad (n=1, 2, 3, \ldots, k-1). \tag{3}$$

Onsager reciprocal relations analogous to those given by equations (3.166) and (3.167) exist between the phenomenological coefficients which occur in the equations above. The Onsager relations must satisfy inequalities similar to inequalities (3.168), (3.169), and (3.170).

Example 3.7. Transform the system of phenomenological equations defining the fluxes $\boldsymbol{J}_q''$ and $\boldsymbol{j}_i$ as functions of the thermodynamic forces $-\nabla \ln T$ and $-\nabla_T(\mu_i'-\mu_k')$ into a system of equations for the fluxes $\boldsymbol{J}_u$ and $\boldsymbol{j}_i$ with $-\nabla \ln T$ and $-\left[T\nabla\left(\frac{\mu_i-\mu_k}{T}\right)-(\boldsymbol{F}_i-\boldsymbol{F}_k)\right]$ as thermodynamic forces.

Give the relationships between the new and old thermodynamic forces and between the phenomenological coefficients for a multicomponent fluid and for the special case of a binary fluid.

Solution. The energy flux $\boldsymbol{J}_u$ and the heat flux $\boldsymbol{J}_q''$ in the entropy balance equation are associated by relation (3.133):

$$\boldsymbol{J}_u=\boldsymbol{J}_q''+\sum_{i=1}^{k}\tilde{h}_i\boldsymbol{j}_i=\boldsymbol{J}_q''+\sum_{i=1}^{k-1}(\tilde{h}_i-\tilde{h}_k)\boldsymbol{j}_i. \tag{1}$$

The diffusion fluxes remain unchanged in both cases.

The phenomenological equations can be written in the first case as

$$J_q'' = L_{qq} X_q + \sum_{i=1}^{k-1} L_{qi} X_i , \tag{2}$$

$$j_n = L_{nq} X_q + \sum_{i=1}^{k-1} L_{ni} X_i \quad (n=1, 2, 3, \ldots, k-1), \tag{3}$$

and, in the second case, in the form

$$J_u = L_{uu} X_q + \sum_{i=1}^{k-1} L_{ui} X_i', \tag{4}$$

$$j_n = L_{nu} X_q + \sum_{i=1}^{k-1} L_{ni}' X_i' \quad (n=1, 2, 3, \ldots, k-1). \tag{5}$$

As is seen, the thermodynamic forces X_q do not change whereas the new and old thermodynamic forces, X_i' and X_i respectively, are related by

$$X_i' = -\left[T\nabla \frac{(\mu_i - \mu_k)}{T} - (F_i - F_k) \right]$$
$$= -\nabla_T(\mu_i' - \mu_k') + (\tilde{h}_i - \tilde{h}_k) \nabla \ln T = X_i - (\tilde{h}_i - \tilde{h}_k) X_q . \tag{6}$$

Insertion of this relation into equation (5) gives

$$j_n = [L_{nu} - \sum_{i=1}^{k-1} L_{ni}' (\tilde{h}_i - \tilde{h}_k)] X_q + \sum_{i=1}^{k-1} L_{ni}' X_i . \tag{7}$$

Comparison of equation (3) and (7) leads to

$$L_{ni}' = L_{ni} \quad (n, i=1, 2, 3, \ldots, k-1), \tag{8}$$

$$L_{nu} = L_{nq} + \sum_{i=1}^{k-1} L_{ni} (\tilde{h}_i - \tilde{h}_k) \quad (n=1, 2, 3, \ldots, k-1). \tag{9}$$

Equations (1), (2), and (3) yield

$$J_u = [L_{qq} + \sum_{n=1}^{k-1} L_{nq} (\tilde{h}_n - \tilde{h}_k)] X_q + [\sum_{i=1}^{k-1} L_{qi} + \sum_{i,n=1}^{k-1} L_{ni} (\tilde{h}_n - \tilde{h}_k)] X_i, \tag{10}$$

whereas the result of combining equations (4) and (6) is

$$J_u = [L_{uu} - \sum_{i=1}^{k-1} L_{ui} (\tilde{h}_i - \tilde{h}_k)] X_q + \sum_{i=1}^{k-1} L_{ui} X_i . \tag{11}$$

Comparing the phenomenological coefficients in equations (10) and (11), we obtain

$$L_{ui} = L_{qi} + \sum_{n=1}^{k-1} L_{ni} (\tilde{h}_n - \tilde{h}_k) \quad (i=1, 2, 3, \ldots, k-1) \tag{12}$$

$$L_{uu} = L_{qq} + 2 \sum_{n=1}^{k-1} L_{qn} (\tilde{h}_n - \tilde{h}_k) + \sum_{i,n=1}^{k-1} L_{ni} (\tilde{h}_n - \tilde{h}_k)(\tilde{h}_i - \tilde{h}_k). \tag{13}$$

If the Onsager reciprocal relations

$$L_{qi} = L_{iq} , \quad L_{ni} = L_{in} \quad (i, n=1, 2, 3, \ldots, k-1) \tag{14}$$

are satisfied, then so are

$$L_{ui} = L_{iu} , \quad L_{ni}' = L_{in}' \quad (i, n=1, 2, 3, \ldots, k-1), \tag{15}$$

and conversely.

For a binary fluid, the phenomenological equations can be written in the first case as

$$J_q'' = L_{qq} X_q + L_{q1} X_1 , \tag{16}$$

$$j_1 = -j_2 = L_{1q} X_q + L_{11} X_1 , \tag{17}$$

and, in the second case, in the form

$$J_u = L_{uu} X_q + L_{u1} X_1' , \tag{18}$$

$$j_1 = L_{1u} X_q + L_{11}' X_1' , \tag{19}$$

where the thermodynamic forces

$$X_1 = -\nabla_T (\mu_1' - \mu_2') , \tag{20}$$

$$X_1' = -\left[T\nabla \frac{(\mu_1 - \mu_2)}{T} - (F_1 - F_2) \right] \tag{21}$$

are related by

$$X_1' = X_1 - (\tilde{h}_1 - \tilde{h}_2) X_q . \tag{22}$$

The phenomenological coefficients are related by:

$$L_{11}' = L_{11}, \tag{23}$$

$$L_{1u} = L_{1q} + (\tilde{h}_1 - \tilde{h}_2) L_{11} , \tag{24}$$

$$L_{u1} = L_{q1} + (\tilde{h}_1 - \tilde{h}_2) L_{11} , \tag{25}$$

$$L_{uu} = L_{qq} + 2L_{q1} (\tilde{h}_1 - \tilde{h}_2) + L_{11} (\tilde{h}_1 - \tilde{h}_2)^2 . \tag{26}$$

8. The Heat of Transport

So-called 'transport quantities' are often used to describe various kinds of transport phenomena in multicomponent fluids. In considerations aimed at introducing these concepts one may start from equation (3.165) for the diffusion flux of component n,

$$-j_n = L_{nq} \nabla \ln T + \sum_{i=1}^{k-1} L_{ni} \nabla_T (\mu_i' - \mu_k') \quad (n = 1, 2, 3, \ldots, k-1) \tag{3.175}$$

after putting it into the form

$$-(j_n + L_{nq} \nabla \ln T) = \sum_{i=1}^{k-1} L_{ni} \nabla_T (\mu_i' - \mu_k') \quad (n = 1, 2, 3, \ldots, k-1) . \tag{3.176}$$

The thermodynamic forces can be determined from the foregoing system of equations:

$$\nabla_T (\mu_i' - \mu_k') = -\sum_{n=1}^{k-1} K_{in} (j_n + L_{nq} \nabla \ln T) \quad (i = 1, 2, 3, \ldots, k-1) . \tag{3.177}$$

Since the matrix of the phenomenological coefficients K_{in} is the inverse of the coefficients L_{ni}, that is,

$$K_{in} = L_{ni}^{-1} , \tag{3.178}$$

the phenomenological coefficients constituting the terms of these matrices are related by

$$\sum_{i=1}^{k-1} L_{ni} K_{im} = \delta_{nm} \quad (n, m = 1, 2, 3, \ldots, k-1), \tag{3.179}$$

$$\sum_{i=1}^{k-1} K_{ni} L_{im} = \delta_{nm} \quad (n, m = 1, 2, 3, \ldots, k-1). \tag{3.180}$$

The first of these relations follows if we substitute equation (3.177) into equation (3.176), replacing the dummy indices n by m in equation (3.177) and noting that the thermodynamic forces $\nabla_T(\mu_i' - \mu_k')$ in equation (3.176) are independent. The second relation is obtained by inserting equation (3.176) into equation (3.177).

The thermodynamic forces defined by equations (3.177) can in turn be put into equation (3.164) for the heat flux in the entropy balance equation

$$-J_q'' = L_{qq} \nabla \ln T + \sum_{i=1}^{k-1} L_{qi} \nabla_T(\mu_i' - \mu_k'),$$

and the result is

$$-J_q'' = (L_{qq} - \sum_{i,n=1}^{k-1} L_{qi} K_{in} L_{nq}) \nabla \ln T - \sum_{i,n=1}^{k-1} L_{qi} K_{in} \boldsymbol{j}_n. \tag{3.181}$$

The heat of transport $Q_n''^*$ of component n is defined as

$$Q_n''^* = \sum_{i=1}^{k-1} L_{qi} K_{in} = \sum_{i=1}^{k-1} L_{qi} L_{ni}^{-1} \quad (n = 1, 2, 3, \ldots, k-1), \tag{3.182}$$

and, on substitution of the Onsager reciprocal relations, also as

$$Q_n''^* = \sum_{i=1}^{k-1} L_{iq} K_{ni} = \sum_{i=1}^{k-1} L_{iq} L_{in}^{-1} \quad (n = 1, 2, 3, \ldots, k-1). \tag{3.183}$$

The heat of transport can thus also be used in the phenomenological equations in order to eliminate the coefficients L_{qi} or L_{iq}. On introduction of the heat of transport, the heat flux in the entropy balance equation is defined by the expression

$$J_q'' = (\sum_{n=1}^{k-1} Q_n''^* L_{nq} - L_{qq}) \nabla \ln T + \sum_{n=1}^{k-1} Q_n''^* \boldsymbol{j}_n. \tag{3.184}$$

In an isothermal system $\nabla \ln T = 0$, whereby

$$J_q'' = \sum_{n=1}^{k-1} Q_n''^* \boldsymbol{j}_n \quad (\nabla T = 0). \tag{3.185}$$

The foregoing equation permits an interpretation of the physical meaning of the heat of transport

$$Q_n''^* = \left(\frac{J_q''}{\boldsymbol{j}_n}\right)_T \quad (\boldsymbol{j}_{i \neq n} = 0). \tag{3.186}$$

The heat of transport $Q_n''^*$ thus is the 'heat carried' by a unit diffusion flux of component n when there is no temperature gradient and no diffusion of the other components.

In the same way (cf. Example 3.8) we introduce the energy of transport, i.e. the internal energy carried by a unit diffusion flux of a component when there is no temperature gradient and no diffusion of the other components. The energy of transport is denoted by U_n^* and is defined by the relationship:

$$J_u = \sum_{n=1}^{k-1} U_n^* j_n \quad (\nabla T = 0). \tag{3.187}$$

Similar relations define other transport quantities, such as the entropy of transport S_n^* of component n:

$$J_s = \sum_{n=1}^{k-1} S_n^* j_n \quad (\nabla T = 0). \tag{3.188}$$

Now, if use is made of equation (3.133), which associates the heat flux in the entropy balance equation with the conduction energy flux,

$$J_q'' = J_u - \sum_{n=1}^{k-1} (\tilde{h}_n - \tilde{h}_k) j_n \tag{3.189}$$

and account is taken of equations (3.185) and (3.187), the heat of transport and the energy of transport are related by

$$Q_n''^* = U_n^* - (\tilde{h}_n - \tilde{h}_k) \quad (\nabla T = 0) \quad (n = 1, 2, 3, \ldots, k-1). \tag{3.190}$$

Likewise, if use is made of relation (3.125), transformed into

$$J_q'' = T \Big[J_s - \sum_{n=1}^{k-1} (\tilde{s}_n - \tilde{s}_k) j_n \Big], \tag{3.191}$$

and the defining equations (3.185) and (3.188) are taken into account, the following relation is obtained for the entropy of transport:

$$S_n^* = \frac{Q_n''^*}{T} + \tilde{s}_n - \tilde{s}_k = \frac{1}{T} (U_n^* - \mu_n + \mu_k) \quad (\nabla T = 0) \quad (n = 1, 2, 3, \ldots, k-1). \tag{3.192}$$

For a binary fluid the heat of transport of the first component is

$$Q_1''^* = \frac{L_{q1}}{L_{11}} = \frac{L_{1q}}{L_{11}} = U_1^* - (\tilde{h}_1 - \tilde{h}_2). \tag{3.193}$$

For a better understanding of the transport quantities introduced above, let us consider the flow of an individual component when there is no temperature gradient and no diffusion of the other components, $j_{i \neq n} = 0$, which corresponds to the interpretation given above for the physical meaning of these quantities.

Suppose that a system N, which is not in equilibrium (Fig. 3.4), has a surface of area A in contact with a region E in which changes in state proceed in equilibrium fashion. This region contains a multicomponent fluid but only component n diffuses from region N to region E. The fluid in equilibrium is kept at a constant pressure p_E by means of a piston loaded with a constant weight and is kept at a constant temperature by means of an additional reservoir R of internal energy from which

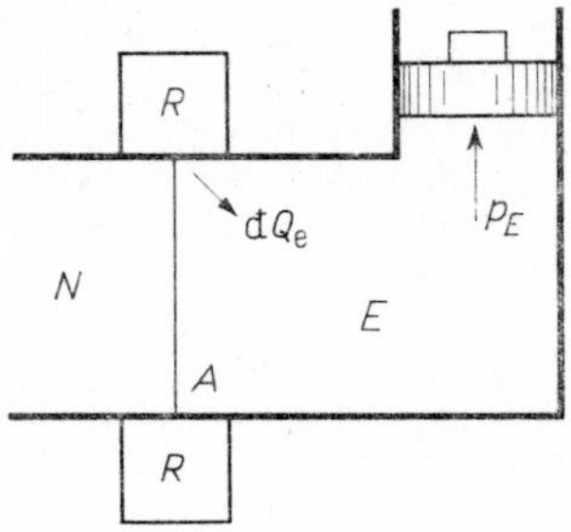

Fig. 3.4 Diagram for interpretation of the heat of transport

heat $đQ_e$ flows into the fluid under an infinitesimal change of state. Application of the first law of thermodynamics to the system under consideration gives

$$J_u A\mathrm{d}t + đQ_e = \mathrm{d}U_E + p_E \mathrm{d}V_E = \tilde{h}_n \mathrm{d}m_n \,. \tag{3.194}$$

Since the diffusion flux of component n is

$$j_n = \frac{\mathrm{d}m_n}{A\mathrm{d}t}, \tag{3.195}$$

the energy of transport of component n is

$$U_n^* = \left(\frac{J_u}{j_n}\right)_T = \tilde{h}_n - \left(\frac{đQ_e}{\mathrm{d}m_n}\right)_T \qquad (j_{n\neq i}=0), \tag{3.196}$$

whereas the heat of transport of component n is

$$Q_n''^* = U_n^* - \tilde{h}_n = -\left(\frac{đQ_e}{\mathrm{d}m_n}\right)_T \qquad (j_{n\neq i}=0). \tag{3.197}$$

The heat of transport also admits interpretation as the flux of heat entering through the surface of contact to maintain isothermal conditions if a unit diffusion flux leaves the region in equilibrium. The value of the heat of transport can be calculated analytically only for some special cases when the energy distribution is known for molecules crossing the surface under consideration. Note that for a non-diffusion flow the partial enthalpy $\tilde{h}_n$ of the component corresponds to this quantity and then satisfies the relation

$$\left(\frac{J_u}{J_n}\right)_T = \tilde{h}_n \,; \tag{3.198}$$

where J_n is the substance flux of component n.

The equations defining the transport quantity contain diffusion fluxes which may be defined in various ways, depending on the units used for the amount of substance and on the reference velocity. The diffusion flux $\boldsymbol{j}_n$ of component n was formed with kilograms as units of amount of substance and the centre-of-mass velocity as the reference velocity for diffusion.

In the special case of heat transport, the diffusion flux of the components may be referred to the velocity of one of the components which is present in the fluid in excess, that is, may be treated as a solvent. The diffusion flux of component i relative to solvent k is, by equation (3.17), defined as

$$\boldsymbol{j}_{k,i}=\rho_i(\boldsymbol{v}_i-\boldsymbol{v}_k) \tag{3.199}$$

where $\boldsymbol{v}_i$ and $\boldsymbol{v}_k$ are the velocities of component i and solvent k, respectively.

The diffusion flux defined above can be evaluated from the previous fluxes referred to the centre-of-mass velocity (3.25) of the mass by means of the expression

$$\boldsymbol{j}_{k,i}=\boldsymbol{j}_i-\frac{\rho_i}{\rho_k}\boldsymbol{j}_k=\boldsymbol{j}_i+\frac{\rho_i}{\rho_k}\sum_{n=1}^{k-1}\boldsymbol{j}_n \quad (i=1,2,3,\ldots,k-1), \tag{3.200}$$

relation (3.26) being taken into account.

The heat transported by diffusion of component i, determined relative to solvent k, can be defined by means of the relation

$$\boldsymbol{J}_q''=\sum_{i=1}^{k-1}Q_{k,i}''^{*}\boldsymbol{j}_{k,i} \quad (\nabla T=0); \tag{3.201}$$

it is then related to the heat of transport $Q_i''^{*}$, defined by equation (3.185), by

$$Q_i''^{*}=Q_{k,i}''^{*}+\sum_{n=1}^{k-1}\frac{\rho_n}{\rho_k}Q_{k,n}''^{*} \quad (i=1,2,3,\ldots,k-1). \tag{3.202}$$

For very dilute solutions the densities ρ_n of the components are low relative to the density ρ_k of the solvent. Since $\rho_n \ll \rho_k$, it may be assumed that $\rho_n/\rho_k \approx 0$. In this case diffusion of a component relative to the solvent coincides practically with the diffusion relative to the centre of mass of all the multicomponent fluid.

Example 3.8. Derive an equation defining the energy of transport U_n^* of component n by starting from the equation defining the conduction energy flux $\boldsymbol{J}_u$.

Solution. On the basis of equations (4) and (6) of Example 3.7, the conduction energy flux is given by

$$-\boldsymbol{J}_u=L_{uu}\nabla\ln T+\sum_{i=1}^{k-1}L_{ui}\left[T\nabla\left(\frac{\mu_i-\mu_k}{T}\right)-\boldsymbol{F}_i+\boldsymbol{F}_k\right]. \tag{1}$$

The diffusion flux of component n becomes

$$-\boldsymbol{j}_n=L_{nu}\nabla\ln T+\sum_{i=1}^{k-1}L_{ni}'\left[T\nabla\left(\frac{\mu_i-\mu_k}{T}\right)-\boldsymbol{F}_i+\boldsymbol{F}_k\right] \quad (n=1,2,3,\ldots,k-1), \tag{2}$$

which can be rewritten as

$$-(j_n+L_{nu}\nabla\ln T)=\sum_{i=1}^{k-1}L'_{ni}\left[T\nabla\left(\frac{\mu_i-\mu_k}{T}\right)-F_i+F_k\right]\qquad(n=1,2,3,\ldots,k-1).\quad(3)$$

The thermodynamic forces can be calculated from equation (3) and found to be

$$T\nabla\left(\frac{\mu_i-\mu_k}{T}\right)-F_i+F_k=-\sum_{n=1}^{k-1}K'_{in}(j_n+L_{nu}\nabla\ln T)\qquad(i=1,2,3,\ldots,k-1),\quad(4)$$

where the phenomenological coefficient matrix K'_{in} is the inverse of the matrix L'_{ni},

$$K'_{in}=L'^{-1}_{ni},\quad(5)$$

and hence the coefficients of these matrices are related by

$$\left.\begin{aligned}&\sum_{i=1}^{k-1}L'_{ni}K'_{im}=\delta_{nm}\\&\sum_{i=1}^{k-1}K'_{ni}L'_{im}=\delta_{nm}\end{aligned}\right\}\qquad(n,m=1,2,3,\ldots,k-1).\quad(7)$$

The thermodynamic forces defined by equation (4) are inserted into equation (1), and then

$$-J_u=(L_{uu}-\sum_{i,n=1}^{k-1}L_{ui}K'_{in}L_{nu})\nabla\ln T-\sum_{i,n=1}^{k-1}L_{ui}K'_{in}j_n.\quad(8)$$

The energy of transport of component n is defined as

$$U_n^*=\sum_{i=1}^{k-1}L_{ui}K'_{in}=\sum_{i=1}^{k-1}L_{ui}L'^{-1}_{ni}\qquad(n=1,2,3,\ldots,k-1),\quad(9)$$

and, when the Onsager reciprocal relations are taken into account, also as

$$U_n^*=\sum_{i=1}^{k-1}L_{iu}K'_{ni}=\sum_{i=1}^{k-1}L_{iu}L'^{-1}_{in}\qquad(n=1,2,3,\ldots,k-1).\quad(10)$$

When the energy of transport is introduced into equation (8), the result is

$$J_u=(\sum_{n=1}^{k-1}U_n^*L_{nu}-L_{uu})\nabla\ln T+\sum_{n=1}^{k-1}U_n^*j_n.\quad(11)$$

In an isothermal system, since $\nabla\ln T=0$,

$$J_u=\sum_{n=1}^{k-1}U_n^*j_n\qquad(\nabla T=0),\quad(12)$$

as previously given by equation (3.187).

9. The Heat and Diffusion Fluxes in a Fluid in Mechanical Equilibrium without Chemical Reactions

In the case of a multicomponent fluid, a state of mechanical equilibrium is attained right at the beginning of a phenomenon and, as follows from the momentum balance equation (3.66), is associated with the pressure gradient being balanced by the mass forces:

$$\nabla p=\rho F=\sum_{i=1}^{k}\rho_iF_i=-\sum_{i=1}^{k}\rho_i\nabla\psi_i.\quad(3.203)$$

The Gibbs–Duhem relation (1.73)

$$s\,\mathrm{d}T - v\,\mathrm{d}p + \sum_{i=1}^{k} x_i\,\mathrm{d}\mu_i = 0 \tag{3.204}$$

at constant temperature yields

$$\nabla p = \sum_{i=1}^{k} \rho_i \nabla_T \mu_i \tag{3.205}$$

which, in combination with the mechanical equilibrium condition (3.203) and with equations (3.142) and (1.36) being taken into account, assumes the form

$$\sum_{i=1}^{k} x_i \nabla_T \mu_i' = 0\,. \tag{3.206}$$

The foregoing relation enables the thermodynamic force $\nabla_T \mu_k'$ to be eliminated from equation (3.149) for the component dissipation function and makes it possible to cast this equation into the form

$$\Psi_1 = -\boldsymbol{J}_q'' \cdot \nabla \ln T - \sum_{i,m=1}^{k-1} \boldsymbol{j}_i \cdot a_{im} \nabla_T \mu_m', \tag{3.207}$$

where

$$a_{im} = \delta_{im} + \frac{x_m}{x_k} \qquad (i, m = 1, 2, 3, \ldots, k-1)\,. \tag{3.208}$$

The chemical potential μ_m of component m is a function of the temperature T, the pressure p, and the mass fractions of $k-1$ components x_j:

$$\mu_m = \mu_m(T, p, x_j) \qquad (j = 1, 2, 3, \ldots, k-1)\,.$$

Hence, the chemical potential gradient is

$$\nabla \mu_m = \sum_{j=1}^{k-1} \left(\frac{\partial \mu_m}{\partial x_j}\right)_{T,p,x_{i\neq j}} \nabla x_j + \left(\frac{\partial \mu_m}{\partial T}\right)_{p,x_i} \nabla T + \left(\frac{\partial \mu_m}{\partial p}\right)_{T,x_i} \nabla p$$

$$= \nabla_{T,p}\,\mu_m - \tilde{s}_m \nabla T + \tilde{v}_m \nabla p\,, \tag{3.209}$$

where the gradient of the chemical potential at constant temperature and constant pressure is equal to

$$\nabla_{T,p}\,\mu_m = \sum_{j=1}^{k-1} \left(\frac{\partial \mu_m}{\partial x_j}\right)_{T,p,x_{i\neq j}} \nabla x_j\,. \tag{3.210}$$

At constant temperature, therefore, the gradient of the total potential can be made dependent on the gradient of the concentrations expressed in terms of the mass fractions as follows:

$$\nabla_T \mu_m' = \sum_{j=1}^{k-1} \left(\frac{\partial \mu_m}{\partial x_j}\right)_{T,p,x_{i\neq j}} \nabla x_j + \tilde{v}_m \nabla p - \boldsymbol{F}_m\,. \tag{3.211}$$

Equation (3.207) can now be put into the form

$$\Psi_1 = -\boldsymbol{J}_q'' \cdot \nabla \ln T - \sum_{i,m=1}^{k-1} \boldsymbol{j}_i \cdot a_{im} \left[\sum_{j=1}^{k-1} \left(\frac{\partial \mu_m}{\partial x_j} \right)_{T,p,xi \neq j} \nabla x_j + \tilde{v}_m \nabla p - \boldsymbol{F}_m \right]. \tag{3.212}$$

The thermodynamic forces can be isolated from the expression above,

$$\boldsymbol{X}_q = -\nabla \ln T, \tag{3.213}$$

$$\boldsymbol{X}_i = -\sum_{m=1}^{k-1} a_{im} \left[\sum_{j=1}^{k-1} \left(\frac{\partial \mu_m}{\partial x_j} \right)_{T,p,xi \neq j} \nabla x_j + \tilde{v}_m \nabla p - \boldsymbol{F}_m \right] \tag{3.214}$$

$(i = 1, 2, 3, \ldots, k-1)$,

and the phenomenological equations can be written for the heat flux in the entropy balance equation,

$$-\boldsymbol{J}_q'' = L_{qq} \nabla \ln T + \sum_{i,m=1}^{k-1} L_{qi} a_{mi} \left[\sum_{j=1}^{k-1} \left(\frac{\partial \mu_m}{\partial x_j} \right)_{T,p,xi \neq j} \nabla x_j + \tilde{v}_m \nabla p - \boldsymbol{F}_m \right] \tag{3.215}$$

and the diffusion flux of component n,

$$-\boldsymbol{j}_n'' = L_{nq} \nabla \ln T + \sum_{i,m=1}^{k-1} L_{ni} a_{mi} \left[\sum_{j=1}^{k-1} \left(\frac{\partial \mu_m}{\partial x_j} \right)_{T,p,xi \neq j} \nabla x_j + \tilde{v}_m \nabla p - \boldsymbol{F}_m \right]$$

$(n = 1, 2, 3, \ldots, k-1)$. (3.216)

These phenomenological equations are simplified considerably if there are no mass forces, $\boldsymbol{F}_m = 0$, since the mechanical equilibrium condition (3.203) then reduces to the absence of a pressure gradient:

$$\nabla p = \rho \boldsymbol{F} = 0. \tag{3.217}$$

In this case the heat flux in the entropy balance equation is of the form

$$-\boldsymbol{J}_q'' = L_{qq} \nabla \ln T + \sum_{i,j,m=1}^{k-1} L_{qi} a_{mi} \left(\frac{\partial \mu_m}{\partial x_j} \right)_{T,p,xi \neq j} \nabla x_j, \tag{3.218}$$

and the diffusion flux of component n is

$$-\boldsymbol{j}_n = L_{nq} \nabla \ln T + \sum_{i,j,m=1}^{k-1} L_{ni} a_{mi} \left(\frac{\partial \mu_m}{\partial x_j} \right)_{T,p,xi \neq j} \nabla x_j$$

$(n = 1, 2, 3, \ldots, k-1)$. (3.219)

By the Onsager reciprocal relations, the matrix of phenomenological coefficients is symmetric:

$$L_{nq} = L_{qn}, \quad L_{ni} = L_{in} \quad (i, n = 1, 2, 3, \ldots, k-1). \tag{3.220}$$

In view of the positive value of the dissipation function, the phenomenological coefficients must satisfy the inequalities

$$\left.\begin{aligned} &L_{qq}>0, \quad L_{ii}>0 \\ &L_{qq}L_{ii}-L_{qi}^2>0 \\ &L_{ii}L_{nn}-L_{ni}^2>0 \end{aligned}\right\} \quad (i, n=1, 2, 3, \ldots, k-1). \quad \begin{aligned} &(3.221) \\ &(3.222) \\ &(3.223) \end{aligned}$$

To simplify the discussion, below we consider a binary fluid ($k=2$) for which

$$a_{mi}=a_{11}=\frac{1}{x_2}=\frac{1}{1-x_1}, \tag{3.224}$$

and the phenomenological equations become:
the heat flux in the entropy balance equation

$$-J_q''=L_{qq}\nabla \ln T+L_{q1}\frac{1}{x_2}\left(\frac{\partial \mu_1}{\partial x_1}\right)_{T,p}\nabla x_1, \tag{3.225}$$

the diffusion flux of component 1 or [by equation (3.26)] the diffusion flux of component 2, taken with the opposite sign,

$$-j_1=j_2=L_{1q}\nabla \ln T+L_{11}\frac{1}{x_2}\left(\frac{\partial \mu_1}{\partial x_1}\right)_{T,p}\nabla x_1. \tag{3.226}$$

The phenomenological coefficients in equations (3.225) and (3.226) are associated by the Onsager reciprocal relations,

$$L_{q1}=L_{1q}. \tag{3.227}$$

The positive value of the dissipation function leads to inequalities pertaining to the phenomenological coefficients:

$$L_{qq}>0, \quad L_{11}>0, \tag{3.228}$$

$$L_{qq}L_{11}-L_{q1}^2>0. \tag{3.229}$$

Heat conduction due only to the temperature gradient is described by Fourier's law

$$J_q''=-\lambda_0 \nabla T \quad (\nabla x_j=0), \tag{2.230}$$

where λ_0 is the thermal conductivity in the absence of concentration gradients.

Comparison of formulae (3.218) and (3.230) yields a relation between the phenomenological coefficient L_{qq} and the empirical coefficient λ_0 in the form

$$L_{qq}=\lambda_0 T. \tag{3.231}$$

It is easily seen that in the case of reasoning in which the entropy source strength is taken as the starting point, a phenomenological coefficient which is more heavily dependent on the temperature is obtained.

Diffusion due only to the concentration gradient is described by an equation defining the diffusion flux of component n, that is,

$$j_n = -\rho \sum_{j=1}^{k-1} D_{nj} \nabla x_j \quad (\nabla T=0) \quad (n=1, 2, 3, \ldots, k-1), \tag{3.232}$$

where a diffusion coefficient given by

$$D_{nj} = \frac{1}{\rho} \sum_{i,m=1}^{k-1} L_{ni} a_{mi} \left(\frac{\partial \mu_m}{\partial x_j}\right)_{T,p,x_i \neq j} \quad (n=1, 2, 3, \ldots, k-1) \tag{3.233}$$

has been introduced.

In an isothermal binary fluid, diffusion is described by Fick's law

$$j_1 = -j_2 = -\rho D \nabla x_1 \quad (\nabla T=0) \tag{3.234}$$

which contains the diffusion coefficient (Tables 3.1 and 3.2)

$$D = D_{11} = L_{11} \frac{1}{\rho x_2} \left(\frac{\partial \mu_1}{\partial x_1}\right)_{T,p}. \tag{3.235}$$

TABLE 3.1

Experimental Values of Diffusion Coefficient D_{AB} for Binary Gases at Pressure of 1 atm [16]

Components A–B	Temperature °C	Diffusion Coefficient D_{AB} cm^2s^{-1}
CO_2–N_2O	0	0.096
CO_2–CO	0	0.139
CO_2–N_2	0	0.144
	15	0.158
	25	0.165
Ar–O_2	20	0.200
H_2–SF_6	25	0.420
H_2–CH_4	25	0.726

Diffusion due only to a temperature gradient is called *thermal diffusion.* In the absence of concentration gradients, equation (3.219) reduces to the form

$$j_n = -L_{nq} \nabla \ln T = -\frac{\rho}{T} D_{Tn} \nabla T \quad (\nabla x_j=0) \quad (n=1, 2, 3, \ldots, k-1), \tag{3.236}$$

where the thermal diffusion coefficient of component n, defined as

$$D_{Tn} = \frac{L_{nq}}{\rho} \quad (n=1, 2, 3, \ldots, k-1), \tag{3.237}$$

has been introduced.

These empirical coefficients enable the diffusion flux of component n to be put into the form

$$-j_n=\rho\left(D_{Tn}\nabla\ln T+\sum_{j=1}^{k-1}D_{nj}\nabla x_j\right)\quad (n=1,2,3,\dots,k-1) \tag{3.238}$$

or as

$$-j_n=\rho\sum_{j=1}^{k-1}D_{nj}(k_{Tj}\nabla\ln T+\nabla x_j)\quad (n=1,2,3,\dots,k-1), \tag{3.239}$$

TABLE 3.2

Experimental Values of Diffusion Coefficient D_{AB} for Binary Fluids [16]

Components A–B	Temperature °C	Molar Fraction z_A	Diffusion Coefficient D_{AB} ($\times 10^5$) cm^2s^{-1}
Chlorobenzene and bromobenzene	10	0.0332	1.007
		0.2642	1.069
		0.5122	1.146
		0.7617	1.226
		0.9652	1.291
	40	0.0332	1.584
		0.2642	1.691
		0.5122	1.806
		0.7617	1.902
		0.9652	1.996
Ethanol and water	25	0.050	1.13
		0.275	0.41
		0.500	0.90
		0.700	1.40
		0.950	2.20
Water and n-butanol	30	0.131	1.24
		0.222	0.92
		0.358	0.56
		0.454	0.437
		0.524	0.267

on introduction of the thermal diffusion ratio k_{Tj}, related to the previous coefficients by

$$D_{Tn}=\sum_{j=1}^{k-1}k_{Tj}D_{nj}\quad (n=1,2,3,\dots,k-1). \tag{3.240}$$

For a binary fluid equations (3.236), (3.238), and (3.239) are written as

$$j_1=-L_{1q}\nabla\ln T=-\rho x_1x_2D'\nabla T\quad (\nabla x_j=0), \tag{3.241}$$

$$-j_1=j_2=\rho(x_1x_2D'\nabla T+D\nabla x_1)=\rho D(k_T\nabla\ln T+\nabla x_1), \tag{3.242}$$

where the thermal diffusion coefficient is usually defined as

$$D'=\frac{D_{T1}}{x_1x_2T}=\frac{L_{1q}}{\rho x_1x_2T}, \tag{3.243}$$

whereas the thermal diffusion ratio is

$$k_T = \frac{D_{T1}}{D} = x_1 x_2 T \left(\frac{D'}{D}\right). \tag{3.244}$$

If k_T is positive, component 1 diffuses to a cooler region, and if k_T is negative, it diffuses to a hotter region.

In addition to the coefficients above, other quantities are also used to describe the phenomenon of thermal diffusion. These are:
the Soret coefficient, applied most often to fluids,

$$s_T = \frac{D'}{D} = \frac{k_T}{x_1 x_2 T} = \frac{L_{1q}}{L_{11}} \frac{1}{x_1 T \left(\frac{\partial \mu_1}{\partial x_1}\right)_{T,p}} \tag{3.245}$$

and the thermal diffusion factor, which in general is independent of concentration for gases,

$$\alpha = T s_T = T \left(\frac{D'}{D}\right). \tag{3.246}$$

The Dufour effect is the flow of heat arising only from a concentration gradient. In the absence of a temperature gradient, as follows from equation (3.218) there may be a heat flux

$$J_q'' = - \sum_{i,j,m=1}^{k-1} L_{qi} a_{mi} \left(\frac{\partial \mu_m}{\partial x_j}\right)_{T,p,x_i \neq j} \nabla x_j \quad (\nabla T = 0), \tag{3.247}$$

and, in a binary fluid, a heat flux

$$J_q'' = -L_{q1} \frac{1}{x_2} \left(\frac{\partial \mu_1}{\partial x_1}\right)_{T,p} \nabla x_1 = -\rho_1 \left(\frac{\partial \mu_1}{\partial x_1}\right)_{T,p} T D'' \nabla x_1 \quad (\nabla T = 0). \tag{3.248}$$

The Dufour coefficient D'' introduced in this equation is related to the phenomenological coefficient L_{q1} by

$$D'' = \frac{L_{q1}}{\rho x_1 x_2 T}; \tag{3.249}$$

when the Onsager reciprocal relations (3.227) are taken into account, this yields

$$D'' = D'. \tag{3.250}$$

The Dufour coefficient is equal to the thermal diffusion coefficient, and hence equation (3.225) can also be expressed in terms of empirical coefficients as

$$-J_q'' = \lambda_0 \nabla T + \rho_1 \left(\frac{\partial \mu_1}{\partial x_1}\right)_{T,p} T D' \nabla x_1. \tag{3.251}$$

Inequalities (3.228) and (3.229) pertaining to the phenomenological coefficients can now also be rewritten in terms of the empirical coefficients

$$\lambda_0 > 0, \quad D > 0, \tag{3.252}$$

$$(D')^2 < \frac{\lambda_0 D}{T\rho x_1^2 x_2 \left(\frac{\partial \mu_1}{\partial x_1}\right)_{T,p}}, \tag{3.253}$$

where the thermodynamic stability condition

$$\left(\frac{\partial \mu_1}{\partial x_1}\right)_{T,p} \geqslant 0 \tag{3.254}$$

has been used in deriving the second and third inequalities.

If a solution of ideal gases contains only two components, then by equation (1.48) the molar fraction of the first component is

$$z_1 = \frac{x_1}{M_1\left(\frac{x_1}{M_1} + \frac{x_2}{M_2}\right)} = \frac{M_2 x_1}{M_2 x_1 + M_1 x_2}, \tag{3.255}$$

and, by equation (1.63), its chemical potential is

$$\mu_1 = f_1(T) + R_1 T \ln \frac{p M_2 x_1}{M_2 x_1 + M_1 x_2}. \tag{3.256}$$

The partial derivative of the chemical potential of an ideal gas at constant temperature and constant pressure is

$$\left(\frac{\partial \mu_1}{\partial x_1}\right)_{T,p} = \frac{R_M T}{x_1 (M_2 x_1 + M_1 x_2)} = \frac{1}{x_1}\frac{M^2 RT}{M_1 M_2}, \tag{3.257}$$

where R_M is the universal gas constant.

The diffusion coefficient for a binary solution of ideal gases, by equation (3.235), is

$$D = L_{11} \frac{R_M T}{\rho} \frac{1}{x_1 x_2 (M_2 x_1 + M_1 x_2)}, \tag{3.258}$$

whereas inequality (3.253) takes the form

$$(D')^2 < \frac{\lambda_0 D (M_2 x_1 + M_1 x_2)}{R_M T^2 \rho x_1 x_2}. \tag{3.259}$$

For a binary solution of ideal gases the phenomenological equation (3.251) gives the heat flux in the entropy balance equation in the form

$$-J_q'' = \lambda_0 \nabla T + D' \frac{R_M T^2 \rho}{M_2 x_1 + M_1 x_2} \nabla x_1. \tag{3.260}$$

Finally, it should be emphasized that the cross-effect coefficients L_{qi} and L_{iq} can be eliminated from the phenomenological equations, by being replaced by heats of transport. In particular, application of the heat of transport $Q_1''^*$, defined by equation (3.193), to a binary fluid which obeys equations (3.225) and (3.226) gives

$$-J_q''=L_{qq}\nabla \ln T+L_{11}Q_1''^*\frac{1}{x_2}\left(\frac{\partial\mu_1}{\partial x_1}\right)_{T,p}\nabla x_1, \tag{3.261}$$

$$-j_1=j_2=L_{11}\left[Q_1''^*\nabla \ln T+\frac{1}{x_2}\left(\frac{\partial\mu_1}{\partial x_1}\right)_{T,p}\nabla x_1\right]. \tag{3.262}$$

By equations (3.193) and (3.245), the Soret coefficient is obtained in the form

$$s_T=\frac{Q_1''^*}{x_1 T\left(\frac{\partial\mu_1}{\partial x_1}\right)_{T,p}}. \tag{3.263}$$

When equation (3.257) is taken into account, the result for ideal gases is

$$s_T=\frac{Q_1''^*(M_2x_1+M_1x_2)}{R_M T^2}. \tag{3.264}$$

The discussion above concerning diffusion phenomena was carried out by using the diffusion flux $\boldsymbol{j}_i$ referred to a unit mass and centre-of-mass velocity, that is, defined by equation (3.25) as

$$\boldsymbol{j}_i=\rho_i(\boldsymbol{v}_i-\boldsymbol{v}). \tag{3.265}$$

As shown by Prigogine, in considerations of diffusion phenomena in mechanical equilibrium, the centre-of-mass velocity which occurs in equation (3.265) can be replaced by any other velocity $\boldsymbol{v}_a$, and the dissipation function does not change. By equation (3.17), the diffusion flux referred to an arbitrary velocity $\boldsymbol{v}_a$ is defined as

$$\boldsymbol{j}_{v_a i}=\rho_i(\boldsymbol{v}_i-\boldsymbol{v}_a). \tag{3.266}$$

The Gibbs–Duhem relation, in combination with the mechanical equilibrium condition yielded relation (3.206) which also states that

$$\sum_{i=1}^{k}\rho_i\nabla_T\mu_i'=0. \tag{3.267}$$

In view of equations (3.144) and (3.265) the dissipation function for diffusion can be written as

$$\Psi_D=-\sum_{i=1}^{k}\boldsymbol{j}_i\cdot\nabla_T\mu_i'=-\sum_{i=1}^{k}\rho_i(\boldsymbol{v}_i-\boldsymbol{v})\cdot\nabla\ \mu_i',\quad (\nabla T=0). \tag{3.268}$$

This expression is equivalent to

$$\Psi_D = -\sum_{i=1}^{k} \boldsymbol{j}_{v_a i} \cdot \nabla_T \mu_i' = -\sum_{i=1}^{k} \rho_i (\boldsymbol{v}_i - \boldsymbol{v}_a) \cdot \nabla_T \mu_i' \quad (\nabla T = 0), \tag{3.269}$$

since the difference between these expressions is zero for the state of mechanical equilibrium in accordance with relation (3.267). When diffusion fluxes are considered relative to various velocities, the thermodynamic forces remain unchanged and only the values of the phenomenological coefficients change. In addition to the centre-of-mass velocity $\boldsymbol{v}$ (3.22), the following are also used as a reference velocity for the diffusion flux: the mean molar velocity $\boldsymbol{v}_M$ (3.23), the mean volume velocity $\boldsymbol{v}_v$ (3.24), and the velocity $\boldsymbol{v}_k$ of the kth component treated as a solvent, since it is then the component which occurs in considerable excess.

In addition to mass diffusion fluxes $\boldsymbol{j}_{v_a i}$, molar diffusion fluxes $\boldsymbol{j}_{Mv_a i}$ (3.20) are often used, as will be examined more closely in Example 3.10.

Example 3.9. Go over from the mass fraction gradient to the molar fraction gradient in the equation defining the mass diffusion flux relative to the centre-of-mass velocity in a binary fluid.

Solution. By equation (3.242), the diffusion flux of the first component is

$$-\boldsymbol{j}_1 = \boldsymbol{j}_2 = \rho D (k_T \nabla \ln T + \nabla x_1). \tag{1}$$

The mass fraction of the component is calculated from the molar fraction, in accordance with relation (1.47):

$$x_1 = \frac{z_1 M_1}{z_1 M_1 + z_2 M_2} = \frac{z_1 M_1}{M}. \tag{2}$$

The gradients of the mass and molar fractions thus are related by

$$\nabla x_1 = \frac{M_1 M_2}{(z_1 M_1 + z_2 M_2)^2} \nabla z_1 = \frac{M_1 M_2}{M^2} \nabla z_1. \tag{3}$$

In view of the relations above,

$$-\boldsymbol{j}_1 = \rho D \left(k_T \nabla \ln T + \frac{M_1 M_2}{M^2} \nabla z_1 \right). \tag{4}$$

Note that the concentration of the solution as a whole is

$$c = \frac{\rho}{M}. \tag{5}$$

and introduce the thermal diffusion ratio (Table 3.3) defined by

$$k_T^z = k_T \frac{M^2}{M_1 M_2} \tag{6}$$

and the diffusion coefficient

$$D^z = \frac{M_1 M_2}{M} D = \frac{c}{\rho} M_1 M_2 D. \tag{7}$$

The result is the diffusion flux in the form

$$-\boldsymbol{j}_1 = c D^z (k_T^z \nabla \ln T + \nabla z_1). \tag{8}$$

A rough value for the thermal diffusion ratio is given by the formula

$$k_T^z \approx 0.25 z_1 z_2 \frac{M_1 - M_2}{M}. \tag{9}$$

TABLE 3.3

Experimental Values of Thermal Diffusion Ratio k_T^z for Gases and Fluids [16]

Components A–B	Temperature K	Molar Fraction z_A	Thermal Diffusion Ratio k_T^z
He–Ne	330	0.20	0.0531
		0.60	0.1004
H_2–N_2	264	0.294	0.0548
		0.775	0.0663
H_2–D_2	327	0.10	0.0145
		0.50	0.0432
		0.90	0.0166
$C_2H_2Cl_4$–n-C_6H_{14}	298	0.5	1.080
$C_2H_4Br_2$–$C_2H_4Cl_2$	298	0.5	0.225
$C_2H_4Cl_2$–CCl_4	298	0.5	0.060
CBr_4–CCl_4	298	0.09	0.129
CCl_4–CH_3OH	313	0.5	−1.230
H_2O–CH_3OH	313	0.5	−0.137
C_6H_6–cyclo-C_6H_{12}	313	0.5	0.100

Example 3.10. Using the molar diffusion flux relative to the solvent, consider the phenomenon of isothermal diffusion in a binary fluid which is in mechanical equilibrium.

Solution. The molar flux of component i relative to solvent k is defined by a relation which follows from equations (3.20) and (3.28):

$$\boldsymbol{j}_{Mk,i} = c_i(\boldsymbol{v}_i - \boldsymbol{v}_k) = \boldsymbol{j}_{Mv_a i} + \frac{z_i}{z_k} \sum_{n=1}^{k-1} \boldsymbol{j}_{Mv_a n} . \tag{1}$$

The chemical potential of component i, relative to the amount of substance in moles, is denoted by

$$\mu_{Mi} = \left(\frac{\partial G}{\partial n_i} \right)_{p,T,n_{j \neq i}} . \tag{2}$$

The dissipation function for the case of isothermal diffusion under consideration is of the form

$$\Psi_D = - \sum_{i=1}^{k} \boldsymbol{j}_{Mk,i} \cdot \nabla_T \mu'_{Mi} \quad (\nabla T = 0), \tag{3}$$

and, when the Gibbs–Duhem relation is taken into account in combination with the mechanical equilibrium condition

$$\sum_{i=1}^{k} z_i \nabla_T \mu'_{Mi} = 0, \tag{4}$$

also of the form

$$\Psi_D = - \sum_{i=1}^{k-1} \boldsymbol{j}_{Mk,i} \cdot \nabla_T \mu'_{Mi} \quad (\nabla T = 0). \tag{5}$$

For a binary fluid calculations yield a dissipation function containing only the product of two factors

$$\Psi_D = -\boldsymbol{j}_{M2,1} \cdot \nabla_T \mu'_{M1} \tag{6}$$

and one phenomenological equation for the diffusion flux of the first component

$$\boldsymbol{j}_{M2,1} = -L_{11} \nabla_T \mu'_{M1}. \tag{7}$$

The isothermal interdiffusion of two components is a phenomenon caused by a single thermodynamic force, and hence no cross-effects occur in this case.

The isothermal gradient of the chemical potential is expressible in terms of the molar concentration gradient or molar fraction gradient as

$$\nabla_T \mu_{M1} = \left(\frac{\partial \mu_1}{\partial z_1}\right)_{T,p} \nabla z_1, \tag{8}$$

and hence

$$\nabla_T \mu'_{M1} = \left(\frac{\partial \mu_1}{\partial z_1}\right)_{T,p} \nabla z_1 + \tilde{V}_{M1} \nabla p - M\boldsymbol{F}_1. \tag{9}$$

In the absence of mass forces in the state of mechanical equilibrium $\nabla p = \boldsymbol{F}_1 = 0$, whereby

$$\boldsymbol{j}_{M2,1} = -L_{11} \left(\frac{\partial \mu_1}{\partial z_1}\right)_{T,p} \nabla z_1. \tag{10}$$

Introducing the diffusion coefficient

$$D^z_{2,1} = L_{11} \frac{1}{c} \left(\frac{\partial \mu_1}{\partial z_1}\right)_{T,p} \tag{11}$$

gives

$$\boldsymbol{j}_{M2,1} = -cD^z_{2,1} \nabla z_1. \tag{12}$$

Since the diffusion flux can also be written as

$$\boldsymbol{j}_{M2,1} = M_1 \boldsymbol{j}_1 + \frac{z_1}{z_2} M_1 \boldsymbol{j}_1 = \frac{M_1}{z_2} \boldsymbol{j}_1, \tag{13}$$

the result is

$$\boldsymbol{j}_{M2,1} = -\frac{M_1}{z_2} \rho D \nabla x_1. \tag{14}$$

As is known, molar fractions are converted to mass fractions as follows

$$z_2 = \frac{x_2}{M_2} M, \tag{15}$$

whereas the molar and mass fractions are related by

$$\nabla z_1 = \frac{M^2}{M_1 M_2} \nabla x_1. \tag{16}$$

In view of the above, comparison of expressions (12) and (14) yields relations between the diffusion coefficients,

$$D^z_{2,1} = \frac{(M_1 M_2)^2}{M^2 x_2} D \tag{17}$$

or

$$D = \frac{z_2 M}{M_1^2 M_2} D^z_{2,1}. \tag{18}$$

10. Heat Conduction and Diffusion in a Fluid in a Stationary State without Chemical Reactions

To determine the thermal diffusion coefficient D' experimentally, a binary fluid in a vessel is examined. Convection can usually be neglected for fluids contained in vessels. Since the concentration gradients arising from thermal diffusion are low, the density of the fluid may be assumed to be constant and the balance equation (3.38) for the amount of substance may be written as

$$\frac{\partial x_1}{\partial t} = -\nabla \cdot \left(\frac{j_1}{\rho}\right), \tag{3.270}$$

where the diffusion flux of component 1 is expressible as a function of the temperature gradient and mass fraction gradient by means of equation (3.242), and thus

$$\frac{\partial x_1}{\partial t} = \nabla \cdot (x_1 x_2 D' \nabla T + D \nabla x_1). \tag{3.271}$$

The temperature distribution is determined from the equation of heat conduction, neglecting the Dufour effect and assuming that the heat conductivity is constant, that is, in conformity with equation (19) of Example 3.4:

$$\frac{\partial T}{\partial t} = a \nabla^2 T. \tag{3.272}$$

In many cases, a linear temperature distribution may be assumed along the normal to the surface on which the temperature difference has been set up, this distribution being independent of the other directions and of the time. Once the temperature distribution is known, the distribution of the mass fractions is found from equation (3.271). If the experiment began with the same concentrations throughout the entire vessel, a gradient of concentration and hence of mass fraction arises with time owing to the temperature difference. Finally, a stationary state is attained in which diffusion flows vanish, i.e. the density of the diffusion flux defined by equation (3.242) becomes zero. In the stationary state, therefore, the Soret coefficient (3.245) can be expressed as

$$s_T = \left(\frac{D'}{D}\right)_{j_1=0} = -\frac{1}{x_1(1-x_1)} \frac{\nabla x_1}{\nabla T}, \tag{3.273}$$

whereas the thermal diffusion ratio (3.244) is

$$k_T = \left(\frac{D_{T1}}{D}\right)_{j_1=0} = x_1 x_2 T s_T = -T \frac{\nabla x_1}{\nabla T}. \tag{3.274}$$

Knowing the Soret coefficient and the diffusion coefficient D, we calculate the thermal diffusion coefficient D'.

The Soret coefficient s_T for solutions of gases and liquids is of the order of 10^{-3}–10^{-5} K^{-1}. For liquids the diffusion coefficient D is of the order of 10^{-5} cm^2/s, and the thermal diffusion coefficient D' is of the order of 10^{-8}–10^{-10} cm²/s·K. For gases the orders of magnitude of these quantities are 10^{-1} cm²/s in the case of D and 10^{-4}–10^{-6} cm²/s·K in the case of D'.

Experiments have shown that the thermal diffusion and Dufour coefficients are equal for gases. The Onsager reciprocal relations were thus verified experimentally. The Dufour effect is much more difficult to detect in liquids. For gases, by equation (3.251) the temperature difference arising from the mass fraction difference in the absence of a heat flux, $J_q''=0$, is

$$\Delta T \approx -\frac{D'}{\lambda_0} T\rho_1 \left(\frac{\partial \mu_1}{\partial x_1}\right)_{T,p} \Delta x_1 \tag{3.275}$$

and, as follows from experimental data, is of the order of 1 K. For liquids the thermal conductivity λ_0 is greater by a factor of $\approx 10^2$, the thermal diffusion coefficient $D'=D''$ is smaller by a factor of $\approx 10^4$, and the density ρ is greater by a factor of $\approx 10^3$ than for gases. The other quantities are more or less of the same order as for gases. In view of this, the temperature difference ΔT produced in a fluid by the mass fraction difference Δx_1 in the absence of a heat flux, $J_q''=0$, is smaller than in gases by a factor of $\approx 10^3$, that is to say, is a hard-to-measure quantity of the order of 10^{-3} K.

The numerical values of empirical coefficients enable inequalities (3.253) and (3.259) to be verified. For gases the left-hand sides of these inequalities are 10^2–10^3 smaller than the right-hand sides, and for liquid solutions, are approximately 10^4–10^5 times smaller.

When considering heat conduction in multicomponent fluids, one can distinguish different heat conductivities, depending on the kind of heat flux and on the conditions imposed on the fluid in question.

In the experiments described above for determining the thermal diffusion coefficient, the fluid is at first homogeneous and the mass fraction gradients are equal to zero:

$$\left.\begin{aligned} &t=0,\\ &\nabla x_i=0. \end{aligned}\right\} \qquad J_q''=-\lambda_0 \nabla T, \qquad J_u=-\lambda \nabla T. \tag{3.276}$$

By equation (3.189), the heat flux and energy flux are related by

$$J_q''=J_u-\sum_{i=1}^{k-1}(\tilde{h}_i-\tilde{h}_k)j_i. \tag{3.277}$$

Near the end of this thermal diffusion experiment, a stationary state is attained in which diffusion flows vanish, and then

$$\left.\begin{aligned} &t=\infty,\\ &j_i=0. \end{aligned}\right\} \qquad J_q''=J_u=-\lambda_\infty \nabla T. \tag{3.278}$$

The heat conductivities λ_0 and λ_∞ are related to the heat flux J_q'' in the entropy balance equation, a quantity which is convenient in theoretical considerations.

The heat conductivities λ and λ_∞ are associated with the conduction energy flux $\boldsymbol{J}_u$, a quantity which is measured directly during experimental investigations.

If equations (3.218) and (3.219) are inserted into equation (3.277), with $\nabla x_j = 0$, then

$$-\boldsymbol{J}_u = \frac{1}{T}\left[L_{qq} + \sum_{i=1}^{k-1} L_{iq}(\tilde{h}_i - \tilde{h}_k)\right]\nabla T. \tag{3.279}$$

By equations (3.276) and (3.279), the heat conductivity is

$$\lambda = \frac{1}{T}\left[L_{qq} + \sum_{i=1}^{k-1} L_{iq}(\tilde{h}_i - \tilde{h}_k)\right] = \lambda_0 + \frac{1}{T}\sum_{i=1}^{k-1} L_{iq}(\tilde{h}_i - \tilde{h}_k), \tag{3.280}$$

and for a binary solution

$$\lambda = \frac{1}{T}\left[L_{qq} + L_{1q}(\tilde{h}_1 - \tilde{h}_2)\right] = \lambda_0 + \frac{1}{T} L_{1q}(\tilde{h}_1 - \tilde{h}_2) \tag{3.281}$$

In the stationary state $\boldsymbol{j}_i = 0$ and hence equation (3.181) yields

$$-\boldsymbol{J}_q'' = -\boldsymbol{J}_u = L_{qq}\nabla \ln T - \sum_{i,n=1}^{k-1} L_{qi} L_{nq} L_{in}^{-1} \nabla \ln T$$

$$= \frac{1}{T}\left(L_{qq} - \sum_{i,n=1}^{k-1} L_{qi} L_{nq} L_{in}^{-1}\right)\nabla T. \tag{3.282}$$

On the basis of equations (3.278) and (3.282) the heat conductivity in the stationary state is

$$\lambda_\infty = \lambda_0 - \frac{1}{T}\sum_{i,n=1}^{k-1} L_{qi} L_{nq} L_{in}^{-1} \tag{3.283}$$

or, on introduction of the heat of transport (3.182), may also be written as

$$\lambda_0 - \lambda_\infty = \frac{1}{T}\sum_{n=1}^{k-1} L_{qn} Q_n''^{*}. \tag{3.284}$$

For binary fluids this last equation becomes

$$\lambda_0 - \lambda_\infty = \frac{1}{T}\frac{L_{q1}^2}{L_{11}} = \frac{1}{T} L_{11}(Q_1''^{*})^2. \tag{3.285}$$

It is easily seen that for one-component fluids all the heat conductivities are the same and are equal to λ.

For binary fluids the phenomenological coefficients in equations (3.281) and (3.285) may be replaced by the empirical diffusion and thermal diffusion coefficients, D and D', and relations (3.235) and (3.243) taken into account, whereupon

$$\lambda - \lambda_0 = D'\rho x_1(1 - x_1)(\tilde{h}_1 - \tilde{h}_2), \tag{3.286}$$

$$\lambda_0 - \lambda_\infty = \frac{(D')^2}{D}\rho x_1^2(1 - x_1)T\left(\frac{\partial \mu_1}{\partial x_1}\right)_{T,p}. \tag{3.287}$$

If the thermodynamic stability condition $\left(\frac{\partial \mu_1}{\partial x_1}\right)_{T,p} \geqslant 0$ and inequality (3.253) are invoked, the result is

$$\lambda_\infty < \lambda_0 . \tag{3.288}$$

The stationary-state heat conductivity λ_∞ is several per cent smaller than the conductivity λ_0 in the absence of concentration gradients. The heat conductivity λ may be greater or smaller than the conductivity λ_0, that is

$$\lambda \lesseqgtr \lambda_0 . \tag{3.289}$$

For ideal gases, when use is made of relation (3.257), equation (3.287) reduces to the form

$$\lambda_0 - \lambda_\infty = \frac{(D')^2}{D} \frac{x_1 x_2 \rho R_M T^2}{M_2 x_1 + M_1 x_2} > 0 . \tag{3.290}$$

In this case inequality (3.288) is immediately obvious.

Example 3.11. Two vessels at the same temperature are connected by a thermally insulated conduit. The system is filled with a solution of hydrogen and neon. The hydrogen mass fraction is $x_{H_2} = 0.8$. Find the difference between the mass fractions of the components in the two vessels in the stationary state when one vessel is kept at $T_{\mathrm{I}} = 280$ K and the other at $T_{\mathrm{II}} = 400$ K.

Solution. The diffusion flows and diffusion fluxes vanish,

$$j_{H_2} = -j_{Ne} = 0 , \tag{1}$$

in the system in question when it is in the stationary state; hence, by equation (3.242), we have

$$\nabla x_{H_2} = -\frac{k_T}{T} \nabla T . \tag{2}$$

Integration of this equation for a one-dimensional system between the temperature limits T_{I} and T_{II} gives

$$x_{H_2 \mathrm{II}} - x_{H_2 \mathrm{I}} = -\int_{T_I}^{T_{II}} \frac{k_T}{T} \mathrm{d}T . \tag{3}$$

Since the thermal diffusion ratio is temperature-dependent, the integral above can be approximated by taking a constant value for k_T for a reference temperature proposed by H. Brown,

$$T_r = \frac{T_{\mathrm{I}} T_{\mathrm{II}}}{T_{\mathrm{II}} - T_{\mathrm{I}}} \ln \frac{T_{\mathrm{II}}}{T_{\mathrm{I}}} ,$$

$$T_r = \frac{280 \times 400}{400 - 280} \ln \frac{400}{280} = 330 \text{ K} , \tag{4}$$

and then the calculations are carried out in conformity with the formula

$$x_{H_2 \mathrm{II}} - x_{H_2 \mathrm{I}} = -k_T \ln \frac{T_{\mathrm{II}}}{T_{\mathrm{I}}} . \tag{5}$$

Table 3.3 gives the values of the thermal diffusion ratio k_T^z corresponding to calculations employing molar fractions. The ratio k_T^z is converted into the thermal diffusion ratio k_T by means of formula (6) from Example 3.9:

$$k_T = \frac{M_{H_2} M_{Ne}}{M^2} k_T^z = M_{H_2} M_{Ne} \left(\frac{x_{H_2}}{M_{H_2}} + \frac{x_{Ne}}{M_{Ne}} \right)^2 k_T^z = \frac{(M_{Ne} x_{H_2} + M_{H_2} x_{Ne})^2}{M_{H_2} M_{Ne}} k_T^z, \tag{6}$$

$$k_T^z = 0.0531, \quad k_T = \frac{(20.2 \times 0.2 + 2 \times 0.8)^2}{2 \times 20.2} 0.0531 = 0.042.$$

Since the thermal diffusion ratio is positive, the hydrogen diffuses into the vessel at lower temperature, that is, into vessel I.

Equation (5) can now be used to calculate the difference between the hydrogen mass fractions in the two vessels in the stationary state:

$$x_{H_2 II} - x_{H_2 I} = -0.042 \ln \frac{400}{280} = -0.015.$$

The molar fraction and the volume fraction differences are the same and equal to

$$z_{H_2 II} - z_{H_2 I} = -0.0531 \ln \frac{400}{280} = -0.019.$$

11. The Rate of Change of Entropy Production in a Multicomponent Fluid

In order to simplify the mathematical notation, let us consider an isotropic, multicomponent fluid in which no electric and magnetic phenomena occur, and in which the internal friction is negligible owing to low viscosity or low flow velocities. In this case there will be phenomena of a scalar nature, arising only from chemical reactions, and phenomena of a vectorial nature, connected with the conduction of heat and transport of substance. By equations (3.139) and (3.140), the entropy source strength is in this case defined by

$$\Phi_s = \sum_{a=1}^{n} J_a X_a = -\sum_{j=1}^{r} J_j \frac{A_j}{T} + \boldsymbol{J}_u \cdot \nabla \left(\frac{1}{T} \right) - \sum_{j=1}^{k} \boldsymbol{j}_i \cdot \left[\nabla \left(\frac{\mu_i}{T} \right) - \frac{\boldsymbol{F}_i}{T} \right]. \tag{3.291}$$

The expression above can be transformed by using the relations

$$A_j = \sum_{i=1}^{k} \nu_{ij} M_i \mu_i, \tag{3.292}$$

$$\boldsymbol{J}_u \cdot \nabla \left(\frac{1}{T} \right) = \nabla \cdot \left(\frac{\boldsymbol{J}_u}{T} \right) - \frac{1}{T} (\nabla \cdot \boldsymbol{J}_u), \tag{3.293}$$

$$\boldsymbol{j}_i \cdot \nabla \left(\frac{\mu_i}{T} \right) = \nabla \cdot \left(\frac{\mu_i}{T} \boldsymbol{j}_i \right) - \frac{\mu_i}{T} (\nabla \cdot \boldsymbol{j}_i) \tag{3.294}$$

and then

$$\Phi_s = \sum_{a=1}^{n} J_a X_a = -\sum_{j=1}^{r} J_j \sum_{i=1}^{k} \nu_{ij} M_i \frac{\mu_i}{T} + \nabla \cdot \left(\frac{\boldsymbol{J}_u}{T} \right) - \frac{1}{T} (\nabla \cdot \boldsymbol{J}_u)$$
$$- \sum_{i=1}^{k} \nabla \cdot \left(\frac{\mu_i}{T} \boldsymbol{j}_i \right) + \sum_{i=1}^{k} \frac{\mu_i}{T} (\nabla \cdot \boldsymbol{j}_i)$$
$$+ \sum_{i=1}^{k} \boldsymbol{j}_i \cdot \frac{\boldsymbol{F}_i}{T}. \tag{3.295}$$

In accordance with the method developed by Glansdorff and Prigogine, we shall consider how the entropy production varies under the influence of the time-variable thermodynamic forces X_a, and with constant generalized flows J_a; that is, we shall consider the quantity

$$\frac{\partial_X P}{\partial t}=\int_V \sum_{a=1}^{n} J_a \frac{\partial X_a}{\partial t} \mathrm{d}V = -\int_V \Big[\sum_{j=1}^{r} J_j \sum_{i=1}^{k} \nu_{ij} M_i \frac{\partial}{\partial t}\left(\frac{\mu_i}{T}\right) + (\nabla \cdot \boldsymbol{J}_u) \frac{\partial}{\partial t}\left(\frac{1}{T}\right) - \sum_{i=1}^{k} (\nabla \cdot \boldsymbol{j}_i) \frac{\partial}{\partial t}\left(\frac{\mu_i}{T}\right) - \sum_{i=1}^{k} \boldsymbol{j}_i \cdot \boldsymbol{F}_i \frac{\partial}{\partial t}\left(\frac{1}{T}\right)\Big] \mathrm{d}V + \int_A \Big[\boldsymbol{J}_u \frac{\partial}{\partial t}\left(\frac{1}{T}\right) - \sum_{i=1}^{k} \boldsymbol{j}_i \frac{\partial}{\partial t}\left(\frac{\mu_i}{T}\right)\Big] \cdot \mathrm{d}\boldsymbol{A}. \tag{3.296}$$

In this transformation the Gauss–Ostrogradsky theorem has been used and the mass forces have been assumed not to vary with time

$$\frac{\partial \boldsymbol{F}_i}{\partial t}=0. \tag{3.297}$$

Next, the further assumption is made that the pressure p also does not vary with time and that the barycentric velocity $\boldsymbol{v}$ is negligible:

$$\frac{\partial p}{\partial t}=0, \qquad \boldsymbol{v}=0. \tag{3.298}$$

In conformity with the assumptions made above,

$$\nabla \cdot \boldsymbol{j}_i = -\rho \frac{\partial x_i}{\partial t} + \sum_{j=1}^{r} \nu_{ij} M_i J_j, \tag{3.299}$$

$$\nabla \cdot \boldsymbol{J}_u - \sum_{i=1}^{k} \boldsymbol{j}_i \cdot \boldsymbol{F}_i = -\rho \sum_{i=1}^{k} \tilde{h}_i \frac{\partial x_i}{\partial t} - \rho c_p \frac{\partial T}{\partial t}, \tag{3.300}$$

$$\frac{\partial}{\partial t}\left(\frac{\mu_i}{T}\right)_p = -\frac{\tilde{h}_i}{T^2}\frac{\partial T}{\partial t} + \frac{1}{T}\sum_{n=1}^{k-1}\left(\frac{\partial \mu_i}{\partial x_n}\right)_{T,\,p,\,x_j \neq n} \frac{\partial x_n}{\partial t}. \tag{3.301}$$

When these relations are taken into account, the volume integral on the right-hand side of equation (3.296) assumes the form

$$\frac{\partial_X P}{\partial t} = -\int_V \left[\frac{\rho c_p}{T^2}\left(\frac{\partial T}{\partial t}\right)^2 + \frac{\rho}{T}\sum_{i,n=1}^{k-1}\left(\frac{\partial(\mu_i-\mu_k)}{\partial x_n}\right)_{T,\,p,\,x_j \neq n} \frac{\partial x_i}{\partial t}\frac{\partial x_n}{\partial t}\right] \mathrm{d}V \leqslant 0, \tag{3.302}$$

which, because of the thermodynamic stability condition

$$c_p>0, \qquad \sum_{i,n=1}^{k-1}\left(\frac{\partial(\mu_i-\mu_k)}{\partial x_n}\right)_{p,\,T,\,x_j \neq n} \frac{\partial x_i}{\partial t}\frac{\partial x_n}{\partial t} \geqslant 0, \tag{3.303}$$

cannot be positive, and is zero only if

$$\frac{\partial T}{\partial t}=0, \quad \frac{\partial x_i}{\partial t}=0,$$

that is to say, in the stationary state.

The surface integral in equation (3.296) is zero: if the system is adiabatic closed, if the temperature and concentrations (mass fractions) of the components do not vary with time over the entire boundary surface delimiting the system from the environment, or if the temperature on the boundary of a closed system does not vary with time. Under these conditions, the rate of entropy production due to time-variable thermodynamic forces, the generalized flows remaining constant, falls off with time,

$$\frac{\partial_X P}{\partial t}=\int\limits_V \sum_{a=1}^{n} J_a \frac{\partial X_a}{\partial t} \mathrm{d}V \leqslant 0, \tag{3.304}$$

until a minimum is reached in the stationary state.

The time derivative of the rate of entropy production with constant thermodynamic forces and variable generalized flows does not have a strictly defined sign and, in the general case, nothing definite can be said about the character of the change in the overall rate of entropy production as the system approaches a stationary state.

If linear phenomenological equations and Onsager reciprocal relations are valid and the phenomenological coefficients are constant, then by equations (2.85) it is found that

$$\frac{\partial P}{\partial t}=2\frac{\partial_X P}{\partial t}\leqslant 0; \tag{3.305}$$

the rate of entropy production in a system with unvarying conditions at the boundaries falls off until a minimum is reached in the stationary state.

12. Uniqueness Conditions for the Solution of Multicomponent Fluid Problems

To solve problems concerning the distribution of the intensive parameters in a multicomponent fluid, one needs a closed system of differential equations describing the phenomenon under consideration as well as initial and boundary conditions ensuring the solution of these equations is unique.

In considering a k-component fluid, one treats as independent variables concentrations x_i of $k-1$ components, the density ρ, the temperature T, and three components v_α, v_β, v_γ of the barycentric velocity, making a total of $k+4$ variables. The equations of state enable the pressure p, the specific internal energy u, and the chemical potential μ_i to be expressed in terms of the aforementioned independent variables. The complete system of equations thus consists of $k+4$ balance equations in the form of k balance equations for the amount of substance of the individual components and of the fluid as a whole, three momentum balance equations for three directions of coordinate axes, and the energy balance equation. The phenomenological equations make it possible to eliminate the fluxes of the vectorial quantities and the rates of scalar quantities.

The initial conditions for fields of intensive parameters z are determined by the distribution of these parameters throughout the entire space occupied by the fluid at the initial time $t=0$:

$$z(x_\alpha, x_\beta, x_\gamma, t) = z(x_\alpha, x_\beta, x_\gamma, 0).$$

In many problems considerations are started with a uniform distribution of the intensive parameters throughout the entire space under study at the beginning of the phenomenon:

$$z(x_\alpha, x_\beta, x_\gamma, 0) = z_0 = \text{const}.$$

For problems of heat conduction and diffusion the temperature and mass fractions (or concentrations) of $k-1$ components are those parameters.

The uniqueness conditions must also encompass the shapes and geometric dimensions of the system in question.

Boundary conditions of the first kind (Dirichlet's) are specified by the distribution of the particular intensive parameter on the external surface of the system at every moment. In the special case, the boundary conditions of the first kind are constant and the intensive parameters do not vary with time over the entire outer surface of the system.

Boundary conditions of the second kind (Neumann's) are specified by the flux distribution on the outer surface of the system at every moment. In the special case the boundary conditions of the second kind are constant, and the fluxes do not vary with time over the entire outer surface of the system.

Boundary conditions of the third kind (Fourier's) consist of a linear combination of boundary conditions of the first and second kinds.

A closed system of differential equations describing a particular phenomenon in a multicomponent fluid, along with the uniqueness conditions for its solution, constitutes a mathematical model of the phenomenon. The solution of the mathematical model gives the distributions of the intensive parameters in the given system, i.e. specifies the state of the system which is not in equilibrium. Analytic solutions of systems of partial differential equations are possible only for bodies of relatively simple shape and uncomplicated boundary conditions. More involved problems can sometimes be solved by numerical or analogue methods.

13. Application of the Methods of Nonequilibrium Thermodynamics to Chemical Reactions

As already stated in Section III.3, cross-effects between chemical reactions and the flow of a viscous fluid are weak enough so as not to have been observed. In view of this, in considering chemical reactions by the methods of nonequilibrium thermodynamics, one may confine oneself to the phenomena occurring in nonviscous fluids. Under these conditions, the expression for the entropy source strength, and hence the dissipation function as well, are limited to scalar terms containing conjugated chemical affinities and chemical reaction rates. The dissipation function (3.148) corresponding to scalar phenomena takes the form

$$\Psi_0 = T\Phi_{s0} = -\sum_{j=1}^{r} A_j J_j \geqslant 0, \tag{3.306}$$

whereas the phenomenological equations for the reaction rates are restricted to

$$J_j = -\sum_{m=1}^{r} L_{jm} A_m = -\sum_{m=1}^{r} L_{jm} \sum_{i=1}^{k} \nu_{im} M_i \mu_i \tag{3.307}$$

$$(j = 1, 2, 3, \ldots, r),$$

where the phenomenological coefficients are linked by the Onsager reciprocal relations:

$$L_{jm} = L_{mj} \quad (j, m = 1, 2, 3, \ldots, r). \tag{3.308}$$

The dissipation function is a quadratic function of the chemical affinities,

$$\Psi_0 = T\Phi_{s0} = \sum_{j,m=1}^{r} L_{jm} A_j A_m \geqslant 0, \tag{3.309}$$

and its positive values give rise to the following inequalities for the phenomenological coefficients:

$$L_{mm} > 0, \quad L_{mm} L_{jj} - L_{mj}^2 > 0 \quad (m, j = 1, 2, 3, \ldots, r). \tag{3.310}$$

The chemical reaction velocity is defined as the ratio of the change in the concentration of one reactant to the time in which that change took place:

$$w = \pm \frac{dc}{dt}. \tag{3.311}$$

The chemical reaction velocity is always designated by a positive number. The plus sign is used to indicate a reaction velocity found from a change in the product concentration, which increases during the reaction. The minus sign is used when the reaction velocity has been determined from the change in the concentration of the parent substrate which decreases during the reaction.

By the law of mass action of Guldberg and Waage, the velocity of a chemical reaction is proportional to the product of the concentrations of the reactants. For a chemical reaction running from left to right the reaction velocity is

$$w' = k' \prod_i c_i'^{\nu_i'}, \tag{3.312}$$

whereas for the reaction in the opposite direction it is

$$w'' = k'' \prod_i c_i''^{\nu_i''}. \tag{3.313}$$

The reaction rate of reaction j thus is

$$J_j = w' - w'' = w'\left(1 - \frac{w''}{w'}\right) = w'\left(1 - \frac{\prod_i c_i^{\nu_i}}{K_c}\right), \tag{3.314}$$

where $K_c = k'/k''$ is the constant of equilibrium calculated from the concentrations.

On the other hand, it is known that for ideal gases the chemical affinity is

$$A_j = \sum_{i=1}^{k} \nu_{ij} M_i \mu_i = -R_M T \ln \frac{\prod_i c_i^{\nu_i}}{K_c}, \tag{3.315}$$

whereby

$$\frac{\prod_i c_i^{\nu_i}}{K_c} = \exp\left(-\frac{A_j}{R_M T}\right) \tag{3.316}$$

and, finally, the reaction rate is

$$J_j = w'\left[1 - \exp\left(-\frac{A_j}{R_M T}\right)\right]. \tag{3.317}$$

The expression in brackets can be expanded in a power series

$$1 - \exp\left(-\frac{A_j}{R_M T}\right) = \frac{A_j}{R_M T} - \frac{1}{2!}\left(\frac{A_j}{R_M T}\right)^2 + \frac{1}{3!}\left(\frac{A_j}{R_M T}\right)^3 + \dots$$

and in the case of a near-equilibrium state, in which

$$\left|\frac{A_j}{R_M T}\right| \ll 1, \tag{3.318}$$

we can confine ourselves to the first term in the series. Only in this case is there a valid linear relationship between the reaction rate and the chemical affinity,

$$J_j = \frac{w'}{R_M T} A_j, \tag{3.319}$$

and the phenomenological coefficient is equal to

$$L_{jj} = \frac{w'}{R_M T}. \tag{3.320}$$

Inequality (3.318) is satisfied only in the immediate proximity of the equilibrium state, this being due to the very strong limitation on the applicability of the linear phenomenological equations of nonequilibrium processes to chemical reactions. This limitation may in some cases be weakened considerably if the reaction proceeds chainwise.

For chain reactions consisting of r consecutive reactions the dissipation function is given by expression (3.306), and in the stationary state, since the rates of the reactions in the chain are equal,

$$J_j = J_m, \tag{3.321}$$

the dissipation function is

$$\Psi_0 = -J_j \sum_{j=1}^{r} A_j. \tag{3.322}$$

In this case it may be that the affinities of the links in the chain satisfy condition (3.318) whereas the affinity of the entire chain reaction $\sum_{j=1}^{r} A_j$ does not.

Henceforth a k-component fluid in which r chemical reactions may occur will also be considered. However, the discussion will be confined to vectorial phenomena of heat conduction and diffusion for which the term for the entropy source strength consisting of the sum of dot products of thermodynamic forces and fluxes conjugated with them takes on the form of equation (3.140)

$$\Phi_{s1} = J_u \cdot \nabla\left(\frac{1}{T}\right) - \frac{1}{T} \sum_{i=1}^{k} j_i \cdot \left[T\nabla\left(\frac{\mu_i}{T}\right) - F_i\right] \geqslant 0. \tag{3.323}$$

On application of the generalized Planck function

$$y_i' = -\frac{\mu_i + \psi_i}{T} = y_i - \frac{\psi_i}{T}, \tag{3.324}$$

this becomes

$$\Phi_{s1} = J_u \cdot \nabla\left(\frac{1}{T}\right) + \sum_{i=1}^{k} j_i \cdot \left[\nabla y_i' + \psi_i \nabla\left(\frac{1}{T}\right)\right]. \tag{3.325}$$

The mechanical equilibrium condition (3.203)

$$\nabla p = -\sum_{i=1}^{k} \rho_i \nabla \psi_i \tag{3.326}$$

can now be combined with the Gibbs–Duhem relation (1.73)

$$s\nabla T - v\nabla p + \sum_{i=1}^{k} x_i \nabla \mu_i = 0 \tag{3.327}$$

into the relation

$$s\nabla T + \sum_{i=1}^{k} x_i \nabla \mu_i' = 0, \tag{3.328}$$

or

$$h'\nabla\left(\frac{1}{T}\right) + \sum_{i=1}^{k} x_i \nabla y_i' = 0. \tag{3.329}$$

To simplify the notation, use has been made of the generalized specific enthalpy of the fluid

$$h' = h + \sum_{i=1}^{k} x_i \psi_i = \sum_{i=1}^{k} x_i \mu_i' + Ts. \tag{3.330}$$

Thus the thermodynamic forces $\nabla(1/T)$ and $\nabla y_i'$, which occur in equation (3.325), are found to be related by:

$$\nabla\left(\frac{1}{T}\right)+\frac{1}{h'}\sum_{i=1}^{k} x_i \nabla y_i'=0. \tag{3.331}$$

Now the term for the entropy source strength in the case of vectorial phenomena becomes

$$\Phi_{s1}=\sum_{i=1}^{k} \boldsymbol{J}_i' \cdot \nabla y_i' \geqslant 0, \tag{3.332}$$

where we have introduced new fluxes, defined as

$$\boldsymbol{J}_i'=\boldsymbol{j}_i-\frac{x_i}{h'}(\boldsymbol{J}_u+\sum_{m=1}^{k}\psi_m \boldsymbol{j}_m) \quad (i=1,2,3,\ldots,k). \tag{3.333}$$

If the potentials ψ_m are identical for all components, as is the case when the effect of the gravitational field is taken into account, the term representing the potential energy flux of the diffusible components vanishes

$$\sum_{m=1}^{k}\psi_m \boldsymbol{j}_m=0, \tag{3.334}$$

since, as is known,

$$\sum_{m=1}^{k}\boldsymbol{j}_m=0. \tag{3.335}$$

The fluxes $\boldsymbol{J}_i'$ are expressible by means of linear phenomenological equations

$$\boldsymbol{J}_i'=\sum_{n=1}^{k} l_{in}^* \nabla y_n' \quad (i=1,2,3,\ldots,k), \tag{3.336}$$

and then the term pertaining to vectorial phenomena in the expression for the entropy source strength is of the form

$$\Phi_{s1}=\sum_{i,n=1}^{k} l_{in}^* \nabla y_n' \cdot \nabla y_i' \geqslant 0. \tag{3.337}$$

The condition that this expression be negative implies that the phenomenological coefficients satisfy the inequalities

$$l_{ii}^*>0; \quad l_{ii}^* l_{nn}^*>\tfrac{1}{4}(l_{in}^*+l_{ni}^*)^2 \quad (i,n=1,2,3,\ldots,k), \tag{3.338}$$

and, when the Onsager reciprocal relations

$$l_{in}^*=l_{ni}^* \quad (i,n=1,2,3,\ldots,k) \tag{3.339}$$

are taken into account, also satisfy

$$l_{ii}^* l_{nn}^*-l_{in}^{*2}>0 \quad (i,n=1,2,3,\ldots,k). \tag{3.340}$$

If the mass forces are limited to gravitational forces, at most, then when equations (3.334) and (3.335) and

$$\sum_{m=1}^{k} x_m = 1$$

are taken into account, it is possible to calculate fluxes with a straightforward physical interpretation:
the conduction energy flux

$$\boldsymbol{J}_u = -h' \sum_{i=1}^{k} \boldsymbol{J}_i' = -h' \sum_{i,n=1}^{k} l_{in}^{*} \nabla y_n', \tag{3.341}$$

the diffusion flux of component i

$$\boldsymbol{j}_i = \boldsymbol{J}_i' - x_i \sum_{n=1}^{k} \boldsymbol{J}_n' = \sum_{n=1}^{k} l_{in}^{*} \nabla y_n' - x_i \sum_{i,j=1}^{k} l_{ji}^{*} \nabla y_j' . \tag{3.342}$$

If the mass forces are neglected, the heat flux in the entropy balance equation is

$$\boldsymbol{J}_q'' = \boldsymbol{J}_u - \sum_{n=1}^{k} \tilde{h}_n \boldsymbol{j}_n = - \sum_{n=1}^{k} \tilde{h}_n \boldsymbol{J}_n'$$

$$= - \sum_{n=1}^{k} \tilde{h}_n \sum_{i=1}^{k} l_{in}^{*} \nabla y_i . \tag{3.343}$$

In the absence of mass forces, the state of mechanical equilibrium entails the absence of a pressure gradient, and then

$$\nabla_p y_i = -\tilde{h}_i \nabla \left(\frac{1}{T} \right) - \frac{1}{T} \sum_{j=1}^{k-1} \left(\frac{\partial \mu_i}{\partial x_j} \right)_{T,\, p,\, x_{n \neq j}} \nabla x_j . \tag{3.344}$$

Hence, when the concentration gradient is omitted, equation (3.343) can be recast into the form of Fourier's law

$$\boldsymbol{J}_q'' = -\frac{1}{T^2} \sum_{n=1}^{k} \tilde{h}_n \sum_{i=1}^{k} l_{in}^{*} \tilde{h}_i \nabla T \tag{3.345}$$

with heat conductivity

$$\Lambda'' = \frac{1}{T^2} \sum_{n=1}^{k} \tilde{h}_n \sum_{i=1}^{k} l_{in}^{*} \tilde{h}_i . \tag{3.346}$$

For a binary fluid in which chemical reactions occur, the heat conductivity is related to the phenomenological coefficients by

$$\Lambda'' = \frac{1}{T^2} \left(l_{11}^{*} \tilde{h}_1^2 + 2 l_{12}^{*} \tilde{h}_1 \tilde{h}_2 + l_{22}^{*} \tilde{h}_2^2 \right) . \tag{3.347}$$

To consider further relationships between phenomenological and empirical coefficients, equation (3.342) for the diffusion flux of component i when chemical reactions occur needs to be put into a form such that the thermodynamic forces be analogous to those in equation (3.219) when no chemical reactions occur. For this purpose use should be made of relation (3.344) which, on insertion into equation (3.342), yields

$$j_n = \Big(\sum_{i=1}^{k} l^*_{ni}\tilde{h}_i - x_n \sum_{i,j=1}^{k} l^*_{ji}\tilde{h}_j\Big)\frac{1}{T}\nabla \ln T$$
$$-\Big(\sum_{i=1}^{k} l^*_{ni} - x_n \sum_{i,n=1}^{k} l^*_{in}\Big)\sum_{j=1}^{k-1}\left(\frac{\partial \mu_i}{\partial x_j}\right)_{T,\,p,\,x_n\neq j}\frac{1}{T}\nabla x_j . \tag{3.348}$$

As shown by comparison of this equation with equation (3.219), the phenomenological coefficients associated with thermal diffusion are related by

$$TL_{nq} = -\sum_{i=1}^{k} l^*_{ni}\tilde{h}_i + x_n \sum_{i,j=1}^{k} l^*_{ij}\tilde{h}_j . \tag{3.349}$$

For a binary system the thermal diffusion coefficient is expressed in terms of the phenomenological coefficients as

$$D' = \frac{L_{1q}}{\rho x_1 x_2 T}$$
$$= \frac{1}{\rho x_1 x_2 T^2}\big[l^*_{22} x_1 \tilde{h}_2 + l^*_{12}(x_1 \tilde{h}_1 - x_2 \tilde{h}_2) - l^*_{11} x_2 \tilde{h}_1\big] . \tag{3.350}$$

The phenomenological coefficients associated with diffusion are related by

$$T\sum_{i,m,j=1}^{k-1} L_{ni}\, a_{im}\left(\frac{\partial \mu_m}{\partial x_j}\right)_{T,\,p,\,x_i\neq j} = \Big(\sum_{i=1}^{k} l^*_{ni} - x_n \sum_{i,n=1}^{k} l^*_{in}\Big)\sum_{j=1}^{k-1}\left(\frac{\partial \mu_i}{\partial x_j}\right)_{T,\,p,\,x_n\neq j} \tag{3.351}$$

The diffusion coefficient for a binary fluid can be written as

$$D = \frac{L_{11}}{\rho x_2}\left(\frac{\partial \mu_1}{\partial x_1}\right)_{T,\,p} = \frac{1}{\rho x_2 T}\big(l^*_{11} x_2^2 - 2 l^*_{12} x_1 x_2 + l^*_{22} x_1^2\big)\left(\frac{\partial \mu_1}{\partial x_1}\right)_{T,\,p} . \tag{3.352}$$

If transformation (3.344) is applied to the conduction energy flux (3.341) in the absence of mass forces and concentration gradients, then

$$J_u = -h\sum_{i,n=1}^{k} l^*_{in}\nabla y_n = -\frac{h}{T^2}\sum_{i,n=1}^{k} l^*_{in}\tilde{h}_n \nabla T \tag{3.353}$$

and the heat conductivity is defined by

$$\Lambda = \frac{h}{T^2}\sum_{i,n=1}^{k} l^*_{in}\tilde{h}_n . \tag{3.354}$$

This heat conductivity for a binary fluid is

$$\Lambda = \frac{h}{T^2}\big[l^*_{11}\tilde{h}_1 + l^*_{12}(\tilde{h}_1 + \tilde{h}_2) + l^*_{22}\tilde{h}_2\big] . \tag{3.355}$$

CHAPTER IV

THERMOELECTRIC AND THERMOMAGNETIC EFFECTS

1. The Fundamental Equations for Unpolarized Media

In examining thermoelectric, thermomagnetic, and galvanomagnetic effects, one needs to take account of electromagnetic forces, hitherto not taken into consideration. It should be remarked that in the general case these forces are not conservative. In proceeding to analyse thermoelectric, thermomagnetic, and galvanomagnetic effects, therefore, the balance equation must be reconsidered with due regard for the electromagnetic forces. Accordingly, we shall discuss a material medium consisting of k components which may have electric charges and is in an electromagnetic field.

It has further been assumed that the medium is unpolarized and the possibility of chemical reactions occurring has been excluded from the considerations. The first fundamental equation for a system under study is the equation for conservation of electric charge, which is a special case of the general balance equation.

If the electric charge per unit mass of solution is denoted by e, the equation can be written in a form emerging from the general conservation equation (3.15)

$$\rho \frac{\mathrm{d}e}{\mathrm{d}t} = -\nabla \cdot \boldsymbol{J}_{\mathrm{e}}, \tag{4.1}$$

where $\boldsymbol{J}_{\mathrm{e}}$ stands for the density of the electric current arising from diffusion of the individual components.

The electric charge e is related to the charge e_i per unit mass of component i by

$$e = \frac{1}{\rho} \sum_{i=1}^{k} \rho_i e_i = \sum_{i=1}^{k} x_i e_i . \tag{4.2}$$

The total electric current density is the sum of the diffusion and convection currents and is expressible by the relation

$$\boldsymbol{J}_{\mathrm{el}} = \sum_{i=1}^{k} \rho_i e_i \boldsymbol{v}_i = \rho e \boldsymbol{v} + \sum_{i=1}^{k} e_i \boldsymbol{j}_i = \rho e \boldsymbol{v} + \boldsymbol{J}_{\mathrm{e}}, \tag{4.3}$$

where $\rho e \boldsymbol{v}$ is the density of the current due to convection, whereas

$$\boldsymbol{J}_{\mathrm{e}} = \sum_{i=1}^{k} e_i \boldsymbol{j}_i \tag{4.4}$$

is the diffusion flux.

As before, $\boldsymbol{v}_i$ and $\boldsymbol{v}$ in equation (4.3) denote the barycentric velocities of the individual components and of the solution as a whole.

The momentum balance equation can be obtained from equation (3.64) in which we must introduce the Lorentz force described by expression (3.48):

$$\rho \frac{\mathrm{d}\boldsymbol{v}}{\mathrm{d}t} = \boldsymbol{\nabla}\cdot\boldsymbol{\sigma} + \sum_{i=1}^{k} \rho_i e_i \left[\boldsymbol{E} + \frac{1}{c} (\boldsymbol{v}_i \times \boldsymbol{B}) \right], \tag{4.5}$$

where $\boldsymbol{E}$ is the electric field strength, $\boldsymbol{B}$ is the magnetic field strength (historically: magnetic induction), and c is the velocity of light. This expression can be transformed by using equations (4.2), (4.3), and (4.4) and it is then put into the form

$$\rho \frac{\mathrm{d}\boldsymbol{v}}{\mathrm{d}t} = \boldsymbol{\nabla}\cdot\boldsymbol{\sigma} + \rho e \left[\boldsymbol{E} + \frac{1}{c} (\boldsymbol{v} \times B) \right] + \frac{1}{c} (\boldsymbol{J}_\mathrm{e} \times B). \tag{4.6}$$

Scalar multiplication of this equation by the centre-of-mass velocity v yields the energy equation

$$\rho \frac{\mathrm{d}(\frac{1}{2}v^2)}{\mathrm{d}t} = \boldsymbol{v}\cdot(\boldsymbol{\nabla}\cdot\boldsymbol{\sigma}) + \rho e \boldsymbol{v}\cdot\boldsymbol{E} + \rho e \frac{1}{c} \boldsymbol{v}\cdot(\boldsymbol{v}\times B) + \frac{1}{c} \boldsymbol{v}\cdot(\boldsymbol{J}_\mathrm{e}\times\boldsymbol{B}).$$

Next, use can be made of the relations

$$\frac{\rho\,\mathrm{d}(\frac{1}{2}v^2)}{\mathrm{d}t} = \frac{\partial(\frac{1}{2}\rho v^2)}{\partial t} + \boldsymbol{\nabla}\cdot(\tfrac{1}{2}\rho v^2 \boldsymbol{v}),$$

$$\boldsymbol{v}\cdot(\boldsymbol{\nabla}\cdot\boldsymbol{\sigma}) = \boldsymbol{\nabla}\cdot(\boldsymbol{v}\cdot\boldsymbol{\sigma}) - \boldsymbol{\sigma} : (\boldsymbol{\nabla}\boldsymbol{v}),$$

$$\boldsymbol{v}\cdot(\boldsymbol{v}\times\boldsymbol{B}) = 0,$$

$$\boldsymbol{v}\cdot(\boldsymbol{J}_\mathrm{e}\times\boldsymbol{B}) = -\boldsymbol{J}_\mathrm{e}\cdot(\boldsymbol{v}\times\boldsymbol{B})$$

and the energy balance equation can be recast into the form

$$\frac{\partial(\frac{1}{2}\rho v^2)}{\partial t}$$

$$= -\boldsymbol{\nabla}\cdot(\tfrac{1}{2}\rho v^2\boldsymbol{v}) + \boldsymbol{\nabla}\cdot(\boldsymbol{v}\cdot\boldsymbol{\sigma}) - \boldsymbol{\sigma} : (\boldsymbol{\nabla}\boldsymbol{v}) + \rho e\boldsymbol{v}\cdot\boldsymbol{E} - \frac{1}{c} \boldsymbol{J}_\mathrm{e}\cdot(\boldsymbol{v}\times\boldsymbol{B}). \tag{4.7}$$

This expression can be transformed further by using Maxwell's equations which relate the electric displacement $\boldsymbol{D}$ and the magnetic field strength $\boldsymbol{B}$ with the electric field strength $\boldsymbol{E}$ and the magnetic displacement (historically: magnetic field strength) $\boldsymbol{H}$:

$$\frac{\partial\boldsymbol{D}}{\partial t} - c(\boldsymbol{\nabla}\times\boldsymbol{H}) = -\boldsymbol{J}_\mathrm{el},$$

$$\frac{\partial\boldsymbol{B}}{\partial t} + c(\boldsymbol{\nabla}\times\boldsymbol{E}) = 0.$$

The relations

$$\boldsymbol{D}=\boldsymbol{E} \quad \text{and} \quad \boldsymbol{H}=\boldsymbol{B}$$

are valid in unpolarized media.

Combinations of Maxwell's equations for unpolarized media enable the Poynting equation to be obtained:

$$\frac{1}{2}\frac{\partial}{\partial t}(E^2+B^2)=-\nabla\cdot c(\boldsymbol{E}\times\boldsymbol{B})-\boldsymbol{J}_{\text{el}}\cdot\boldsymbol{E}. \tag{4.8}$$

The meaning of the terms in this equation is as follows: $\frac{1}{2}(E^2+B^2)$ is the electromagnetic energy density, $c(\boldsymbol{E}\times\boldsymbol{B})$ is the Poynting vector, or the electromagnetic energy flux, and $\boldsymbol{J}_{\text{el}}\cdot\boldsymbol{E}$ is the work of the electromagnetic field.

Adding equations (4.7) and (4.8) memberwise, we obtain a new form for the energy balance equation, also containing the electromagnetic energy:

$$\frac{\partial}{\partial t}[\tfrac{1}{2}(\rho v^2+E^2+B^2)]=-\nabla\cdot[\tfrac{1}{2}\rho v^2\boldsymbol{v}-\boldsymbol{\sigma}\cdot\boldsymbol{v}+c(\boldsymbol{E}\times\boldsymbol{B})]$$
$$-\boldsymbol{\sigma}:(\nabla\boldsymbol{v})-\boldsymbol{J}_{\text{e}}\cdot\left[\boldsymbol{E}+\frac{1}{c}(\boldsymbol{v}\times\boldsymbol{B})\right]. \tag{4.9}$$

It follows from this equation that the sum of the kinetic and electromagnetic energy is not conserved, but the energy $\boldsymbol{J}_{\text{e}}\cdot\left[\boldsymbol{E}+\frac{1}{c}(\boldsymbol{v}\times\boldsymbol{B})\right]+\boldsymbol{\sigma}:(\nabla\boldsymbol{v})$ is converted to another form.

The total energy e_{t} of the system must satisfy the general balance equation

$$\frac{\partial e_{\text{t}}}{\partial t}=-\nabla\cdot\boldsymbol{J}_{\text{et}}, \tag{4.10}$$

where $\boldsymbol{J}_{\text{et}}$ denotes the total energy flux.

The quantities e_{t} and $\boldsymbol{J}_{\text{et}}$ are described by:

$$e_{\text{t}}=\rho u+\tfrac{1}{2}(\rho v^2+E^2+B^2),$$

where u is the specific internal energy, and

$$\boldsymbol{J}_{\text{et}}=\boldsymbol{J}_u+[\tfrac{1}{2}\rho v^2\boldsymbol{v}+\rho u\boldsymbol{v}-\boldsymbol{\sigma}\cdot\boldsymbol{v}+c(\boldsymbol{E}\times\boldsymbol{B})].$$

If equations (4.9) and (4.10) are subtracted from each other memberwise and if the expressions describing e_{t} and $\boldsymbol{J}_{\text{et}}$ are used, the result is

$$\frac{\partial(\rho u)}{\partial t}=-\nabla\cdot(\rho u\boldsymbol{v}+\boldsymbol{J}_u)+\boldsymbol{\sigma}:(\nabla\boldsymbol{v})+\boldsymbol{J}_{\text{e}}\cdot\left[\boldsymbol{E}+\frac{1}{c}(\boldsymbol{v}\times\boldsymbol{B})\right] \tag{4.11}$$

or

$$\rho\frac{\mathrm{d}u}{\mathrm{d}t}=-\nabla\cdot\boldsymbol{J}_u+\boldsymbol{\sigma}:(\nabla\boldsymbol{v})+\boldsymbol{J}_{\text{e}}\cdot\left[\boldsymbol{E}+\frac{1}{c}(\boldsymbol{v}\times\boldsymbol{B})\right]$$
$$=-\nabla\cdot\boldsymbol{J}_u-\rho p\frac{\mathrm{d}v}{\mathrm{d}t}+\boldsymbol{\tau}:(\nabla\boldsymbol{v})+\boldsymbol{J}_{\text{e}}\cdot\left[\boldsymbol{E}+\frac{1}{c}(\boldsymbol{v}\times\boldsymbol{B})\right]. \tag{4.12}$$

The last two terms in equation (4.12) stand for the energy converted into internal energy as a result of dissipation, and the term $\boldsymbol{J}_e \cdot \left[\boldsymbol{E} + \frac{1}{c}(\boldsymbol{v} \times \boldsymbol{B})\right]$ is the dissipated electromagnetic energy (per unit volume and time).

The next equation describing the system under consideration is the entropy balance equation.

For an unpolarized medium the specific entropy is a function only of the internal energy, specific volume, and the fractions of the individual components, and hence can be described by the Gibbs relation

$$T\frac{\mathrm{d}s}{\mathrm{d}t} = \frac{\mathrm{d}u}{\mathrm{d}t} + p\frac{\mathrm{d}v}{\mathrm{d}t} - \sum_{i=1}^{k} \mu_i \frac{\mathrm{d}x_i}{\mathrm{d}t}.$$

In the absence of chemical reactions, the conservation equation for the amount of substance is of the form

$$\rho\frac{\mathrm{d}x_i}{\mathrm{d}t} = -\boldsymbol{\nabla}\cdot\boldsymbol{j}_i \qquad (i=1, 2, 3, \ldots, k).$$

Combination of the two equations above with expression (4.12) leads to

$$\rho\frac{\mathrm{d}s}{\mathrm{d}t} = -\boldsymbol{\nabla}\cdot\left(\frac{\boldsymbol{J}_u - \sum_{i=1}^{k}\mu_i\boldsymbol{j}_i}{T}\right) - \frac{1}{T^2}\boldsymbol{J}_u\cdot\boldsymbol{\nabla}T - \frac{1}{T}\sum_{i=1}^{k}\boldsymbol{j}_i\cdot\left\{T\boldsymbol{\nabla}\left(\frac{\mu_i}{T}\right)\right.$$
$$\left. - e_i\left[\boldsymbol{E} + \frac{1}{c}(\boldsymbol{v}\times\boldsymbol{B})\right]\right\} + \frac{1}{T}\boldsymbol{\tau}:(\boldsymbol{\nabla}\boldsymbol{v}) = -\boldsymbol{\nabla}\cdot\boldsymbol{J}_s + \Phi_s, \tag{4.13}$$

where Φ_s is the entropy source strength.

The conduction entropy flux is described by

$$\boldsymbol{J}_s = \frac{1}{T}\left(\boldsymbol{J}_u - \sum_{i=1}^{k}\mu_i\boldsymbol{j}_i\right),$$

whereas the entropy source strength is equal to

$$\Phi_s = \boldsymbol{J}_u\cdot\boldsymbol{\nabla}\left(\frac{1}{T}\right) - \frac{1}{T}\sum_{i=1}^{k}\boldsymbol{j}_i\cdot\left\{T\boldsymbol{\nabla}\left(\frac{\mu_i}{T}\right) - e_i\left[\boldsymbol{E} + \frac{1}{c}(\boldsymbol{v}\times\boldsymbol{B})\right]\right\}$$
$$+ \frac{1}{T}\boldsymbol{\tau}:(\boldsymbol{\nabla}\boldsymbol{v}). \tag{4.14}$$

Similarly, the dissipation function can be written as

$$\Psi = T\Phi_s = -\frac{1}{T}\boldsymbol{J}_u\cdot\boldsymbol{\nabla}T$$
$$- \sum_{i=1}^{k}\boldsymbol{j}_i\cdot\left\{T\boldsymbol{\nabla}\cdot\left(\frac{\mu_i}{T}\right) - e_i\left[\boldsymbol{E} + \frac{1}{c}(\boldsymbol{v}\times\boldsymbol{B})\right]\right\} + \boldsymbol{\tau}:(\boldsymbol{\nabla}\boldsymbol{v}). \tag{4.15}$$

Use of the expression describing the entropy flux enables the dissipation function to be written in another form convenient for applications, viz.:

$$\Psi = -\boldsymbol{J}_s \cdot \nabla T - \sum_{i=1}^{k} \boldsymbol{j}_i \cdot \left\{ \nabla \mu_i - e_i \left[\boldsymbol{E} + \frac{1}{c} (\boldsymbol{v} \times \boldsymbol{B}) \right] \right\} + \boldsymbol{\tau} : (\nabla \boldsymbol{v}) . \tag{4.16}$$

In the sequel, phenomena in which viscosity plays no role will be considered, thus allowing the last term in equation (4.16) to be neglected.

Setting $\boldsymbol{v} = \boldsymbol{v}_i - \frac{\overset{\circ}{\boldsymbol{j}}_i}{\rho_i}$ and $\boldsymbol{j}_i \cdot (\boldsymbol{j}_i \times \boldsymbol{B}) = 0$, we can write the dissipation function as

$$\Psi = -\boldsymbol{J}_s \cdot \nabla T - \sum_{i=1}^{k} \boldsymbol{j}_i \cdot \left\{ \nabla \mu_i - e_i \left[\boldsymbol{E} + \frac{1}{c} (\boldsymbol{v}_i \times \boldsymbol{B}) \right] \right\} . \tag{4.17}$$

If the system is in a state of mechanical equilibrium, equations (3.203) and (3.205) can be used to obtain

$$\sum_{i=1}^{k} \rho_i (\nabla_T \mu_i - \boldsymbol{F}_i) = \sum_{i=1}^{k} \rho_i \left\{ \nabla_T \mu_i - e_i \left[\boldsymbol{E} + \frac{1}{c} (\boldsymbol{v}_i \times \boldsymbol{B}) \right] \right\} = 0 , \tag{4.18}$$

which, on insertion of

$$\nabla_T \mu_i = \nabla \mu_i + \tilde{s}_i \nabla T ,$$

can be recast into the form

$$\sum_{i=1}^{k} \rho_i \left\{ \nabla \mu_i - e_i \left[\boldsymbol{E} + \frac{1}{c} (\boldsymbol{v}_i \times \boldsymbol{B}) \right] \right\} = -\rho s \nabla T . \tag{4.19}$$

The equations describing the dissipation function can also be written in terms of an arbitrary reference velocity $\boldsymbol{v}_a$ and the corresponding diffusion fluxes $\boldsymbol{j}_{vai}$. If, as before, viscosity effects are neglected and systems in mechanical equilibrium are considered, then – because $\boldsymbol{j}_{vai} = \rho_i (\boldsymbol{v}_i - \boldsymbol{v}_a)$ – equations (4.17) and (4.19) yield

$$\Psi = -[\boldsymbol{J}_s + \rho s (\boldsymbol{v} - \boldsymbol{v}_a)] \cdot \nabla T - \sum_{i=1}^{k} \boldsymbol{j}_{vai} \cdot \left\{ \nabla \mu_i - e_i \left[\boldsymbol{E} + \frac{1}{c} (\boldsymbol{v}_i \times \boldsymbol{B}) \right] \right\} . \tag{4.20}$$

For a metal conductor it is convenient to choose a crystal lattice as a reference frame and then assume that its barycentric velocity, which is the reference velocity in calculations of the diffusion flux, is zero, $\boldsymbol{v}_a = 0$. In this case the electron flux $\boldsymbol{j}_e$ is equal to $\boldsymbol{j}_{vai=0} = \rho_i (\boldsymbol{v}_i - \boldsymbol{v}_a) = \rho_i \boldsymbol{v}_i$, whereas the charge of a unit mass of electron is $e_i = e_e$.

Using the relation $\boldsymbol{j}_e \cdot (\boldsymbol{j}_e \times \boldsymbol{B}) = 0$, which enables $\boldsymbol{v}_i$ in equation (4.20) to be replaced by $\boldsymbol{v}_a$, and making the aforementioned assumptions, we rewrite equation (4.20) as

$$\Psi = -(\boldsymbol{J}_s + \rho s \boldsymbol{v}) \cdot \nabla T - \boldsymbol{j}_e \cdot (\nabla \mu_e - e_e \boldsymbol{E}) , \tag{4.21}$$

where μ_e is the electron chemical potential.

In the given case of system consisting of electrons and a crystal lattice the following relations are obeyed:

for the electron flux $\boldsymbol{j}_e = \rho_e \boldsymbol{v}_e$

for the crystal lattice $\boldsymbol{j}_i = 0$

and subsequently $\rho s \boldsymbol{v} = \sum_{i=1}^{2} \rho_i \tilde{s}_i \boldsymbol{v} = -\sum_{i=1}^{2} \rho_i \tilde{s}_i (\boldsymbol{v}_i - \boldsymbol{v} + \boldsymbol{v}_i - \boldsymbol{v}_a)$ since $\boldsymbol{v}_a = 0$. Application of formulae (3.17) and (3.25) yields

$$\rho s \boldsymbol{v} = -(\tilde{s}_e \boldsymbol{j}_e - \sum_{i=1}^{2} \tilde{s}_i \boldsymbol{j}_i).$$

When these results are taken into account, equation (4.21) can be transformed into

$$\Psi = -(\boldsymbol{J}_s + \tilde{s}_e \boldsymbol{j}_e - \sum_{i=1}^{2} \tilde{s}_i \boldsymbol{j}_i) \cdot \nabla T - \boldsymbol{j}_e \cdot (\nabla \mu_e - e_e \boldsymbol{E}). \tag{4.22}$$

If $\boldsymbol{J}_s$ is replaced by $\boldsymbol{J}_q''$ in accordance with equation (3.125), then

$$\Psi = -\left(\frac{\boldsymbol{J}_q''}{T} + \tilde{s}_e \boldsymbol{j}_e\right) \cdot \nabla T - \boldsymbol{j}_e \cdot (\nabla \mu_e - e_e \boldsymbol{E}). \tag{4.23}$$

Relation (4.3) implies that

$$\boldsymbol{J}_{el} = \sum_{i=1}^{2} \rho_i e_i \boldsymbol{v}_i = e_e \boldsymbol{j}_e,$$

which makes it possible to obtain one more form of the expression for the dissipation function:

$$\Psi = -\left(\frac{\boldsymbol{J}_q''}{T} + \tilde{s}_e \boldsymbol{j}_e\right) \cdot \nabla T - \boldsymbol{J}_{el} \cdot \left(\nabla \frac{\mu_e}{e_e} - \boldsymbol{E}\right) = -\frac{\boldsymbol{J}_q''}{T} \cdot \nabla T - \boldsymbol{J}_{el} \cdot \left(\nabla_T \frac{\mu_e}{e_e} - \boldsymbol{E}\right). \tag{4.24}$$

The entropy source strength is equal to

$$\Phi_s = \left(\boldsymbol{J}_q'' + \frac{1}{T} \tilde{s}_e \boldsymbol{j}_e\right) \cdot \nabla \left(\frac{1}{T}\right) - \boldsymbol{J}_{el} \frac{1}{T} \cdot \left(\nabla \frac{\mu_e}{e_e} - \boldsymbol{E}\right). \tag{4.25}$$

It is sometimes more convenient to operate with the internal energy flux $\boldsymbol{J}_u$ which is related to $\boldsymbol{J}_q''$ by

$$\boldsymbol{J}_u = \boldsymbol{J}_q'' + \sum_{i=1}^{k} \tilde{h}_i \boldsymbol{j}_i.$$

Introducing this into expression (4.24) gives

$$\Psi = -\frac{\boldsymbol{J}_u}{T} \cdot \nabla T - \boldsymbol{J}_{el} \cdot \left[T \nabla \left(\frac{\mu_e}{e_e T}\right) - \boldsymbol{E}\right] \tag{4.26}$$

and

$$\Phi_s = \boldsymbol{J}_u \cdot \nabla \left(\frac{1}{T}\right) - \boldsymbol{J}_{el} \cdot \left[\nabla \left(\frac{\mu_e}{e_e T}\right) - \frac{\boldsymbol{E}}{T}\right]. \tag{4.27}$$

2. Analysis of Phenomena in an Electrical Conductor

In view of the range of topics which will be considered further on, this analysis of phenomena occurring in a solid electrical conductor will initially be confined to cases of a one-dimensional flow of heat and electric current. Moreover, we shall also consider phenomena which occur in the absence of a magnetic field, neglecting viscosity. In that case the energy balance equation (4.12) can be simplified to

$$\rho \frac{du}{dt} = -\nabla \cdot \boldsymbol{J}_u + \boldsymbol{J}_e \cdot \boldsymbol{E} = -\nabla \cdot \boldsymbol{J}_u + e_e \boldsymbol{j}_e \cdot \boldsymbol{E}, \tag{4.28}$$

since the other terms are zero. It should be added that in the model adopted for the phenomenon the current is carried by electron diffusion, hence $\boldsymbol{J}_e = \boldsymbol{J}_{el}$, and furthermore $\boldsymbol{j}_e$ is the electron flux relative to the crystal lattice of the conductor.

In some discussions it is more convenient to operate with the heat flux $\boldsymbol{J}_q''$, described [in accordance with equation (3.133)] by the relation

$$\boldsymbol{J}_q'' = \boldsymbol{J}_u - \tilde{h}_e \boldsymbol{j}_e,$$

where $\tilde{h}_e$ is the partial electron enthalpy.

On introduction of $\boldsymbol{J}_q''$ the energy balance equation (4.28) becomes

$$\rho \frac{du}{dt} = -\nabla \cdot \boldsymbol{J}_q'' - \nabla \cdot (\tilde{h}_e \boldsymbol{j}_e) + e_e \boldsymbol{j}_e \cdot \boldsymbol{E} \tag{4.29}$$

or

$$\rho \frac{du}{dt} = -\nabla \cdot \boldsymbol{J}_q'' - \tilde{h}_e (\nabla \cdot \boldsymbol{j}_e) - \boldsymbol{j}_e \cdot \nabla \tilde{h}_e + e_e \boldsymbol{j}_e \cdot \boldsymbol{E}. \tag{4.30}$$

The dissipation function is described by equation (4.24) if the fluxes $\boldsymbol{J}_q''$ and $\boldsymbol{J}_{el}$ are used, and by equation (4.26) if $\boldsymbol{J}_u$ and $\boldsymbol{J}_{el}$ are used. It is more convenient to adapt these relations to the electron flux $\boldsymbol{j}_e$, and then

$$\Psi = -\frac{\boldsymbol{J}_q''}{T} \cdot \nabla T - \boldsymbol{j}_e \cdot (\nabla_T \mu_e - e_e \boldsymbol{E}) = -\boldsymbol{J}_q'' \cdot \nabla (\ln T) - \boldsymbol{j}_e \cdot \nabla_T \mu_e' \tag{4.31}$$

and

$$\Psi = -\frac{\boldsymbol{J}_u}{T} \cdot \nabla T - \boldsymbol{j}_e \cdot \left[T \nabla \left(\frac{\mu_e}{T} \right) - e_e \boldsymbol{E} \right]$$

$$= -\boldsymbol{J}_u \cdot \nabla (\ln T) - \boldsymbol{j}_e \cdot \left[T \nabla \left(\frac{\mu_e}{T} \right) - e_e \boldsymbol{E} \right]. \tag{4.32}$$

In equation (4.31), as in equation (4.24), use is made of the previous notation

$$\nabla_T \mu_e = \nabla \mu_e + \tilde{s}_e \nabla T,$$

where $\tilde{s}_e$ is the partial electron entropy. Moreover, the symbol μ_e' has been used to denote the total value of the electrochemical potential, being the sum of the chemical potential μ_e of the electrons and the electric potential φ. By definition, in the absence of a magnetic field the electric potential is described by the relation

$$E = -\nabla\varphi, \tag{4.33}$$

where, moreover, in the given case

$$\nabla_T \mu'_e = \nabla_T \mu_e + e_e \nabla\varphi = \nabla_T \mu_e - e_e E, \tag{4.34}$$

in conformity with the definition of the electric potential φ.

In further considerations use will be made of relation (4.31) describing the dissipation function.

The phenomenological equations can then be written in the form

$$-J''_q = L_{qq} \nabla(\ln T) + L_{qe} \nabla_T \mu'_e, \tag{4.35}$$

$$-j_e = L_{eq} \nabla(\ln T) + L_{ee} \nabla_T \mu'_e. \tag{4.36}$$

By the Onsager principle, the phenomenological coefficients L_{qe} and L_{eq} are related by

$$L_{qe} = L_{eq}, \tag{4.37}$$

and since the dissipation function cannot be negative, the following inequalities must be satisfied:

$$L_{qq} > 0, \quad L_{ee} > 0, \quad L_{qq} L_{ee} - L_{qe}^2 > 0. \tag{4.38}$$

By the definition given in Section III.8, the heat of transport of electrons $Q_e''^*$ is described by the equation

$$Q_e''^* = \left(\frac{J''_q}{j_e}\right)_{\nabla T = 0} = \frac{L_{qe}}{L_{ee}} \tag{4.39}$$

and is the ratio of the heat flux to the electron flux for a zero temperature gradient.

The electron transport entropy can be defined by means of the equation

$$S_e^* = \frac{Q_e''^*}{T} + \tilde{s}_e, \tag{4.40}$$

where, as before, $\tilde{s}_e$ is the partial electron entropy.

It should be added that the value of the heat of transport of electrons depends on the heat or energy flux appearing in relation (4.39) and that there are other definitions for the heat of transport, based on other quantities J and j. A similar remark applies to the concept of the entropy of transport. In any event, the defining equations are always the same, that is, are in the form of relations (4.39) and (4.40), except that appropriate values of the fluxes J and j appear in them and the appropriate value of Q appears in relation (4.40).

The literature on thermoelectric effects makes use of the concept of the energy of transport of electrons which also depends on the energy flux on which it is based. The definition employed most frequently is

$$U_e^* = \left(\frac{J_u}{j_e}\right)_{\nabla T = 0}. \tag{4.41}$$

The relation (3.133) between J_q'' and J_u can be used to put equation (4.41) into the form

$$U_e^* = Q_e''^* + \tilde{h}_e . \tag{4.42}$$

By relations (4.40) and (4.42), the relation between U_e^* and the entropy of transport S_e^* is

$$S_e^* = \frac{U_e^*}{T} - \frac{\tilde{h}_e}{T} + \tilde{s}_e ,$$

whence

$$U_e^* = TS_e^* + (\tilde{h}_e - T\tilde{s}_e) = TS_e^* + \mu_e . \tag{4.43}$$

The individual heat, electric, and thermoelectric effects may be analysed on the basis of the equation derived, suitably modified.

Such an analysis will now be presented.

a. *Heat Conduction.* Pure conduction of heat takes place when there is no conduction of electricity. In this case $j_e = 0$ and equations (4.35) and (4.36) give rise to the relation

$$-J_q'' = \frac{L_{qq} L_{ee} - L_{qe}^2}{T L_{ee}} \nabla T , \tag{4.44}$$

use having been made here of the Onsager reciprocal relation (4.37). The quantity

$$\frac{L_{qq} L_{ee} - L_{qe}^2}{T L_{ee}} = \lambda_\infty \tag{4.45}$$

is the thermal conductivity in the absence of a flow of electricity.

Note that λ_∞ characterizes heat conduction in the presence of a potential gradient $\nabla_T \mu_e'$ which exerts an additional influence on the heat conduction. Hence, λ_∞ describes heat conduction in a stationary state, after enough time has elapsed for the distributions of both the temperature and the potential μ_e' to have become stable.

The equation for heat flow in a solid conductor without a flow of electricity can thus be written as

$$-(J_q'')_{j_e=0} = \lambda_\infty \nabla T . \tag{4.46}$$

Apart from λ_∞, we can introduce yet another coefficient, λ_0, describing heat conduction when no electric field appears in the system, that is, when $\nabla_T \mu_e' = 0$.

It follows from equation (4.35) that the heat equation then is of the form

$$-(J_q'')_{\nabla_T \mu_e = 0} = L_{qq} \nabla (\ln T) = \frac{L_{qq}}{T} \nabla T = \lambda_0 \nabla T ,$$

whereby

$$\lambda_0 = \frac{L_{qq}}{T} . \tag{4.47}$$

In practice, it should be noted, it is λ_∞ in principle which is measured since a gradient of the potential μ_e' appears in the conductor very quickly once a temperature gradient has been applied to it. Therefore, λ_∞ is of greater practical importance than is λ_0 since it is closer to physical reality than is the latter.

The difference between these thermal conductivities, in accordance with relations (4.45) and (4.47), is

$$\lambda_0 - \lambda_\infty = \frac{1}{T}\frac{L_{qe}^2}{L_{ee}} = \frac{1}{T}(Q_e''^*)^2 L_{ee}. \tag{4.48}$$

By inequality (4.38), the relation

$$\lambda_0 - \lambda_\infty > 0,$$

must be satisfied, and hence

$$\lambda_0 > \lambda_\infty > 0.$$

b. *Electrical Conduction*. Pure electrical conduction occurs when there is no temperature gradient in the conductor, that is, when

$$\nabla T = 0,$$

and then, by equation (4.36),

$$-\boldsymbol{j}_e = L_{ee} \nabla_T \mu_e'. \tag{4.49}$$

Electrical conductivity may be defined as

$$\sigma = -\frac{\boldsymbol{j}_e e_e}{\nabla \varphi}, \tag{4.50}$$

where φ is the electric potential.

Since the relation

$$\nabla_T \mu_e' = e_e \nabla \varphi \tag{4.51}$$

holds for a homogeneous isothermal electrical conductor, we have

$$\sigma = e_{ee}^2 L_{ee}. \tag{4.52}$$

c. *The Thomson Effect*. Relations (4.39) and (4.40), describing the heat of transport and the entropy of transport, can be used together with expressions (4.45) and (4.52) to transform equations (4.35) and (4.36) into

$$\boldsymbol{J}_q'' = -\lambda_\infty \nabla T + Q_e''^* \boldsymbol{j}_e, \tag{4.53}$$

$$-\boldsymbol{j}_e = Q_e''^* \frac{\sigma}{T e_e^2} \nabla T + \frac{\sigma}{e_e^2} \nabla_T \mu_e'. \tag{4.54}$$

The energy balance equation (4.30) allows itself to be simplified further since a conductor of electricity must obey the relation $\nabla \cdot \boldsymbol{j}_e = 0$, and thus

$$\rho \frac{\mathrm{d}u}{\mathrm{d}t} = -\nabla \cdot \boldsymbol{J}_q'' - \boldsymbol{j}_e \cdot \nabla \tilde{h}_e - e_e \boldsymbol{j}_e \cdot \nabla \varphi, \tag{4.55}$$

use having been made here of relation (4.33).

The gradient of the partial electron enthalpy $\tilde{h}_e$ can be written as follows (omitting the partial derivative of the enthalpy with respect to the pressure since the pressure is constant in the system):

$$\nabla \tilde{h}_e = c_{pe} \nabla T + (\nabla \tilde{h}_e)_{T,p}, \tag{4.56}$$

where $(\nabla \tilde{h}_e)_{T,p}$ denotes the component of $\nabla \tilde{h}_e$ at constant p and T, whereas c_{pe} is the electron specific heat. The quantity $(\nabla \tilde{h}_e)_{T,p}$ is zero for a homogeneous conductor and appears at the junction between different conductors. Accordingly, once expression (4.56) has been taken into account the energy balance equation for a homogeneous conductor becomes

$$\rho \frac{du}{dt} = -\nabla \cdot J_q'' - e_e j_e \cdot \nabla \varphi - j_e \cdot c_{pe} \nabla T. \tag{4.57}$$

Equation (4.54) implies that in the case under study

$$\nabla_T \mu_e' = -\frac{j_e}{\sigma} e_e^2 - \frac{Q_e''^*}{T} \nabla T = e_e \nabla \varphi. \tag{4.58}$$

Expressions (4.53) and (4.58) can be used to transform relation (4.57) into

$$\rho \frac{du}{dt} = \nabla \cdot (\lambda_\infty \nabla T) - \nabla \cdot (Q_e''^* j_e) + \frac{e_e^2 j_e^2}{\sigma} + j_e \cdot \frac{Q_e''^*}{T} \nabla T - j_e \cdot c_{pe} \nabla T. \tag{4.59}$$

The fact that $\nabla \cdot j_e = 0$ leads to

$$\nabla \cdot (Q_e''^* j_e) = j_e \cdot \nabla Q_e''^* + Q_e''^* \nabla \cdot j_e = j_e \cdot \nabla Q_e''^*,$$

and

$$-j_e \cdot \nabla Q_e''^* + j_e \cdot \frac{Q_e''^*}{T} \nabla T = j_e \cdot T \frac{Q_e''^* \nabla T - T \nabla Q_e''^*}{T^2} = -j_e \cdot T \nabla \frac{Q_e''^*}{T},$$

whereby one may also write

$$\rho \frac{du}{dt} = \nabla \cdot (\lambda_\infty \nabla T) + \frac{e_e^2 j_e^2}{\sigma} - j_e \cdot \left(T \nabla \frac{Q_e''^*}{T} + c_{pe} \nabla T \right). \tag{4.60}$$

It thus follows from equation (4.60) that the change in the internal energy of an element of a non-isothermal conductor is due to three causes:

(1) heat conduction characterized by the thermal conductivity λ_∞,

(2) electrical conduction, described by the conductivity σ,

(3) an additional effect expressed by the third term on the right-hand side of equation (4.60).

The last effect is known as Thomson heat. The existence of this effect was first pointed out by Thomson in 1856.

Since

$$T \left(\frac{\partial \tilde{s}_e}{\partial T} \right)_p = c_{pe} \quad \text{and} \quad \frac{Q_e''^*}{T} = S_e^* - \tilde{s}_e,$$

or introducing the notation

$$\tau = -\frac{T}{e_e}\left(\frac{\partial S_e^*}{\partial T}\right)_p \tag{4.61}$$

we have

$$T\nabla\frac{Q_e''^*}{T} + c_{pe}\nabla T = -\tau e_e \nabla T\,,$$

which enables the energy balance equation to be recast into the form

$$\rho\frac{\mathrm{d}u}{\mathrm{d}t} = \nabla\cdot(\lambda_\infty \nabla T) + \frac{e_e^2 j_e^2}{\sigma} + \tau e_e \boldsymbol{j}_e\cdot\nabla T\,. \tag{4.62}$$

The last term on the right-hand side of equation (4.62) is the Thomson heat, expressed as a function of the coefficient τ.

As emerges from equation (4.62), the Thomson heat is proportional to both the temperature gradient ∇T and the electron flux $\boldsymbol{j}_e$, and hence to the electric current density $\boldsymbol{J}_{el} = e_e \boldsymbol{j}_e$. The sign of the Thomson heat may thus vary as it depends on many factors. For example, in the case of a copper conductor heat flows in when current flows in the direction of the temperature rise, whereas the opposite effect is observed in an iron conductor. Thomson heat does not occur at all in a lead conductor.

It is to be noted that the generation or absorption of Thomson heat is reversible in both the physical and the thermodynamic sense. Reversal of the flow of electricity changes the direction of heat flow without altering its magnitude and, moreover, a coefficient characterizing the Thomson effect does not appear in the expression for the entropy source strength.

Thomson heat can be found experimentally from the energy balance equation. Knowing the thermal and electrical characteristics of the conductor, that is, the values of the thermal and electrical conductivities, and knowing the values of the electrical current and the temperature differences, one can calculate the Joule heat and the heat conducted in accordance with Fourier's law. The heat exchanged by the given conductor with the environment can be measured and then the Thomson heat is calculated as the item which completes the energy balance. In the steady state the entire heat exchanged by the conductor is equal to the algebraic sum of the Fourier heat, the Joule heat, and the Thomson heat. It should be added that, on account of its small value, the Thomson coefficient τ cannot be evaluated very accurately by experiment, and the measurement must be performed with extreme care since the Thomson effect is determined indirectly as a quantity which emerges from the addition and subtraction of several different quantities measured separately.

Example 4.1. Use the energy flux $\boldsymbol{J}_u$ to calculate the quantities characterizing effects which occur in a solid conductor during the flow of electricity and heat.

Solution. If the quantity $\boldsymbol{J}_u$ is introduced (along with $\boldsymbol{j}_e$) in order to describe effects occurring in a solid conductor, the dissipation function is of the form [equation (4.26)]

$$\Psi = -\boldsymbol{J}_u\cdot\nabla(\ln T) - \boldsymbol{j}_e\cdot\left[T\nabla\left(\frac{\mu_e}{T}\right) - e_e E\right], \tag{1}$$

whereas the energy balance equation, by equation (4.12) and with simplifying assumptions, can be written as

$$\rho \frac{\mathrm{d}u}{\mathrm{d}t} = -\nabla \cdot \boldsymbol{J}_u + \boldsymbol{j}_e \cdot e_e \boldsymbol{E}. \tag{2}$$

The phenomenological equations are expressed by the relations

$$-\boldsymbol{J}_u = L^u_{qq} \nabla (\ln T) + L^u_{qe} \left[T\nabla \left(\frac{\mu_e}{T} \right) - e_e \boldsymbol{E} \right], \tag{3}$$

$$-\boldsymbol{j}_e = L^u_{eq} \nabla (\ln T) + L^u_{ee} \left[T\nabla \left(\frac{\mu_e}{T} \right) - e_e \boldsymbol{E} \right], \tag{4}$$

where the symbols L^u_{qq}, L^u_{qe}, L^u_{eq}, and L^u_{ee} have been used to denote the new phenomenological coefficients so as to distinguish them from the old ones.

By the Onsager principle

$$L^u_{qe} = L^u_{eq}. \tag{5}$$

The heat of transport of electrons is now defined as

$$Q^u_e = \left(\frac{\boldsymbol{J}_u}{\boldsymbol{j}_e} \right)_{\nabla T = 0} = \frac{L^u_{qe}}{L^u_{ee}}, \tag{6}$$

and hence is equal to the transport energy U^*_e specified by equation (4.41).

Similarly, the entropy of transport is

$$S^u_e = \frac{Q^u_e}{T} + \tilde{s}_e. \tag{7}$$

a. *Heat Conduction*. Pure conduction of heat occurs when $\boldsymbol{j}_e = 0$, and hence equations (3) and (4) imply the relation

$$-\boldsymbol{J}_u = \frac{L^u_{qq} L^u_{ee} - (L^u_{eq})^2}{T L^u_{ee}} \nabla T.$$

If there is no flow of electricity ($\boldsymbol{j}_e = 0$), then $\boldsymbol{J}_u = \boldsymbol{J}''_q$ and subsequently

$$\frac{L^u_{qq} L^u_{ee} - (L^u_{eq})^2}{T L^u_{ee}} = \lambda_\infty \tag{8}$$

and

$$-(\boldsymbol{J}_u)_{\boldsymbol{j}_e = 0} = \lambda_\infty \nabla T. \tag{9}$$

b. *Conduction of Electricity*. The condition for 'pure' conduction of electricity is that $\nabla T = 0$, and then

$$-\boldsymbol{j}_e = L^u_{ee} (\nabla \mu_e - e_e \boldsymbol{E}). \tag{10}$$

The electrical conductivity is

$$\sigma = -\left(\frac{\boldsymbol{j}_e e^2_e}{\nabla \mu_e - e_e \boldsymbol{E}} \right)_{\nabla T = 0},$$

or

$$\sigma = e^2_e L^u_{ee}. \tag{11}$$

c. *The Thomson effect*. The quantities Q''^*_e and Q^u_e (U^*_e) are related by

$$Q^u_e = \left(\frac{\boldsymbol{J}_u}{\boldsymbol{j}_e} \right)_{\nabla T = 0} = \left(\frac{\boldsymbol{J}''_q + \tilde{h}_e \boldsymbol{j}_e}{\boldsymbol{j}_e} \right)_{\nabla T = 0} = \left(\frac{\boldsymbol{J}''_q}{\boldsymbol{j}_e} + \tilde{h}_e \right)_{\nabla T = 0},$$

and thus

$$Q_e^u = Q_e^{\prime\prime *} + \tilde{h}_e . \tag{12}$$

Similarly,

$$S_e^u = S_e^* + \frac{\tilde{h}_e}{T} . \tag{13}$$

Using the expressions for σ and Q_e^u, we can obtain the relations

$$J_u = -\lambda_\infty \nabla T + Q_e^u j_e , \tag{14}$$

$$-j_e = (Q_e^u - \tilde{h}_e) \frac{\sigma}{Te_e^2} \nabla T + \frac{\sigma}{e_e^2} \nabla_T \mu_e' \tag{15}$$

or

$$-j_e = (Q_e^u - \mu_e) \frac{\sigma}{Te_e^2} \nabla T + \frac{\sigma}{e_e^2} \nabla \mu_e' . \tag{16}$$

The energy balance equation (2) can also be recast into the form

$$\rho \frac{du}{dt} = -\nabla \cdot J_u - e_e j_e \cdot \nabla \varphi , \tag{17}$$

use having been made here of relation (4.33).

If equation (14) is used to eliminate J_u and then the value

$$\nabla_T \mu_e' = e_e \nabla \varphi = -\frac{j_e}{\sigma} e_e^2 - \frac{Q_e^u - \tilde{h}_e}{T} \nabla T$$

is substituted from expression (15), the result is a new form of the energy balance equation,

$$\rho \frac{du}{dt} = \nabla \cdot (\lambda_\infty \nabla T) + \frac{e_e^2 j_e^2}{\sigma} + \frac{j_e (Q_e^u - \tilde{h}_e)}{T} \cdot \nabla T - \nabla \cdot (Q_e^u j_e) .$$

Since

$$\nabla \cdot (Q_e^u j_e) = j_e \cdot \nabla Q_e^u + Q_e^u (\nabla \cdot j_e) = j_e \cdot \nabla Q_e^u ,$$

transformations similar to those preceding formula (4.60) yield

$$\rho \frac{du}{dt} = \nabla \cdot (\lambda_\infty \nabla T) + \frac{e_e^2 j_e^2}{\sigma} - j_e \cdot \left(T \nabla \frac{Q_e^u}{T} + \nabla \tilde{h}_e \right) . \tag{18}$$

On introduction of the coefficient τ defined by formula (4.61), this equation leads to relation (4.62) derived previously. The coefficient τ can also be expressed as a function of the heat of transport Q_e^u, viz.

$$-\tau e_e \nabla T = T \nabla \frac{Q_e^u - \tilde{h}_e}{T} + c_{pe} \nabla T \tag{19}$$

or, on using formulae (4.62) and (13), as

$$\tau e_e = -T \left(\frac{\partial S_e^u}{\partial T} \right)_p - \frac{\tilde{h}_e}{T} . \tag{20}$$

Example 4.2. Using the energy flux $J_q^{\prime\prime\prime} = TJ_s$, calculate the quantities characterizing the effects which occur in a solid conductor during the flow of electricity and heat.

Solution. The flux $J_q^{\prime\prime\prime}$ is related to the quantities $J_q^{\prime\prime}$ and J_u used earlier by

$$J_q^{\prime\prime\prime} = J_u - j_e \mu_e , \tag{1}$$

$$J_q^{\prime\prime\prime} = J_q^{\prime\prime} + j_e T \tilde{s}_e , \tag{2}$$

which follows immediately from the definition of the entropy flux J_s.

It can be easily shown that the dissipation function is now given by the formula

$$\Psi = -J_q''' \cdot \nabla(\ln T) - j_e \cdot (\nabla\mu_e - e_e E). \tag{3}$$

The energy balance equation can be obtained, for instance, from expression (2), Example 4.1, upon taking formula (1) into account. This leads to

$$\rho \frac{du}{dt} = -\nabla \cdot J_q''' - \nabla \cdot (\mu_e j_e) + e_e j_e \cdot E. \tag{4}$$

Expanding the expression

$$\nabla \cdot (\mu_e j_e) = \mu_e (\nabla \cdot j_e) + j_e \cdot \nabla\mu_e = j_e \cdot \nabla\mu_e,$$

where $\nabla \cdot j_e = 0$ for a conductor, we get another form of the energy equation:

$$\rho \frac{du}{dt} = -\nabla \cdot J_q''' - j_e \cdot \nabla\mu_e + e_e \cdot j_e E. \tag{5}$$

The phenomenological equations are expressed by the relations

$$-J_q''' = L_{qq}''' \nabla(\ln T) + L_{qe}''' (\nabla\mu_e - e_e E), \tag{6}$$

$$-j_e = L_{eq}''' \nabla(\ln T) + L_{ee}''' (\nabla\mu_e - e_e E). \tag{7}$$

The symbols L_{qq}''', L_{qe}''', L_{eq}''', and L_{ee}''' are the phenomenological coefficients relative to the new thermodynamic flows J_q''' and j_e.

By the Onsager reciprocal principle,

$$L_{qe}''' = L_{eq}'''. \tag{8}$$

The heat of transport of electrons is defined as

$$Q_e''' = \left(\frac{J_q'''}{j_e}\right)_{\nabla T = 0} = \frac{L_{eq}'''}{L_{ee}'''} \tag{9}$$

and the entropy of transport as

$$S_e''' = \frac{Q_e'''}{T} + \tilde{s}_e. \tag{10}$$

a. *Heat Conduction.* The conduction of heat without a flow of electricity occurs when $j_e = 0$. Hence, equations (6) and (7) imply the relation

$$-J_q''' = \frac{L_{qq}''' L_{ee}''' - (L_{eq}''')^2}{T L_{ee}'''} \nabla T. \tag{11}$$

Since diffusion does not take place in this case,

$$_q''' = J_q'' = J_u$$

and subsequently

$$\frac{L_{qq}''' L_{ee}''' - (L_{eq}''')^2}{T L_{ee}'''} = \lambda_\infty \tag{12}$$

and

$$-(J_q''')_{j_e = 0} = \lambda_\infty \nabla T \tag{13}$$

b. *Conduction of Electricity*. The condition for 'pure' conduction of electricity is $\nabla T=0$, and then

$$-\boldsymbol{j}_e=L'''_{ee}(\nabla\mu_e-e_e\boldsymbol{E}) \tag{14}$$

and the electrical conductivity is

$$\sigma=-\left(\frac{j_e e_e^2}{\nabla\mu_e-e_e E}\right)_{\nabla T=0}=e_e^2 L'''_{ee}\,. \tag{15}$$

c. *The Thomson Effect*. The heats of transport $Q_e''^*$ and Q_e''' are related by

$$Q_e'''=\left(\frac{J_q'''}{j_e}\right)_{\nabla T=0}=\left(\frac{J_q''+T\tilde{s}_e j_e}{j_e}\right)_{\nabla T=0}=\left(\frac{J_q''}{j_e}+T\tilde{s}_e\right)_{\nabla T=0},$$

or

$$Q_e'''=Q_e''^*+T\tilde{s}_e\,. \tag{16}$$

Similarly, the entropy of transport satisfies the formula

$$S_e'''=S_e^*+\tilde{s}_e\,,$$

which follows from the definition of this quantity.

The expressions for σ and Q_e''' can be used to get the relations

$$J_q'''=-\lambda_\infty\nabla T+Q_e'''j_e\,, \tag{17}$$

$$-j_e=Q_e''\frac{\sigma}{Te_e^2}\nabla T+\frac{\sigma}{e_e^2}\nabla\mu_e'\,. \tag{18}$$

The energy equation (5) can also be written as

$$\rho\frac{du}{dt}=-\nabla\cdot J_q'''-j_e\cdot\nabla\mu_e-e_e\,j_e\cdot\nabla\varphi\,. \tag{19}$$

If the value of J_q''' from equation (17) and the value of

$$e_e\nabla\varphi=\nabla_T\,\mu_e'=\nabla\mu_e'+\tilde{s}_e\nabla T=-\frac{e_e^2 j_e}{\sigma}-\frac{Q_e'''}{T}\nabla T+\tilde{s}_e\nabla T$$

from equation (18) are inserted into expression (19), the result is

$$\rho\frac{du}{dt}=\nabla\cdot(\lambda_\infty\nabla T)-\nabla\cdot(Q_e'''j_e)-j_e\cdot\nabla\mu_e+\frac{j_e^2e_e^2}{\sigma}+\frac{j_e(Q_e'''-T\tilde{s}_e)}{T}\cdot\nabla T,$$

whereby

$$\nabla\cdot(Q_e'''j_e)=j_e\cdot\nabla Q_e'''+Q_e'''\nabla\cdot j_e=j_e\cdot\nabla Q_e'''\quad(\text{since }\nabla\cdot j_e=0)$$

and finally

$$\rho\frac{du}{dt}=\nabla\cdot(\lambda_\infty\nabla T)+\frac{e_e^2 j_e^2}{\sigma}-j_e\cdot\left(T\nabla\frac{Q_e'''}{T}+\tilde{s}_e\nabla T+\nabla\mu_e\right). \tag{20}$$

When the definition of τ is taken into account formula (4.62) is obtained from this relation. As a function of the heat of transport Q_e''', the coefficient τ is

$$-\tau e_e\nabla T=T\nabla\frac{Q_e'''}{T}$$

or

$$\tau e_e=-T\left[\left(\frac{\partial S_e'''}{\partial T}\right)_p-\left(\frac{\partial\tilde{s}_e}{\partial T}\right)_p\right], \tag{21}$$

which follows from the definition of τ and the relation between S_e^* and S_e'''.

Example 4.3. Evaluate the quantities characterizing the effects which occur in a solid conductor during the flow of electricity and flow of heat. To do so, make use of the energy flux $\boldsymbol{J}^u$ defined as

$$\boldsymbol{J}^u = \boldsymbol{J}_u + \boldsymbol{j}_e(\mu_e' - \mu_e).$$

This quantity is used by some authors and is often seen in publications concerning thermoelectric effects.

Solution. The dissipation function is expressed by the formula

$$\Psi = -\boldsymbol{J}^u \cdot \nabla(\ln T) - \boldsymbol{j}_e \cdot \left(T\nabla \frac{\mu_e'}{T}\right), \tag{1}$$

which can be obtained, for instance, from formula (4.32) and the relation between $\boldsymbol{J}^u$ and $\boldsymbol{J}_u$.

The energy balance equation, which is found by inserting the formula for $\boldsymbol{J}^u$ into the relevant relation, is of the form

$$\rho \frac{du}{dt} = -\nabla \cdot \boldsymbol{J}^u. \tag{2}$$

The phenomenological equations are given by the expressions

$$-\boldsymbol{J}^u = L_{qq}'^u \nabla(\ln T) + L_{qe}'^u \, T\nabla \frac{\mu_e'}{T}, \tag{3}$$

$$-\boldsymbol{j}_e = L_{eq}'^u \nabla(\ln T) + L_{ee}'^u \, T\nabla \frac{\mu_e'}{T} \tag{4}$$

with suitable phenomenological coefficients, corresponding to the flux $\boldsymbol{J}^u$.

The Onsager reciprocal principle is expressed by the equality

$$L_{qe}'^u = L_{eq}'^u. \tag{5}$$

The heat of transport of electrons is given by

$$Q_e'^u = \left(\frac{\boldsymbol{J}^u}{\boldsymbol{j}_e}\right)_{\nabla T = 0} = \frac{L_{qe}'^u}{L_{ee}'^u} \tag{6}$$

and, similarly, the entropy of transport is

$$S_e'^u = \frac{Q_e'^u}{T} + \tilde{s}_e. \tag{7}$$

a. *Heat Conduction.* The conduction of heat in the pure form occurs when $\boldsymbol{j}_e = 0$, and thus (3) and (4) lead to

$$-\boldsymbol{J}^u = \frac{L_{qq}'^u L_{ee}'^u - (L_{ee}'^u)^2}{TL_{ee}'^u} \nabla T. \tag{8}$$

Since $\boldsymbol{J}_q'' = \boldsymbol{J}_u = \boldsymbol{J}^u$ in the absence of diffusion, we have

$$\frac{L_{qq}'^u L_{ee}'^u - (L_{eq}'^u)^2}{TL_{ee}'^u} = \lambda_\infty \tag{9}$$

and

$$-(\boldsymbol{J}^u)_{\boldsymbol{j}_e = 0} = \lambda_\infty \nabla T. \tag{10}$$

b. *The Conduction of Electricity.* The condition for 'pure' conduction of electricity is that $\nabla T = 0$, and then the relation

$$-\boldsymbol{j}_e = L_{ee}'^u \, T\nabla \frac{\mu_e'}{T} \tag{11}$$

follows from equation (4). The electrical conductivity is

$$\sigma = -\left(\frac{j_e e_e^2}{\nabla \mu'_e}\right)_{\nabla T=0} = e_e^2 L'^u_{ee}. \tag{12}$$

c. *The Thomson Effect*. The heats of transport Q'^u_e and Q''^*_e are associated by a relation which follows from the relationship between J^u and J''_q:

$$Q'^u_e = \left(\frac{J^u}{j_e}\right)_{\nabla T=0} = \left[\frac{J''_q + \tilde{h}_e j_e + j_e(\mu'_e - \mu_e)}{j_e}\right]_{\nabla T=0}, \tag{13}$$

$$Q'^u_e = Q''^*_e + T\tilde{s}_e + \mu'_e.$$

The corresponding relation for the entropy of transport is

$$S'^u_e = S^*_e + \tilde{s}_e + \frac{\mu'_e}{T}. \tag{14}$$

If the expressions for σ and Q'^u_e are used, the relations describing J^u and j_e can be rewritten as

$$J^u = -\lambda_\infty \nabla T + Q'^u_e j_e \tag{15}$$

$$-j_e = (Q'^u_e - \mu'_e)\frac{\sigma}{Te_e^2}\nabla T + \frac{\sigma}{e_e^2}\nabla \mu'_e. \tag{16}$$

The energy balance equation (2) can be transformed by means of equation (15), viz.

$$\rho\frac{du}{dt} = \nabla \cdot (\lambda_\infty \nabla T) - \nabla \cdot (Q'^u_e j_e)$$

and then

$$\nabla \cdot (Q'^u_e j_e) = j_e \cdot \nabla Q'^u_e + Q'^u_e(\nabla \cdot j_e) = j_e \cdot \nabla Q'^u_e \quad (\text{since } \nabla \cdot j_e = 0).$$

Finally, therefore,

$$\rho\frac{du}{dt} = \nabla \cdot (\lambda_\infty \nabla T) - j_e \cdot \nabla Q'^u_e.$$

The coefficient τ allows itself to be recast into

$$-\tau e_e \nabla T = \nabla Q'^u_e + \frac{j_e e_e^2}{\sigma},$$

or

$$\tau = -\frac{\nabla Q'^u_e}{e_e \nabla T} - \frac{j_e e_e}{\sigma \nabla T}. \tag{17}$$

3. Thermoelectric Effects

In addition to the aforementioned effects others – known as thermoelectric effects – also take place in a circuit consisting of different conductors of electricity. These are the Seebeck effect and the Peltier effect.

a. *The Seebeck Effect*. In 1826 the German physicist, Thomas Johann Seebeck (1770–1831) discovered this effect, which is named after him; in it a potential difference (EMF) is set up in a circuit consisting of two dissimilar conductors with junctions at different temperatures. This effect is utilized in practice first and foremost in thermoelectric thermometers. And, more recently, new potential uses have offered themselves in thermoelectric generators for the direct conversion of heat into electricity; this will be dealt with in greater detail later in this chapter.

It follows from equation (4.54) that if no current is flowing ($j_e = 0$), then

$$\nabla_T \mu_e' = -\frac{Q_e''^*}{T} \nabla T = -(S_e^* - \tilde{s}_e) \nabla T, \tag{4.63}$$

which means that a gradient of electric potential associated with the heat of electron transport is set up in a non-isothermal conductor. This thermoelectric potential, as it is called, is utilized in a circuit consisting of two dissimilar conductors A and B. The value of the thermoelectric potential is characterized by the Seebeck coefficient, written as ε_{AB} for a pair of conductors A and B. In addition to ε_{AB} which describes the behaviour of a circuit consisting of conductors A and B, use is also made of Seebeck coefficients for a single material, A or B.

This latter coefficient is defined as

$$\varepsilon_A = -\lim_{\Delta T \to 0} \left(\frac{\Delta \mu_e'}{e_e \Delta T} \right)_{j_e = 0} = \left(\frac{\nabla \mu_e'}{e_e \nabla T} \right)_{j_e = 0} \tag{4.64}$$

and hence is the limit of the ratio of the increment in the value of electrochemical potential to the temperature difference as that difference tends to zero in the absence of a flow of electric current.

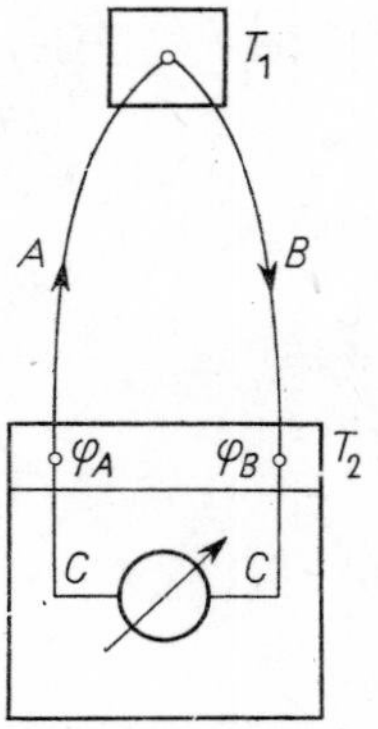

Fig. 4.1 Generation of thermoelectric force

The relation describing $\nabla_T \mu_e'$ and equations (4.40) and (4.63) are used to obtain the expression

$$-\nabla \mu_e' = -\nabla_T \mu_e' + \tilde{s}_e \nabla T = (S_e^* - \tilde{s}_e) \nabla T + \tilde{s}_e \nabla T = S_e^* \nabla T$$

whence for a particular conductor A

$$\varepsilon_A = -\frac{\nabla \mu_{eA}'}{e_e \nabla T} = \frac{1}{e_e} S_{eA}^*. \tag{4.65}$$

In the case of a system consisting of two different conductors, the electrical circuit in Fig. 4.1 must be considered in order to determine the Seebeck coefficient. This circuit consist of conductors of type A and B whose junction is in a thermostat at temperature T_1. The other ends of A and B, which are in a thermostat at temperature T_2, are connected by a conductor C with an instrument measuring the electromotive force of the circuit. The instrument and the conductor C are also in the thermostat at temperature T_2, and it is assumed that there is no voltage drop in the instrument. This arrangement enables the EMF developed in the circuit to be measured as a function of the difference between temperatures T_1 and T_2.

The Seebeck coefficient for the thermocouple AB is defined as

$$\varepsilon_{AB} = -\lim_{\Delta T\to 0} \frac{\phi}{\Delta T} = -\lim_{\Delta T\to 0} \frac{\Delta\varphi}{\Delta T} = -\lim_{\Delta T\to 0} \frac{\varphi_B - \varphi_A}{\Delta T}, \qquad (4.66)$$

where ϕ is the EMF developed by the thermocouple (equal to the potential difference across the terminals of the measuring instrument shown in Fig. 4.1), whereas $\Delta T = T_1 - T_2$.

It should be mentioned that the definition of ε_{AB}, in conformity with equation (4.66), corresponds to the actual direction in which the current flows, that is, if ε_{AB} is positive current flows in conductor A from a cooler spot to a warmer spot and then in conductor B from a warmer spot to a cooler spot. Hence the relation

$$\varepsilon_{AB} = -\varepsilon_{BA}$$

is valid.

The potential difference $\phi = \varphi_B - \varphi_A$ is measured on the conductors C, across the measuring instrument; both of these conductors are at the same constant temperature T_2 and hence the chemical potential of the electrons at the junctions of these conductors with the instrument is the same. This means that the difference of the electric potentials ϕ is equal to the difference of the electrochemical potentials of the electrons, that is,

$$e_e \phi = \Delta\mu'_e = \mu'_{eB} - \mu'_{eA},$$

where μ'_{eA} and μ'_{eB} are the electrochemical potentials of electrons in conductors A and B at temperature T_2.

Accordingly, the Seebeck coefficient can also be written as

$$\varepsilon_{AB} \underset{\Delta T\to 0,\, j_e=0}{=} -\frac{\Delta\mu'_e}{e_e \Delta T} = -\left(\frac{\nabla\mu'_{eB}}{e_e \nabla T} - \frac{\nabla\mu'_{eA}}{e_e \nabla T}\right)$$

and then, if use is made of relation (4.65), as

$$\varepsilon_{AB} = \varepsilon_B - \varepsilon_A. \qquad (4.67)$$

Thus, ε is additive, this being implied by the additivity of the electrostatic and electrochemical potentials. Further consequences also follow from this property. If the

Seebeck coefficients are known for the pairs AX and BX and are ε_{AX} and ε_{BX}, then

$$\varepsilon_{AB}=\varepsilon_B-\varepsilon_A=(\varepsilon_B-\varepsilon_X)-(\varepsilon_A-\varepsilon_X)=\varepsilon_{XB}-\varepsilon_{XA}, \tag{4.68}$$

all of the quantities being taken for the same temperature.

Moreover, the value of ε_{AB} does not change if an additional material X is introduced between the ends of conductors A and B, provided that the junctions AX and BX are at the same temperature.

Relation (4.65) allows the expression for ε_{AB} to be cast into the form

$$\varepsilon_{AB}=\frac{1}{e_e}(S^*_{eB}-S^*_{eA}), \tag{4.69}$$

which shows the connection between this coefficient and the entropy of electron transport for conductors A and B.

Furthermore, comparison of expressions (4.61) and (4.65) yields an additional relation between the Seebeck and Thomson coefficients

$$\tau=-T\left(\frac{\partial\varepsilon}{\partial T}\right)_p. \tag{4.70}$$

To return to the generation of a thermoelectric potential, it should be noted that a continuous change of electrochemical potential occurs along the thermocouple circuit and is also continuous at the junction of two dissimilar conductors. However, the chemical potential experiences a jump at the junction owing to a change in the concentration of the free electron flux in the two conductors. As a result of this change in concentration there also is a jump in the electrostatic potential, this being known as the contact potential difference. Of course, the electrostatic potential difference is equal to the chemical potential difference (with account for the algebraic signs of these differences) so that the continuity of the values of the electrochemical potential be preserved. Thus, the thermoelectric potential grows continuously along the entire thermocouple circuit to attain a particular value determined by the properties of the conductor materials and the temperature difference between the junctions.

If the temperature difference ΔT is finite, the thermocouple EMF can be evaluated from the relation

$$\phi=\int_{T_1}^{T_2}\varepsilon_{AB}\,\mathrm{d}T; \tag{4.71}$$

if ε_{AB} is constant and independent of the temperature, then

$$\phi=\varepsilon_{AB}\Delta T. \tag{4.72}$$

b. *The Peltier Effect*. This effect, first discovered by Peltier in 1834, consists in heat being evolved or absorbed at the junction of two dissimilar conductors. The direction in which heat flows at the junction of the conductors depends on the direction in which current is flowing and can be reversed by changing the direction of the current. A fixed temperature exists at the junction, and hence the two conductors have the same temperature where they meet. Moreover, other conditions are also satisfied at the junction, viz. the j_e have the same value in both conductors and the equality

$$\mu'_{eA} = \mu'_{eB}$$

holds, in conformity with the reasoning given earlier.

The essence of the Peltier effect thus lies in the difference in the flux of energy flowing through the two kinds of conductors. The energy flux inserted into the relevant equations must comprise the energy transported by heat conduction and diffusion and the energy associated with the flow of current as it overcomes the contact potential difference at the junction of the two conductors.

Such a flux, for instance, is the quantity J_q''' introduced in Example 4.2 and defined as

$$J_q''' = J_u - j_e \mu_e$$

or

$$J_q''' = J_q'' + j_e T \tilde{s}_e, \tag{4.73}$$

where

$$J_q''' = T J_s .$$

The energy balance equation written with the aid of the flux J_q''' is of the form [equation (5) in Example (4.2)]

$$\rho \frac{du}{dt} = -\nabla \cdot J_q''' - j_e \cdot \nabla \mu_e + j_e \cdot e_e E$$

or

$$\rho \frac{du}{dt} = -\nabla \cdot J_q''' - j_e \cdot \nabla \mu'_e .$$

If this equation is applied to a very small element of the circuit, containing the junction AB, the second term on the right-hand side is zero since continuity of the electrochemical potential μ'_e is preserved at the junction. Accordingly, the change in the internal energy of the element is equal to the divergence of J_q''', and hence J_q''' does indeed fully describe the exchange of energy which occurs at the junction of the conductors.

The Peltier heat thus is

$$Q_p = (J_{qB}''' - J_{qA}''')_{\nabla T=0} . \tag{4.74}$$

Combination of relations (4.73) and (4.74) yields

$$Q_p = (J_{qB}'' - J_{qA}'')_{\nabla T=0} + j_e T (\tilde{s}_{eB} - \tilde{s}_{eA}) . \tag{4.75}$$

By the definition of the heat of transport, it follows that [equation (4.53)]

$$(J_{qA}'')_{\nabla T=0} = j_e Q_{eA}''^{*} = j_e T (S_{eA}^{*} - \tilde{s}_{eA}) , \tag{4.76}$$

$$(J_{qB}'')_{\nabla T=0} = j_e Q_{eB}''^{*} = j_e T (S_{eB}^{*} - \tilde{s}_{eB}) . \tag{4.77}$$

When relations (4.76) and (4.77) are substituted into expression (4.75), the result is

$$Q_p = j_e T (S^*_{eB} - S^*_{eA}). \tag{4.78}$$

The quantity

$$\Pi_{AB} = \frac{T}{e_e} (S^*_{eB} - S^*_{eA}) \tag{4.79}$$

is called the Peltier coefficient. With the aid of this coefficient, the Peltier heat can be written as

$$Q_p = \Pi_{AB} e_e j_e = \Pi_{AB} J_e. \tag{4.80}$$

The coefficient Π is also specified for an individual conductor as follows:

$$\Pi_A = \frac{T}{e_e} S^*_{eA}, \tag{4.81}$$

$$\Pi_B = \frac{T}{e_e} S^*_{eB}. \tag{4.82}$$

Then

$$\Pi_{AB} = \Pi_B - \Pi_A. \tag{4.83}$$

It should be emphasized that the Seebeck and Peltier effects are both reversible in the thermodynamic sense since they do not entail an irreversible growth of entropy inside the system. With respect to the Peltier effect the reversibility does, of course, require the temperature difference between the junction of the conductors and the environment exchanging heat with it to be infinitesimal. In any case, the physical essence of the Peltier effect does not entail its thermodynamic irreversibility.

Example 4.4. Using the energy flux $\boldsymbol{J}_u$, derive relations describing the coefficients which characterize the thermoelectric effects.

Solution. By definition, the Seebeck coefficient is

$$\varepsilon_A = -\left(\frac{\nabla \mu'_{eA}}{e_e \nabla T}\right)_{\nabla T \to 0,\, j_e = 0}. \tag{1}$$

The relation [equation (16) in Example 4.1]

$$-j_e = (Q^u_e - \mu_e)\frac{\sigma}{T e_e^2} \nabla T + \frac{\sigma}{e_e^2} \nabla \mu'_e$$

holds in the case under consideration, and hence

$$\varepsilon_A = \frac{Q^u_{eA} - \mu_{eA}}{T e_e} = \frac{1}{e_e}\left(S^u_{eA} - \tilde{S}_{eA} - \frac{\mu_{eA}}{T}\right) \tag{2}$$

or

$$\varepsilon_A = \frac{1}{e_e}\left(S^u_{eA} - \frac{\tilde{h}_{eA}}{T}\right). \tag{3}$$

The Seebeck coefficient for the circuit consisting of conductors A and B is, by relation (4.67), equal to

$$\varepsilon_{AB}=\varepsilon_B-\varepsilon_A=\frac{1}{e_e}\left[(S^u_{eB}-S^u_{eA})-\frac{\tilde{h}_{eB}-\tilde{h}_{eA}}{T}\right]. \quad (4)$$

Comparison of relation (20) from Example 4.1 with relation (4) confirms the validity of formula (4.70).

The Peltier coefficient can be calculated by using relation (13) from Example 4.1, viz.

$$S^u_e=S^*_e+\frac{\tilde{h}_e}{T}$$

and equations (4.81) and (4.82). This yields the relations

$$\Pi_A=\frac{T}{e_e}S^u_{eA}-\frac{\tilde{h}_{eA}}{e_e}, \quad (5)$$

$$\Pi_B=\frac{T}{e_e}S^u_{eB}-\frac{\tilde{h}_{eB}}{e_e} \quad (6)$$

and

$$\Pi_{AB}=\frac{T}{e_e}(S^u_{eB}-S^u_{eA})-\frac{1}{e_e}(\tilde{h}_{eB}-\tilde{h}_{eA}). \quad (7)$$

Example 4.5. Use the heat flux $J_q'''=TJ_s$ to derive relations describing the thermoelectric coefficients.

Solution. By definition, the Seebeck effect is

$$\varepsilon_A=-\left(\frac{\nabla\mu'_{eA}}{e_e\nabla T}\right)_{\nabla T\to 0,\, j_e=0}.$$

Relation (18) from Example 4.2 yields

$$\varepsilon_A=\frac{Q'''_{eA}}{Te_e}=(S'''_{eA}-\tilde{s}_{eA})\frac{1}{e_e}. \quad (1)$$

The Seebeck coefficient for the circuit of conductors A and B is

$$\varepsilon_{AB}=\varepsilon_B-\varepsilon_A=\frac{1}{e_e}[(S'''_{eB}-S'''_{eA})-(\tilde{s}_{eB}-\tilde{s}_{eA})]. \quad (2)$$

Comparison of relation (21) from Example 4.2 with relation (2) confirms the validity of formula (4.70).

The Peltier coefficient can be calculated by using equations (4.81) and (4.82) and equation (10) from Example 4.2, that is, $S'''_e=\frac{Q'''_e}{T}+\tilde{s}_e$. This gives the relations

$$\Pi_A=\frac{T}{e_e}(S'''_{eA}-\tilde{s}_{eA}), \quad (3)$$

$$\Pi_B=\frac{T}{e_e}(S'''_{eB}-\tilde{s}_{eB}) \quad (4)$$

and

$$\Pi_B=\frac{T}{e_e}[(S'''_{eB}-S'''_{eA})-(\tilde{s}_{eB}-\tilde{s}_{eA})]. \quad (5)$$

Example 4.6. Derive relations describing the thermoelectric coefficients by making use of the energy flux J^u, defined as

$$J^u = J_u + j_e(\mu_e' - \mu_e).$$

Solution. By definition, the Seebeck coefficient is

$$\varepsilon_A = -\left(\frac{\nabla \mu_{eA}'}{e_e \nabla T}\right)_{\nabla T \to 0,\, j_e = 0} .$$

With the aid of relation (16) from Example 4.3 it is found that

$$\varepsilon_A = \frac{Q_{eA}'^u - \mu_{eA}'}{Te_e} = \frac{1}{e_e}\left[S_{eA}^u - \tilde{s}_{eA} - \frac{\mu_{eA}'}{T}\right]. \tag{1}$$

The Seebeck coefficient for a circuit of two conductors A and B is

$$\varepsilon_{AB} = \varepsilon_B - \varepsilon_A = \frac{1}{e_e}\left[(S_{eB}'^u - S_{eA}'^u) - (\tilde{s}_{eB} - \tilde{s}_{eA})\right], \tag{2}$$

since the relation

$$\mu_{eA}' = \mu_{eB}'$$

holds at the junction AB.

The Peltier coefficient can be found by employing equations (4.81) and (4.82) as well as equation (14) from Example 4.3,

$$S_e'^u = S_e^* + \tilde{s}_e + \frac{\mu_e'}{T} .$$

This leads to the relations

$$\Pi_A = \frac{T}{e_e}\left(S_{eA}'^u - \tilde{s}_{eA} - \frac{\mu_{eA}'}{T}\right), \tag{3}$$

$$\Pi_B = \frac{T}{e_e}\left(S_{eB}'^u - \tilde{s}_{eB} - \frac{\mu_{eB}'}{T}\right) \tag{4}$$

and

$$\Pi_{AB} = \frac{T}{e_e}\left[(S_{eB}'^u - S_{eA}'^u) - (\tilde{s}_{eB} - \tilde{s}_{eA})\right], \tag{5}$$

since once again the condition

$$\mu_{eA}' = \mu_{eB}'$$

is satisfied at the junction of conductors A and B.

4. The Absolute Value of the Entropy of Transport of Electrons

The analysis and the relations above enable analytic relationships to be sought between the various thermoelectric effects.

Such a relation exists between the Thomson and Seebeck coefficients. This is equation (4.70),

$$\tau = -T\left(\frac{\partial \varepsilon}{\partial T}\right)_p ,$$

which, for a pair of conductors A and B, can be written as

$$\tau_A = -T\left(\frac{\partial \varepsilon_A}{\partial T}\right)_p,$$

$$\tau_B = -T\left(\frac{\partial \varepsilon_B}{\partial T}\right)_p,$$

$$\tau_B - \tau_A = -T\left(\frac{\partial \varepsilon_{AB}}{\partial T}\right)_p. \tag{4.84}$$

Similarly, formulae (4.81) to (4.83), (4.65), and (4.69) lead to a relation between the Seebeck and Peltier coefficients, viz.

$$\Pi_{AB} = T\varepsilon_{AB} \tag{4.85}$$

or

$$\Pi_A = T\varepsilon_A,$$

$$\Pi_B = T\varepsilon_B.$$

Equations (4.84) and (4.85) were derived by Thomson back in 1854, of course without aid of the methods of nonequilibrium thermodynamics which was not yet known at the time. These relations are referred to in the literature as the *Thomson equations* describing thermoelectric effects.

The Thomson equations imply that all thermoelectric effects can be described in terms of the entropy of transport, since, apart from formulae (4.84) and (4.85), relations (4.65) and (4.79) represent the relation between the Seebeck and Peltier coefficients and the entropy of transport.

The entropy of transport hence is the parameter which uniquely characterizes the thermoelectric properties of a conductor, and thus knowledge of its absolute value is essential. The magnitude can be found from equation (4.61) which, on integration, can be rewritten as

$$S_e^* = S_{0e}^* - e_e \int_0^T \frac{\tau}{T}\,dT, \tag{4.86}$$

where S_{0e}^* is the value of the entropy of electron transport at 0 K. The integral containing the Thomson coefficient can be evaluated experimentally, and experiments show that

$$\lim_{T\to 0} \tau = 0.$$

Hence the integral in expression (4.86) converges and, moreover, the value of the entropy of electron transport at 0 K is finite, regardless of the kind of conductor.

Equations (4.69) and (4.86) yield

$$\varepsilon_{AB} = \frac{1}{e_e}(S_{eB}^* - S_{eA}^*) = \frac{1}{e_e}(S_{0eB}^* - S_{0eA}^*) - \int_0^T \frac{\tau_B - \tau_A}{T}\,dT. \tag{4.87}$$

By the third law of thermodynamics, which is confirmed by the results of experiments,

$$\lim_{T\to 0} \varepsilon_{AB}=0,$$

which leads to

$$S^*_{0eA}=S^*_{0eB}. \tag{4.88}$$

The entropy of electron transport at absolute zero thus does not depend on the kind of conductor. Accordingly, it may be assumed that

$$S^*_{0e}=0 \tag{4.89}$$

for any conductor.

This assumption enables a conventional absolute value to be calculated for the entropy of transport for any conductor:

$$S^*_e=-e_e\int_0^T \frac{\tau}{T}\,dT, \tag{4.90}$$

which follows directly from formulae (4.86) and (4.89).

If the integral in equation (4.90) is evaluated for some conductor, the absolute value of the entropy of transport for that conductor can also be calculated. Once the Seebeck coefficient is known (its determination requires knowledge of the thermoelectric force of the circuit) relation (4.87) can be used to calculate the entropy of transport for the other conductor.

The calculations can be carried out for various temperatures so as to find the temperature dependence of the entropy of transport.

TABLE 4.1
Absolute Value of Entropy of Electron Transport at 298 K

Metal	S^*_e $\frac{\text{J}}{\text{mole K}}$
Cu	−0.188
Ag	−0.21
Pt	0.21
Bi	7.1

Next, relation (4.81) enables us to find the Peltier coefficient

$$\Pi=T\varepsilon=\frac{TS^*_e}{e_e}$$

and the magnitude of the Seebeck coefficient.

The method described above thus makes it possible to determine the complete thermoelectric characteristics of any material and the absolute value of the thermoelectric coefficients.

Table 4.1 gives the absolute values of the entropy of electron transport for some metals.

5. Thermoelectric Generators

Thermoelectric effects may be exploited in a device for converting heat into electricity. The device may take heat from the environment and direct conversion of heat into electricity thus occurs inside it. A device of this kind is called a *thermoelectric generator*, and is in fact a heat engine since it functions in conformity with a closed thermodynamic cycle, the working medium which is subject to thermodynamic processes being a stream of electrons.

The diagram of a thermoelectric generator is given in Fig. 4.2. The generator consists of two elements made of materials A and B with electrical resistance R_A and R_B, respectively. Elements A and B are connected at one end by a strip of a different material (e.g. copper) and the strip and its junctions with elements A and B are at a constant temperature T. In accordance with what has been shown above, since the connections are isothermal, the presence of the connecting strip does not affect the magnitude of the thermoelectric effects.

The other ends of elements A and B are at the same temperature T_0 and are connected to an electrical load, indicated in the drawing as a symbolical electrical resistance R_e. Heat Q is delivered to the junction at temperature T, whereas heat Q_0 is with-

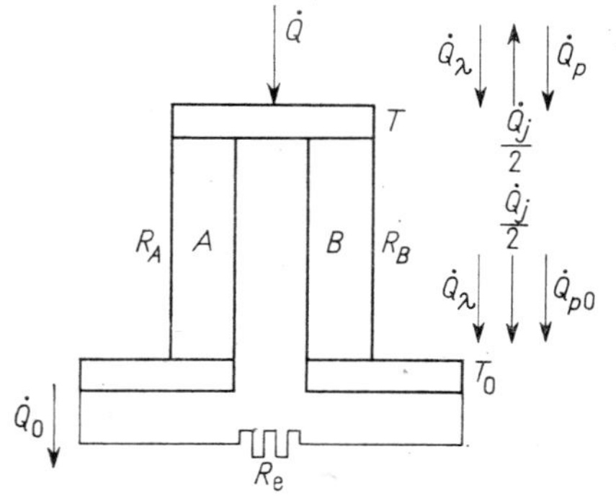

Fig. 4.2 Diagram of thermoelectric generator

drawn from the ends at temperature T_0. Moreover, the lateral surfaces of elements A and B are assumed to be insulated, so that no heat exchange with the environment occurs at these surfaces.

The energy balance of this device is

$$P=\dot{Q}-\dot{Q}_0, \tag{4.91}$$

where P is the power of the electric current produced by the generator.

The heat flows $\dot{Q}$ and $\dot{Q}_0$ consist of the following quantities:

(a) the flow due to the Peltier effect,

(b) the flow of Joule heat evolved in the conductor, and

(c) the flow of heat conducted by the material of conductors A and B as a result of the temperature difference $T-T_0$.

Thomson heat would also have to be taken into account in a more accurate analysis.

Henceforth the reasoning will be carried out for the case when all the physical parameters of materials A and B and the coefficients are constant and independent of the temperature. Moreover, the Thomson heat drops out since $\tau=0$ by equation (4.70).

It can be proved that if the lateral surface of A and B are insulated, Joule heat flows into the hot and cold ends of elements A and B in equal amounts, whereas the heat conducted can be calculated by the formulae which hold for a flat wall. Denoting the flow of Joule heat by $\dot{Q}_j$, of the conducted heat by $\dot{Q}_\lambda$, and the Peltier heat by $\dot{Q}_p$ (at temperature T) and Q_{p0} (at temperature T_0), we can write relations for $\dot{Q}$ and $\dot{Q}_0$ (cf. Fig. 4.2):

$$\dot{Q}=\dot{Q}_p+\dot{Q}_\lambda-\tfrac{1}{2}\dot{Q}_j\,, \tag{4.92}$$

$$|\dot{Q}_0|=\dot{Q}_{p0}+\dot{Q}_\lambda+\tfrac{1}{2}\dot{Q}_j\,. \tag{4.93}$$

The generator power is

$$P=\dot{Q}-|\dot{Q}_0|=\dot{Q}_p-\dot{Q}_{p0}-\dot{Q}_j\,. \tag{4.94}$$

The generator efficiency is defined as the ratio

$$\eta=\frac{P}{\dot{Q}}=\frac{\dot{Q}_p-\dot{Q}_{p0}-\dot{Q}_j}{\dot{Q}_p+\dot{Q}_\lambda-\tfrac{1}{2}\dot{Q}_j}\,. \tag{4.95}$$

Formula (4.95) can be transformed further, viz.

$$\dot{Q}_p=\Pi_{AB}I=TI\varepsilon_{AB}\,,$$

where I is the current flowing in the generator circuit.

Similarly

$$\dot{Q}_{p0}=\Pi_{AB}I=T_0I\varepsilon_{AB}\,.$$

The Joule heat flow can be evaluated by the formula

$$\dot{Q}_j=I^2(R_A+R_B)=\left(\frac{\rho_A}{A_A}+\frac{\rho_B}{A_B}\right)lI^2,$$

where ρ_A and ρ_B denote the resistivities of materials A and B, whereas A_A and A_B are the cross-sectional areas of elements A and B and l is their length.

The conducted heat flow is

$$\dot{Q}_\lambda=(\lambda_A A_A+\lambda_B A_B)\frac{T-T_0}{l}\,,$$

where λ_A and λ_B are the thermal conductivities of materials A and B (these are the thermal conductivities λ_∞).

Substitution of the expressions for $\dot{Q}_p$, $\dot{Q}_{p0}$, $\dot{Q}_j$, and $\dot{Q}_\lambda$ into formula (4.95) yields

$$\eta = \frac{\varepsilon_{AB}(T-T_0)I - \left(\frac{\rho_A}{A_A}+\frac{\rho_B}{A_B}\right) lI^2}{\varepsilon_{AB} TI - \frac{1}{2}\left(\frac{\rho_A}{A_A}+\frac{\rho_B}{A_B}\right) lI^2 + (\lambda_A A_A + \lambda_B A_B)\frac{T-T_0}{l}}. \tag{4.96}$$

The numerator of expression (4.96) is equal to P and can be rewritten as

$$P = I^2 R_e = I V_e,$$

where V_e is the potential drop across the resistor R_e.

The expression for the efficiency can be optimized, a distinction being made between two basic cases, viz.:

(a) achievement of maximum power,

(b) achievement of maximum efficiency.

Maximum power can be obtained if the electrical resistances R_A, R_B, and R_e are related by

$$\frac{R_e}{R_A+R_B} = 1. \tag{4.97}$$

The power of an engine satisfying this condition is

$$P = \frac{\varepsilon_{AB}^2 (T-T_0)^2}{4R_e} = \frac{\varepsilon_{AB}^2 (\Delta T)^2}{4R_e};$$

relation (4.97) can be satisfied for various geometrical dimensions of elements A and B, this corresponding to various generator efficiencies. The highest generator efficiency, however, is ensured by the condition (with the assumption that the two elements are of identical length)

$$\frac{A_A}{A_B} = \left(\frac{\rho_A \lambda_B}{\rho_B \lambda_A}\right)^{\frac{1}{2}}. \tag{4.98}$$

Simultaneous satisfaction of conditions (4.97) and (4.98) ensures maximum work at the given temperatures T and T_0 and material parameters and at the same time ensures the highest efficiency.

This efficiency is given by

$$\eta = \frac{\Delta T}{2T - \frac{1}{2}\Delta T + \frac{4}{Z}}, \tag{4.99}$$

where the figure of merit Z is equal to

$$Z = \frac{\varepsilon_{AB}^2}{[(\lambda_A \rho_A)^{\frac{1}{2}} + (\lambda_B \rho_B)^{\frac{1}{2}}]^2}. \tag{4.100}$$

If λ, ρ, and ε do not depend on the temperature, then Z is also a constant independent of the temperature.

If the generator is calculated by the maximum efficiency condition, the best ratio of resistances is

$$\frac{R_e}{R_A+R_B}=\sqrt{1+\frac{Z}{2}(T+T_0)}=\sqrt{1+Z\bar{T}}. \tag{4.101}$$

The optimal dimensions are characterized by relation (4.98) which held in the previous case.

When the conditions expressed by equations (4.98) and (4.101) are satisfied simultaneously, the generator has the maximum efficiency for the given conditions:

$$\eta=\frac{\Delta T}{T}\frac{\sqrt{1+Z\bar{T}}-1}{\sqrt{1+Z\bar{T}}+\frac{T_0}{T}}=\frac{\Delta T}{T}\frac{\sqrt{1+Z\bar{T}}-1}{\sqrt{1+Z\bar{T}}+1-\frac{\Delta T}{T}}. \tag{4.102}$$

This efficiency is expressible as the product of the Carnot cycle efficiency $\eta_c=\Delta T/T$ and the coefficient α which is equal to

$$\alpha=\frac{\sqrt{1+Z\bar{T}}-1}{\sqrt{1+Z\bar{T}}+\frac{T_0}{T}},$$

that is,

$$\eta=\eta_c\alpha.$$

The coefficient α is the exergetic efficiency of the thermocouple, and hence is a measure of the losses due to the irreversibility of the thermodynamic processes occurring

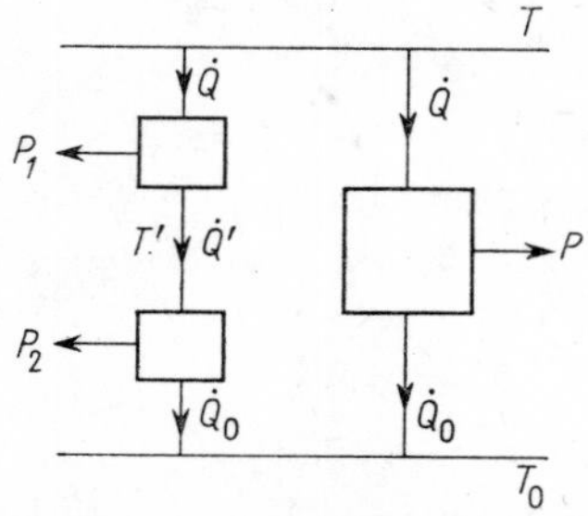

Fig. 4.3 Cascade thermoelectric generator

n the generator. It should be noted that the greater Z is, the closer the value of α is to 1 and the closer the generator efficiency is to the Carnot cycle efficiency. The generator efficiency can be increased if temperature stages are used as shown in Fig. 4.3 where the temperature drop has been divided into two parts. The heat rejected by the first stage is absorbed in the second stage which gives off heat $\dot{Q}_0$ to the environment.

The efficiency of this device can be calculated as follows:
efficiency of stage I

$$\eta_1 = \frac{P_1}{\dot{Q}},$$

efficiency of stage II

$$\eta_2 = \frac{P_2}{\dot{Q}'},$$

where $\dot{Q}' = \dot{Q} - P_1$.

The efficiency of the system is

$$\eta_0 = \frac{P_1 + P_2}{\dot{Q}} = \frac{P_1}{\dot{Q}} + \frac{P_2}{\dot{Q}}.$$

Since

$$\frac{P_2}{\dot{Q}'} = \frac{P_2}{\dot{Q} - P_1} = \eta_2, \quad \text{we have} \quad P_2 = \eta_2 \dot{Q} - P_1 \eta_2$$

and

$$\frac{P_2}{\dot{Q}} = \eta_2 - \frac{P_1}{\dot{Q}} \eta_2 .$$

Insertion of this relation into the expression for η_0 yields

$$\eta_0 = \eta_1 + \eta_2 (1 - \eta_1) = 1 - (1 - \eta_1)(1 - \eta_2).$$

This result can be generalized to more stages and the formula obtained is

$$\eta_0 = 1 - \prod_i (1 - \eta_i). \tag{4.103}$$

If the efficiency of every stage is optimized individually, the efficiency of the device is greater than that of a single-stage device operating over the whole temperature drop.

In the extreme case, the number of stages may theoretically be increased to infinity. If each of these stages is optimized so as to achieve maximum efficiency, the stage efficiency is

$$\eta_{st} = \frac{\mathrm{d}T}{T} \alpha_0 ,$$

where $\alpha_0 = \dfrac{\sqrt{1+ZT}-1}{\sqrt{1+ZT}+1}$ is the limiting value of α as $\Delta T \to 0$.

Such a device enables maximum efficiency to be obtained under given conditions. It can be calculated as

$$\eta_{st} = \frac{P}{\dot{Q}} = \frac{\mathrm{d}\dot{Q}}{\dot{Q}} = \alpha_0 \frac{\mathrm{d}T}{T}, \quad \text{since} \quad P = \mathrm{d}\dot{Q}.$$

Integration of this equation leads to the result

$$\int_{\dot{Q}}^{\dot{Q}_0} \frac{\mathrm{d}\dot{Q}}{\dot{Q}} = \int_T^{T_0} \alpha_0 \frac{\mathrm{d}T}{T}; \qquad \frac{\dot{Q}}{\dot{Q}_0} = \exp\left(\int_T^{T_0} \alpha_0 \frac{\mathrm{d}T}{T}\right)$$

and further

$$\eta = 1 - \frac{\dot{Q}_0}{\dot{Q}} = 1 - \exp\left(-\int_{T_0}^{T} \alpha_0 \frac{\mathrm{d}T}{T}\right).$$

Introducing the notation

$$\bar{\alpha} = \frac{\displaystyle\int_{T_0}^{T} \alpha_0 \frac{\mathrm{d}T}{T}}{\ln \dfrac{T}{T_0}},$$

yields

$$\eta = 1 - \left(\frac{T_0}{T}\right)^{\bar{\alpha}}, \tag{4.104}$$

since

$$\frac{\dot{Q}_0}{\dot{Q}} = \left(\frac{T_0}{T}\right)^{\bar{\alpha}},$$

where $\bar{\alpha}$ is the mean value of α in the interval (T, T_0).

The stage efficiency

$$\eta_{\mathrm{st}} = \frac{\mathrm{d}T}{T} \alpha_0$$

is the product of the efficiency $\mathrm{d}T/T$ of a Carnot cycle with a very small temperature difference and the efficiency α_0 of a thermocouple with a very small temperature drop ΔT. As follows from the formula defining α, with the coefficient Z and the temperature T remaining constant the value of α increases as ΔT grows. This fact indicates an analogy between the coefficient α and the efficiency of a multi-stage turbine. If α_0 is treated as a quantity analogous to the efficiency of a stage with a small pressure drop, then α may be said to be analogous to the internal efficiency which also grows as the pressure drop increases provided that the process obeys the equation $pv^n = \mathrm{const}$ $(n<k)$.

6. Thermoelectric Refrigerators

Thermoelectric effects can also be put to work in a refrigerator, a block diagram of which is given in Fig. 4.4. It is similar in arrangement to a thermoelectric generator. The principal parts are two elements made of materials A and B. Direct current from an external source (a battery in the drawing) is applied to the refrigerator and then heat is absorbed at one junction and given off at the other. If the temperature T is equal to the ambient temperature, the second junction will be cooled to a tem-

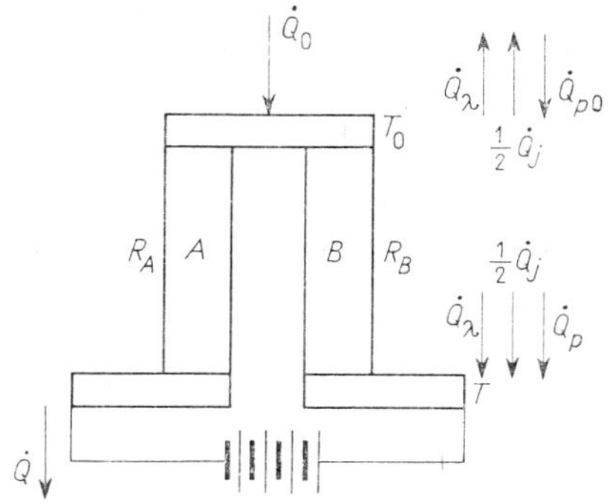

Fig. 4.4 Block diagram of thermoelectric refrigerator

perature T_0, below the ambient temperature. Assuming as before that the lateral surfaces of elements A and B are insulated, and that the physical parameters and the Seebeck coefficient are constant regardless of the temperature, then, neglecting the Thomson heat, we can carry out an energy analysis of the refrigerator.

Its energy balance is of the form

$$P+\dot{Q}_0=|\dot{Q}|,$$

where $\dot{Q}_0$ is the supplied heat flow, and $\dot{Q}$ is the extracted heat flow from the refrigerator; P is the power of the electric current supplied to the refrigerator.

The flows $\dot{Q}$ and $\dot{Q}_0$ can be written, just as for the generator, as functions of the Peltier effect, Joule heat, and conducted heat:

$$|\dot{Q}|=\dot{Q}_p+\tfrac{1}{2}\dot{Q}_j-\dot{Q}_\lambda,$$

$$\dot{Q}_0=\dot{Q}_{p0}-\tfrac{1}{2}\dot{Q}_j-\dot{Q}_\lambda.$$

The power supplied to the refrigerator is

$$P=\dot{Q}_p-\dot{Q}_{p0}+\dot{Q}_j,$$

and on the other hand

$$P=IV_e,$$

where I is the supply current, and V_e is the potential drop in the refrigerator.

The quantities $\dot{Q}_p$, $\dot{Q}_{p0}$, $\dot{Q}_\lambda$, and $\dot{Q}_j$ are described by the same relations as given in the discussion of the thermoelectric generator.

The coefficient of performance is equal to

$$\varepsilon_r=\frac{\dot{Q}_0}{P}=\frac{\dot{Q}_{p0}-\frac{1}{2}\dot{Q}_j-\dot{Q}_\lambda}{\dot{Q}_p-\dot{Q}_{p0}+\dot{Q}_j}, \tag{4.105}$$

and substitution of the values of $\dot{Q}_p$, $\dot{Q}_{p0}$, $\dot{Q}_j$, and $\dot{Q}_\lambda$ leads to the expression

$$\varepsilon_r = \frac{\varepsilon_{AB} T I - \frac{1}{2}\left(\frac{\rho_A}{A_A}+\frac{\rho_B}{A_B}\right) l I^2 - (\lambda_A A_A + \lambda_B A_B)\frac{T-T_0}{l}}{\varepsilon_{AB}(T-T_0) I + \left(\frac{\rho_A}{A_A}+\frac{\rho_B}{A_B}\right) l I^2},$$

the notation being the same as in the case of the generator.

The refrigerator is also optimizable for

(a) maximum refrigerator power,

(b) maximum coefficient of performance.

The condition for maximum refrigerating power (maximum $\dot{Q}_0$) is given by the value of the current

$$I = \frac{\varepsilon_{AB} T_0}{\left(\frac{\rho_A}{A_A}+\frac{\rho_B}{A_B}\right) l} \tag{4.106}$$

and the geometrical dimensions according to relation (4.98)

$$\frac{A_A}{A_B} = \left(\frac{\lambda_B \rho_A}{\lambda_A \rho_B}\right)^{\frac{1}{2}},$$

hence just as for the thermoelectric generator.

The relations above result when the maximum of the numerator is sought in the expression for ε_r.

The maximum value of $\dot{Q}_0$ obtained on substituting $\dot{Q}_{p0}$, $\dot{Q}_j$, and $\dot{Q}_\lambda$ into the expression for $\dot{Q}_0$ and on taking account of conditions (4.98) and (4.106) is

$$\dot{Q}_{0\,\max} = \frac{\varepsilon_{AB}^2 T_0^2}{2\left(\frac{\rho_A}{A_A}+\frac{\rho_B}{A_B}\right) l} - \frac{[(\lambda_A \rho_A)^{\frac{1}{2}} + (\lambda_B \rho_B)^{\frac{1}{2}}]^2}{\left(\frac{\rho_A}{A_A}+\frac{\rho_B}{A_B}\right) l}(T-T_0).$$

The maximum temperature difference can be achieved when $\dot{Q}_0 = 0$, which leads to

$$(T-T_0)_{\max} = \Delta T_{\max} = \frac{\varepsilon_{AB}^2 T_0^2}{2[(\rho_A \lambda_A)^{\frac{1}{2}} + (\rho_B \lambda_B)^{\frac{1}{2}}]^2}$$

or, on introduction of the parameter Z described by equation (4.100), to

$$\Delta T_{\max} = \frac{Z T_0^2}{2},$$

to which $\varepsilon_r = 0$ clearly corresponds.

The condition for maximum coefficient of performance is given by the value of the current

$$I = \frac{\varepsilon_{AB}(T-T_0)}{\left(\frac{\rho_A}{A_A}+\frac{\rho_B}{A_B}\right) l \sqrt{1+\frac{Z}{2}(T+T_0)} - 1},$$

the maximum value of ε_r being

$$\varepsilon_r = \frac{T_0}{T-T_0}\,\frac{\sqrt{1+\frac{1}{2}Z(T+T_0)}-\dfrac{T}{T_0}}{\sqrt{1+\frac{1}{2}Z(T+T_0)}+1} = \frac{T_0}{T-T_0}\,\frac{\sqrt{1+Z\bar{T}}-\dfrac{T}{T_0}}{\sqrt{1+Z\bar{T}}+1}\,. \tag{4.107}$$

As before, therefore, ε_r can be written as the product of the coefficient of performance of the inverse Carnot cycle,

$$\varepsilon_c = \frac{T_0}{T-T_0}$$

and the coefficient

$$\alpha_r = \frac{\sqrt{1+Z\bar{T}}-\dfrac{T}{T_0}}{\sqrt{1+Z\bar{T}}+1}\,,$$

that is,

$$\varepsilon_r = \varepsilon_c\,\alpha_r\,.$$

The coefficient α_r is the exergetic efficiency of the refrigerator, and hence is also a measure of the losses due to the irreversibility of the refrigerator cycle. Its value is less than 1 and increases along with Z.

It is worth noting that the values of α_r and α approach each other as $T \to T_0$ (obviously, with the same value of $Z\bar{T}$), that is,

$$\alpha_{r0} = \alpha_0\,.$$

On the other hand, α_r decreases as ΔT increases, wherein an analogy can be seen between α_r and the efficiency of the turbocompressor.

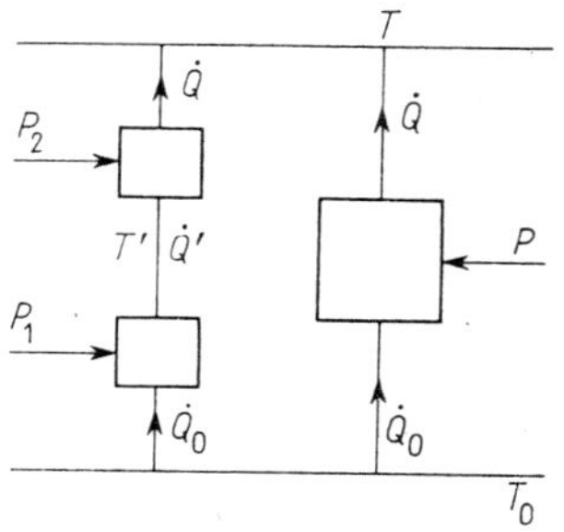

Fig. 4.5 Cascade thermoelectric refrigerator

A thermoelectric refrigerator can also be made as a multi-stage device. This is particularly necessary when considerable temperature differences are involved since the value ΔT_{max} cannot be exceeded in one stage.

By way of example let us consider the two-stage refrigerator depicted in Fig. 4.5. The coefficient of performance is

$$\varepsilon_1 = \frac{\dot{Q}_0}{P_1}$$

for the first stage, and

$$\varepsilon_2 = \frac{\dot{Q}'}{P_2}$$

for the second.

The coefficient of performance for the entire system is

$$\varepsilon_0 = \frac{\dot{Q}_0}{P_1 + P_2} = \frac{1}{\dfrac{P_1}{\dot{Q}_0} + \dfrac{P_2}{\dot{Q}_0}} . \tag{4.108}$$

Next we have

$$\frac{P_2}{\dot{Q}'} = \frac{P_2}{\dot{Q}_0 + P_1} = \frac{1}{\varepsilon_2}$$

and further

$$P_2 = \frac{1}{\varepsilon_2}(\dot{Q}_0 + P_1) \quad \text{and} \quad \frac{P_2}{\dot{Q}_0} = \frac{1}{\varepsilon_2}\left(1 + \frac{P_1}{\dot{Q}_0}\right) = \frac{1}{\varepsilon_2}\left(1 + \frac{1}{\varepsilon_1}\right).$$

When these relations are inserted into equation (4.108) the result is

$$\varepsilon_0 = \frac{1}{\dfrac{1}{\varepsilon_1} + \dfrac{1}{\varepsilon_2}\left(1 + \dfrac{1}{\varepsilon_1}\right)} = \frac{1}{\left(1 + \dfrac{1}{\varepsilon_1}\right)\left(1 + \dfrac{1}{\varepsilon_2}\right) - 1} .$$

This result can be generalized to more stages

$$\varepsilon_0 = \frac{1}{\prod_i \left(1 + \dfrac{1}{\varepsilon_i}\right) - 1} . \tag{4.109}$$

In the case of a system with an infinite number of individually optimized stages the coefficient of performance of a small stage is

$$\varepsilon_d = \frac{T}{\mathrm{d}T} \frac{\sqrt{1+ZT} - 1}{\sqrt{1+ZT} + 1} = \frac{\dot{Q}'}{\mathrm{d}\dot{Q}'} = \frac{T}{\mathrm{d}T} \alpha_0 ,$$

that is,

$$\frac{\mathrm{d}\dot{Q}'}{\dot{Q}'} = \frac{1}{\alpha_0} \frac{\mathrm{d}T}{T} .$$

Integrating this expression gives

$$\ln \frac{\dot{Q}_0}{|\dot{Q}|} = \int_T^{T_0} \frac{1}{\alpha_0} \frac{\mathrm{d}T}{T}$$

and the coefficient of performance of the multi-stage system is

$$\varepsilon_r = \frac{1}{\exp\left(\int_T^{T_0} \frac{1}{\alpha_0} \frac{\mathrm{d}T}{T}\right) - 1} .$$

If the value of the integral is replaced, as before, by the mean value of α_0,

$$\int_T^{T_0} \frac{1}{\alpha_0} \frac{\mathrm{d}T}{T} = \frac{1}{\bar{\alpha}_r} \ln \frac{T_0}{T} ,$$

we can write

$$\frac{\dot{Q}_0}{T} = \left(\frac{T_0}{T}\right)^{1/\bar{\alpha}_r}$$

and the expression for the coefficient of performance of the multi-stage system takes the form

$$\varepsilon_r = \frac{\dot{Q}_0}{|\dot{Q}| - \dot{Q}_0} = \frac{1}{\frac{|\dot{Q}|}{\dot{Q}_0} - 1} = \frac{1}{\left(\frac{T}{T_0}\right)^{1/\bar{\alpha}_r} - 1} \tag{4.110}$$

7. Problems in the Development of Practical Thermoelectric Generators and Refrigerators

The construction of practical thermoelectric generators and refrigerators with satisfactory characteristics requires suitable materials first and foremost. From the point of view of suitability for use in thermoelectric devices these materials can be characterized by a *figure of merit*

$$Z_A = \frac{\varepsilon_A^2}{\rho_A \lambda_A} \tag{4.111}$$

or the *Joffe coefficient*

$$\theta_A = \frac{T\varepsilon_A^2}{\rho_A \lambda_A} . \tag{4.112}$$

Both of these contain the Seebeck coefficient, electrical resistivity, and thermal conductivity. The figure of merit is purely a material parameter, dependent exclusively on the kind of material, whereas the Joffe coefficient contains the temperature as well.

Thus both parameters, Z_A and θ_A, in a way characterize irreversible phenomena which occur in the conductor material, viz. heat conduction and Joule heating. To be more precise, they could be viewed as a measure of the ratio of the quantity characterizing the reversible thermoelectric phenomena (Seebeck effect) to the quantities characterizing irreversible phenomena.

As follows from the discussion on generators and refrigerators, the greater Z_A (or θ_A), the greater the efficiency and the coefficient of performance. Thus, the endeavour to develop economical thermoelectric devices involves a search for materials possessing a high Z, the higher the better. A thorough analysis shows that the properties of various materials depend on the concentration of charge carriers, as indicated in Fig. 4.6. Insulators have a low number of free charge carriers, and this is responsible for their high electrical resistivity and low thermal conductivity. They are characterized by a large Seebeck coefficient, it is true, but their Z is not very large.

Metals, in turn, have a high concentration of free charge carriers, large thermal and electrical conductivities, but relatively small Seebeck coefficients. In effect, their Z is also not very large.

The largest values of Z are displayed by semiconductors which are characterized by not very high concentrations of free charge carriers. Their thermal and electrical conductivities attain values intermediate between those of insulators and metals and the same is true of the Seebeck coefficient. The semiconductors used in thermoelectric devices may be of two kinds, viz. type p or type n. Those of type p are charac-

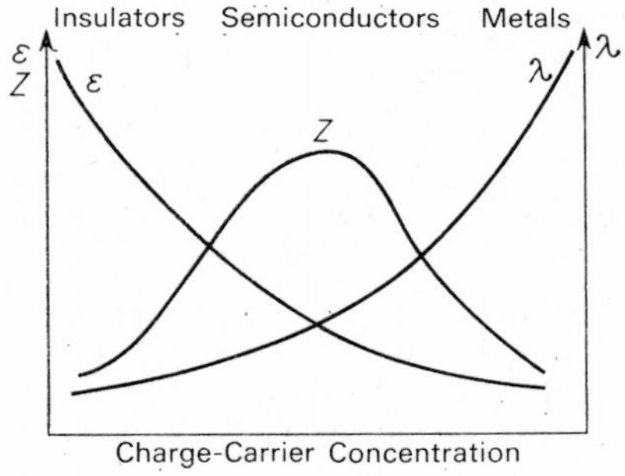

Fig. 4.6 Thermoelectric parameters of electrical conductors

terized by a deficit of free electrons and n-type semiconductors, by an excess. Accordingly, negative carriers (electrons) are the free charge carriers in n-type semiconductors, and positive carriers are in p-type. Positive carriers are due to the absence of electrons at various points of the crystal lattice. These defect electrons in atoms can be treated as holes which move along the semiconductor, just as free electrons do but in the opposite direction.

The effect of an excess or deficit of electrons can be increased by adding suitable impurities to semiconducting material. It should be pointed out that these impurities also have an important influence on the figure of merit.

The figures of merit for some semiconducting materials used in thermoelectric refrigerators and generators are listed in Table 4.2.

TABLE 4.2
Figures of Merit of Typical Semiconductors

Semiconductors	Type	Temperature °C	Figure of Merit $10^3 ZK^{-1}$
Bi_2Te_3	*p*	25	2.5
Bi_2Te_3	*n*	25	2.5
PbTe	*n*	450	1.3
ZnSb	*p*	175	1.4
GeTe	*p*	450	1.7
MnTe	*p*	900	0.4
$AgSbTe_2$	*p*	400	1.8
InAs	*n*	700	0.7

In the case of thermoelectric generators there is an additional problem relating to attainment of maximum efficiency. Semiconducting materials enable high figures of merit to be obtained only in a not very large temperature interval. Thus, if a generator is to operate with a considerable temperature difference, it should be constructed with several elements arranged in a cascade and made of different materials. This complicates the system somewhat but it does permit every stage to be optimized individually and relatively high efficiencies to be achieved.

Other major problems which may crop up are: the semiconducting elements may corrode at higher temperatures upon contact with air, the impurities may diffuse under the influence of the temperature difference, and the technology for making mechanically strong junctions but with the smallest possible contact resistances may present difficulties.

It should also be noted that a thermoelectric refrigerator should have a D. C. power supply. Similarly, a thermoelectric generator produces low-voltage direct current.

The further development and construction of thermoelectric generators and refrigerators depends first of all on the advances made in semiconducting materials. Efforts are being made to produce materials which have the highest possible values of Z and, for applications in thermoelectric generators, also withstand high temperatures.

8. Electromagnetic Effects

If a solid conductor carrying an electric current and displaying a temperature difference is in an external magnetic field $\boldsymbol{B}$, new effects which hold out the promise of practical applications occur in it. These effects will be considered here for a homogeneous, isotropic conductor carrying an electric current and in a uniform magnetic field. Moreover, the magnetic field is assumed to be in the direction of the z-axis whereas the flow of current and heat is assumed to occur in the xy-plane.

The analysis thus does not comprise effects occurring at the interface between two different conductors or effects associated with a chemical reaction taking place.

Note that even in the case of an isotropic conductor the application of a magnetic field causes anisotropies which disappear when that field is removed. An isotropic conductor in a magnetic field will thus be characterized by some parameters such as thermal conductivity and electrical resistivity which are tensorial.

The dissipation function in the case under consideration is described by the same equation as was derived earlier, that is, equation (4.26). On the other hand, equations (4.53) and (4.54) now assume a different form. Relation (4.54) can be transformed into

$$-\nabla_T\left(\frac{\mu_e'}{e_e}\right)=\frac{e_e j_e}{\sigma}+\frac{Q_e''^*}{e_e T}\nabla T=\nabla\left(\frac{\mu_e'}{e_e}\right)+\frac{\tilde{s}_e}{e_e}\nabla T$$

$$=\nabla\left(\frac{\mu_e}{e_e}\right)-E+\frac{\tilde{s}_e}{e_e}\nabla T.$$

If the electrical conductivity σ is replaced by its inverse (electrical resistivity), the transformed expression (4.54) can be rewritten as

$$E-\nabla\left(\frac{\mu_e}{e_e}\right)=e_e j_e \rho+\left(\frac{Q_e''^*}{e_e T}+\frac{\tilde{s}_e}{e_e}\right)\nabla T.$$

Next, the scalar quantities ρ, $Q_e''^*$, and λ should be replaced by tensors of rank two since in the presence of a magnetic field the electrical resistance, heat of transport, and thermal conductivity become tensorial quantities of rank two.

Finally, therefore, relations (4.53) and (4.54) become

$$E-\nabla\left(\frac{\mu_e}{e_e}\right)=e_e\,\boldsymbol{\rho}\cdot j_e+\frac{\mathbf{Q}_e''^*}{e_e T}+\frac{\tilde{s}_e}{e_e}\cdot\nabla T, \tag{4.113}$$

$$J_q''=\mathbf{Q}_e''^*\cdot j_e-\boldsymbol{\lambda}\cdot\nabla T. \tag{4.114}$$

In accordance with the results of the previous section, it can be said that the heat of transport is related to the Seebeck and the Peltier coefficients and that the tensor $Q_e''^*$ can be replaced by the tensors $\boldsymbol{\varepsilon}$ and Π. Moreover, it is convenient to substitute the flux J_s for the flux J_q'',

$$J_s=\frac{J_q''}{T}+\tilde{s}_e\,j_e.$$

Then equations (4.113) and (4.114) can be cast into the form

$$E-\nabla\left(\frac{\mu_e}{e_e}\right)=e_e\,\boldsymbol{\rho}\cdot j_e+\boldsymbol{\varepsilon}\cdot\nabla T, \tag{4.115}$$

$$J_s=\frac{\Pi e_e}{T}\cdot j_e-\frac{\boldsymbol{\lambda}}{T}\cdot\nabla T. \tag{4.116}$$

In view of the assumption made above limiting the effect under consideration to the xy-plane and in view of the fact that the system under study becomes isotropic upon removal of the magnetic field, the tensors $\boldsymbol{\rho}$, $\boldsymbol{\lambda}$, $\boldsymbol{\varepsilon}$, and $\boldsymbol{\Pi}$ can be written in matrix form:

$$\boldsymbol{\rho}=\begin{vmatrix} \rho_{xx} & \rho_{xy} \\ -\rho_{xy} & \rho_{xx} \end{vmatrix}, \tag{4.117}$$

$$\boldsymbol{\lambda}=\begin{vmatrix} \lambda_{xx} & \lambda_{xy} \\ -\lambda_{xy} & \lambda_{xx} \end{vmatrix}, \tag{4.118}$$

$$\boldsymbol{\varepsilon}=\begin{vmatrix} \varepsilon_{xx} & \varepsilon_{xy} \\ -\varepsilon_{xy} & \varepsilon_{xx} \end{vmatrix}, \tag{4.119}$$

$$\boldsymbol{\Pi}=\begin{vmatrix} \Pi_{xx} & \Pi_{xy} \\ -\Pi_{xy} & \Pi_{xx} \end{vmatrix}. \tag{4.120}$$

The validity of expressions (4.117) to (4.120) can be demonstrated with the example of tensor $\boldsymbol{\rho}$. By equation (4.113) we have $\nabla T=0$ in the isothermal conductor and

$$\boldsymbol{E}-\nabla\left(\frac{\mu_{\mathrm{e}}}{e_{\mathrm{e}}}\right)=e_{\mathrm{e}}\,\boldsymbol{\rho}\cdot\boldsymbol{j}_{\mathrm{e}}\,.$$

This expression should be invariant under rotation about the z-axis along which the external magnetic field acts; the form of equation (4.117) follows from this.

In the case under analysis, a rotation through an angle π about the x-axis causes no changes in the general relations since the x-axis is perpendicular to the z-axis. Accordingly, the following relations which express the Onsager–Casimir principle are satisfied:

$$\rho_{xx}(\boldsymbol{B})=\rho_{xx}(-\boldsymbol{B}),\qquad \rho_{xy}(\boldsymbol{B})=-\rho_{xy}(-\boldsymbol{B}), \tag{4.121}$$

$$\lambda_{xx}(\boldsymbol{B})=\lambda_{xx}(-\boldsymbol{B}),\qquad \lambda_{xy}(\boldsymbol{B})=-\lambda_{xy}(-\boldsymbol{B}), \tag{4.122}$$

$$\varepsilon_{xx}(\boldsymbol{B})=\varepsilon_{xx}(-\boldsymbol{B}),\qquad \varepsilon_{xy}(\boldsymbol{B})=-\varepsilon_{xy}(-\boldsymbol{B}), \tag{4.123}$$

$$\Pi_{xx}(\boldsymbol{B})=\Pi_{xx}(-\boldsymbol{B}),\qquad \Pi_{xy}(\boldsymbol{B})=-\Pi_{xy}(-\boldsymbol{B}). \tag{4.124}$$

The consequences stemming from the Onsager reciprocal relations for a conductor which is not in a magnetic field can be used to write the Onsager–Casimir principle in the given case as

$$T\varepsilon_{xx}(\boldsymbol{B})=\Pi_{xx}(\boldsymbol{B})=\Pi_{xx}(-\boldsymbol{B}), \tag{4.125}$$

$$T\varepsilon_{xy}(\boldsymbol{B})=\Pi_{yx}(-\boldsymbol{B})=-\Pi_{xy}(-\boldsymbol{B})=\Pi_{xy}(\boldsymbol{B}) \tag{4.126}$$

or, more generally, as

$$T\boldsymbol{\varepsilon}=\boldsymbol{\Pi}. \tag{4.127}$$

These relations make it possible to analyse a number of effects associated with the flow of heat and electric current in a conductor in a magnetic field.

The relations derived here can also be obtained by basing the discussion directly on the phenomenological equations. The dissipation function, in conformity with equations (4.26) and (4.31), is of the form

$$\Psi = -\boldsymbol{J}_q'' \cdot \nabla(\ln T) - \boldsymbol{j}_e \cdot (\nabla_T \mu_e - e_e \boldsymbol{E}).$$

Accordingly, the energy flux $\boldsymbol{J}_q''$ and the diffusion flux $\boldsymbol{j}_e$ should be used as the flows in phenomenological equations whereas $\nabla(\ln T)$ and $(\nabla_T \mu_e - e_e \boldsymbol{E})$ would be the thermodynamic forces. Since this choice is somewhat arbitrary, for convenience in further applications it is best to take $\boldsymbol{J}_q''$ and $(e_e \boldsymbol{E} - \nabla_T \mu_e)$ for the flows and $\nabla(\ln T)$ and $\boldsymbol{j}_e$ then are thermodynamic forces. Bearing in mind that the effects under consideration occur in the xy-plane, we can write the phenomenological equations as:

$$(e_e E - \nabla_T \mu_e)_x = L_{11} j_{ex} + L_{12} j_{ey} + L_{13} \nabla_x(\ln T) + L_{14} \nabla_y(\ln T), \qquad (4.128)$$

$$(e_e E - \nabla_T \mu_e)_y = L_{21} j_{ex} + L_{22} j_{ey} + L_{23} \nabla_x(\ln T) + L_{24} \nabla_y(\ln T), \qquad (4.129)$$

$$-J_{qx}'' = L_{31} j_{ex} + L_{32} j_{ey} + L_{33} \nabla_x(\ln T) + L_{34} \nabla_y(\ln T), \qquad (4.130)$$

$$-J_{qy}'' = L_{41} j_{ex} + L_{42} j_{ey} + L_{43} \nabla_x(\ln T) + L_{44} \nabla_y(\ln T). \qquad (4.131)$$

In view of the assumption that the system is isotropic (upon removal of the magnetic field), symmetry with respect to the x- and y-axes must be preserved and this leads to the relations

$$L_{11} = L_{22}, \quad L_{12} = -L_{21},$$

$$L_{13} = L_{24}, \quad L_{14} = -L_{23},$$

$$L_{31} = L_{42}, \quad L_{32} = -L_{41},$$

$$L_{33} = L_{44}, \quad L_{34} = -L_{43}.$$

Thus equations (4.128) to (4.131) can be cast into the form

$$(e_e E - \nabla_T \mu_e)_x = L_{11} j_{ex} + L_{12} j_{ey} + L_{13} \nabla_x(\ln T) + L_{14} \nabla_y(\ln T), \qquad (4.132)$$

$$(e_e E - \nabla_T \mu_e)_y = -L_{12} j_{ex} + L_{11} j_{ey} - L_{14} \nabla_x(\ln T) + L_{13} \nabla_y(\ln T), \qquad (4.133)$$

$$-J_{qx}'' = L_{31} j_{ex} + L_{32} j_{ey} + L_{33} \nabla_x(\ln T) + L_{34} \nabla_y(\ln T), \qquad (4.134)$$

$$-J_{qy}'' = -L_{32} j_{ex} + L_{31} j_{ey} - L_{34} \nabla_x(\ln T) + L_{33} \nabla_y(\ln T). \qquad (4.135)$$

The Onsager–Casimir reciprocal relations now become

$$L_{12}(\boldsymbol{B}) = L_{21}(-\boldsymbol{B}) = -L_{12}(-\boldsymbol{B}), \qquad (4.136)$$

$$L_{13}(\boldsymbol{B}) = -L_{31}(-\boldsymbol{B}) = -L_{31}(\boldsymbol{B}), \qquad (4.137)$$

$$L_{14}(\boldsymbol{B}) = -L_{41}(-\boldsymbol{B}) = L_{32}(-\boldsymbol{B}) = -L_{32}(\boldsymbol{B}), \qquad (4.138)$$

$$L_{34}(\boldsymbol{B}) = L_{43}(-\boldsymbol{B}) = -L_{34}(-\boldsymbol{B}), \qquad (4.139)$$

use having been made here of the fact that the coefficients L_{11}, L_{13}, L_{31}, and L_{33}

are even functions of $\boldsymbol{B}$, whereas L_{12}, L_{14}, L_{32}, and L_{34} are odd functions of $\boldsymbol{B}$. Use of the symmetry conditions reduces the number of phenomenological coefficients from 16 in equations (4.128) to (4.131) down to 8 in equations (4.132) to (4.135).

The equations (4.136) to (4.139) describing the Onsager–Casimir reciprocal relations make it possible to consider two more coefficients. In the end the following relations containing six different phenomenological coefficients are obtained:

$$(e_e \boldsymbol{E} - \nabla_T \mu_e)_x = L_{11} j_{ex} + L_{12} j_{ey} + L_{13} \nabla_x(\ln T) + L_{14} \nabla_y(\ln T), \qquad (4.140)$$

$$(e_e \boldsymbol{E} - \nabla_T \mu_e)_y = -L_{12} j_{ex} + L_{11} j_{ey} - L_{14} \nabla_x(\ln T) + L_{13} \nabla_y(\ln T), \qquad (4.141)$$

$$-J''_{qx} = -L_{13} j_{ex} - L_{14} j_{ey} + L_{33} \nabla_x(\ln T) + L_{34} \nabla_y(\ln T), \qquad (4.142)$$

$$-J''_{qy} = L_{14} j_{ex} - L_{13} j_{ey} - L_{34} \nabla_x(\ln T) + L_{33} \nabla_y(\ln T). \qquad (4.143)$$

Analysis of these equations makes it possible to find relations describing galvanomagnetic and thermomagnetic effects.

9. Classification of Thermomagnetic and Galvanomagnetic Effects

Thermomagnetic and galvanomagnetic effects are sometimes classified as follows:

(a) If the original current (thermodynamic force) is perpendicular to the effect produced, the effect is called transverse.

(b) If the original current is parallel to the effect produced, the effect is called longitudinal.

Moreover, a distinction is made between isothermal effects when there is no temperature gradient perpendicular to the direction of the original current, and adiabatic effects when there is no flow of heat in the direction perpendicular to the original current.

Finally, an effect is called galvanomagnetic if it is evoked by an electric current, and thermomagnetic if it is due to the flow of heat.

In the case at hand, that is, when an isotropic conductor is in a magnetic field in the direction of the z-axis whereas the flow of current and heat occurs in the xy-plane, the following thermomagnetic and galvanomagnetic effects may be observed.

9.1. GALVANOMAGNETIC EFFECTS

(a) *Transverse Effects*

1. The *isothermal Hall effect* consists in a potential gradient being set up in the direction transverse to the direction in which current flows under isothermal conditions.

This effect is characterized by the coefficient

$$R_i^t = \frac{\dfrac{E_y}{e_e} - \dfrac{1}{e_e^2}\dfrac{\partial \mu_e}{\partial y}}{j_{ex}} = \frac{E_y - \dfrac{1}{e_e}\dfrac{\partial \mu_e}{\partial y}}{J_{el\,x}} \qquad (4.144)$$

under the conditions

$$J_{\text{el}\,y}=e_{\text{e}}j_{\text{e}y}=0 \quad \text{and} \quad \nabla T=0.$$

It follows from equations (4.115) and (4.117) that

$$R_{\text{i}}^{\text{t}}=-\rho_{xy}. \tag{4.145}$$

2. The *adiabatic Hall effect* is the same kind of effect as before, except that it takes place under adiabatic conditions.

This effect is characterized by the same kind of relation as before,

$$R_{\text{a}}^{\text{t}}=\frac{E_y-\dfrac{1}{e_{\text{e}}}\dfrac{\partial\mu_{\text{e}}}{\partial y}}{J_{\text{el}\,x}}$$

under the conditions

$$J_{\text{el}\,y}=e_{\text{e}}j_{\text{e}y}=0, \quad \frac{\partial T}{\partial x}=0, \quad J_{sy}=0.$$

Equations (4.115), (4.116), and (4.118) to (4.120) lead to

$$R_{\text{a}}^{\text{t}}=-\rho_{xy}-\frac{\varepsilon_{xx}\Pi_{xy}}{\lambda_{xx}}. \tag{4.146}$$

3. The *Ettingshausen effect* consists in a temperature gradient being set up in the direction transverse to the flow of current.

This effect is characterized by the coefficient

$$P^{\text{t}}=\frac{\dfrac{\partial T}{\partial y}}{J_{\text{el}\,x}}=\frac{\nabla_y T}{J_{\text{el}\,x}} \tag{4.147}$$

under the conditions

$$J_{\text{el}\,y}=e_{\text{e}}j_{\text{e}y}=0, \quad J_{sy}=0, \quad \frac{\partial T}{\partial x}=0.$$

From equations (4.115), (4.118), and (4.120) it follows that

$$P^{\text{t}}=-\frac{\Pi_{xx}}{\lambda_{xx}}. \tag{4.148}$$

(b) Longitudinal Effects

1. *Isothermal electrical resistivity* is characterized by the coefficient

$$R_{\text{i}}^{\text{l}}=\frac{E_x-\dfrac{1}{e_{\text{e}}}\dfrac{\partial\mu_{\text{e}}}{\partial x}}{J_{\text{el}\,x}} \tag{4.149}$$

under the conditions

$$J_{\mathrm{el}\,y}=0, \qquad \nabla T=0.$$

The relation

$$R_{\mathrm{i}}^{\mathrm{l}}=\rho_{xx} \tag{4.150}$$

follows from equations (4.115) and (4.117).

2. *Adiabatic electrical resistivity* is characterized by a coefficient defined in the same way as before,

$$R_{\mathrm{a}}^{\mathrm{l}}=\frac{E_x-\dfrac{1}{e_{\mathrm{e}}}\dfrac{\partial \mu_{\mathrm{e}}}{\partial x}}{J_{\mathrm{el}\,x}},$$

but the conditions are changed, viz.:

$$J_{\mathrm{el}\,y}=0, \qquad J_{sy}=0, \qquad \frac{\partial T}{\partial x}=0.$$

Equations (4.115) and (4.118) to (4.120) lead to the relation

$$R_{\mathrm{a}}^{\mathrm{l}}=\rho_{xx}-\frac{\varepsilon_{xy}\Pi_{xy}}{\lambda_{xx}}. \tag{4.151}$$

9.2. THERMOMAGNETIC EFFECTS

(*a*) *Transverse Effects*

1. The *Righi–Leduc effect* is one in which a temperature gradient arises in the direction transverse to the applied temperature gradient.

The coefficient characterizing this effect is described by the formula

$$S^{\mathrm{t}}=\frac{\dfrac{\partial T}{\partial y}}{\dfrac{\partial T}{\partial x}}=\frac{\nabla_y T}{\nabla_x T} \tag{4.152}$$

under the conditions

$$J_{\mathrm{el}}=0, \qquad J_{sy}=0.$$

From equations (4.116) and (4.118) it follows that

$$S^{\mathrm{t}}=\frac{\lambda_{xy}}{\lambda_{xx}}. \tag{4.153}$$

2. The *isothermal Nernst effect* consists in a potential gradient being set up transverse to the temperature gradient when isothermal conditions are maintained.

The coefficient describing the isothermal Nernst effect is defined as

$$Q_{\mathrm{i}}^{\mathrm{t}}=\frac{E_y-\dfrac{1}{e_{\mathrm{e}}}\dfrac{\partial\mu_{\mathrm{e}}}{\partial y}}{\nabla_x T}, \tag{4.154}$$

under the conditions

$$\boldsymbol{J}_{\mathrm{el}}=0, \quad \frac{\partial T}{\partial y}=0.$$

It follows from equations (4.115) and (4.119) that

$$Q_{\mathrm{i}}^{\mathrm{t}}=-\varepsilon_{xy}. \tag{4.155}$$

3. The *adiabatic Nernst effect* is the same kind of effect as above, except that adiabatic conditions are maintained.

The coefficient describing this effect is defined as

$$Q_{\mathrm{a}}^{\mathrm{t}}=\frac{E_y-\dfrac{1}{e_{\mathrm{e}}}\dfrac{\partial\mu_{\mathrm{e}}}{\partial y}}{\nabla_x T}$$

with the conditions

$$\boldsymbol{J}_{\mathrm{el}}=0, \quad J_{sy}=0.$$

Equations (4.115) to (4.120) lead to the relation

$$Q_{\mathrm{a}}^{\mathrm{t}}=-\varepsilon_{xy}+\frac{\varepsilon_{xx}\lambda_{xy}}{\lambda_{xx}}. \tag{4.156}$$

(*b*) *Longitudinal Effects*

1. *Isothermal thermal conductivity* is defined by the coefficient

$$\lambda_{\mathrm{i}}=-\frac{TJ_{sx}}{\nabla_x T} \tag{4.157}$$

under the conditions

$$\boldsymbol{J}_{\mathrm{el}}=0, \quad \frac{\partial T}{\partial y}=0.$$

Equations (4.116) and (4.118) give the relation

$$\lambda_{\mathrm{i}}=\lambda_{xx}. \tag{4.158}$$

2. *Adiabatic thermal conductivity* is defined as the preceding variety, but adiabatic conditions must be satisfied. It is described by the formula

$$\lambda_{\mathrm{a}}=-\frac{TJ_{sx}}{\nabla_x T}$$

under the conditions

$$J_{el}=0, \quad J_{sy}=0.$$

Thus it follows from equations (4.116) and (4.118) that

$$\lambda_a=\lambda_{xx}+\frac{\lambda_{xy}^2}{\lambda_{xx}}. \tag{4.159}$$

3. The *isothermal Ettingshausen–Nernst effect* consists in a potential gradient being set up in the same direction as the applied temperature gradient.

The coefficient for this effect is

$$Q_i^l=\frac{E_x-\dfrac{1}{e_e}\dfrac{\partial\mu_e}{\partial x}}{\nabla_x T} \tag{4.160}$$

under the conditions

$$J_{el}=0, \quad \frac{\partial T}{\partial y}=0.$$

Equations (4.115) and (4.119) lead to the relation

$$Q_i^l=\varepsilon_{xx}. \tag{4.161}$$

4. The *adiabatic Ettingshausen–Nernst effect* is defined as the effect above, except that adiabatic conditions are satisfied.

This effect is thus defined as

$$Q_a^l=\frac{E_x-\dfrac{1}{e_e}\dfrac{\partial\mu_e}{\partial x}}{\nabla_x T}$$

under the conditions

$$J_{el}=0, \quad J_{sy}=0.$$

A relation describing the Ettingshausen–Nernst coefficient follows from equations (4.115), (4.116), (4.118), and (4.119):

$$Q_a^l=\varepsilon_{xx}+\frac{\varepsilon_{xy}\lambda_{xy}}{\lambda_{xx}}. \tag{4.162}$$

In the relations above a superscript l denotes longitudinal effects and t transverse effects. Similarly, a subscript i denotes isothermal effects and a adiabatic effects.

The formulae given above for the various effects enable us to establish the relations between the coefficients for longitudinal and transverse effects. These relations, called *Heulinger equations*, are of the following form:

$$R_a^t - R_i^t = Q_i^l P^t, \tag{4.163}$$

$$R_a^l - R_i^l = -Q_i^t P^t, \tag{4.164}$$

$$Q_a^t - Q_i^t = Q_i^l S^t, \tag{4.165}$$

$$Q_a^l - Q_i^l = -Q_i^t S^t, \tag{4.166}$$

$$\lambda_a - \lambda_i = \lambda_i (S^t)^2. \tag{4.167}$$

Eight different coefficients characterizing thermomagnetic and galvanomagnetic effects appear in equations (4.117) to (4.120). Only seven of these eight values have been used in the 12 such effects described. The five Heulinger equations thus are auxiliary relations describing effects which occur during a flow of electric current and a flow of heat in a conductor immersed in a magnetic field.

Another important relation which can be obtained is a consequence of the Onsager relations (4.125) and (4.126). Formulae (4.148), (4.155), and (4.158) can be used to get the expression

$$TQ_i^t = \lambda_i P^t, \tag{4.168}$$

which is known as the *Bridgman relation*. This relation ties together the isothermal Nernst effect and the Ettingshausen effect.

10. The Potential Applications of Nernst and Ettingshausen Effects in Generators and Refrigerators

Thermoelectric effects can be employed in devices for the direct conversion of heat to electricity. Best suited for this purpose are the Nernst effect, which permits construction of a generator for producing electricity at the cost of heat supplied to it, and the Ettingshausen effect, which makes it possible to build a refrigerator to which electricity is supplied directly, without necessity of any intermediate link for converting electricity into mechanical energy.

For a closer analysis of these potentialities, the essence of the Nernst and Ettingshausen effects should be re-examined.

Figure 4.7 depicts an arrangement in which the Nernst effect occurs. The directions of the magnetic field, heat flow, and poles of the source of current are indicated, with the assumption that the Nernst effect is positive in the situation shown. In the case of a negative Nernst effect the current would flow in the opposite direction. As can be seen from the drawing, the Nernst effect can be used to produce electricity by supplying heat from an external source and extracting it. The difference between the heat supplied and extracted is converted into electricity.

Figure 4.8 represents an arrangement in which the Ettingshausen effect occurs. As before, the directions of the magnetic field, flow of current, and of the temperature difference are indicated. It is assumed that in the situation depicted the Ettingshausen effect is positive. The warm and cool ends would be interchanged if the effect were negative.

It follows from the drawing that the Ettingshausen effect could be used for refrigeration. Passing an electric current through the arrangement produces a temperature below the ambient at the cool end.

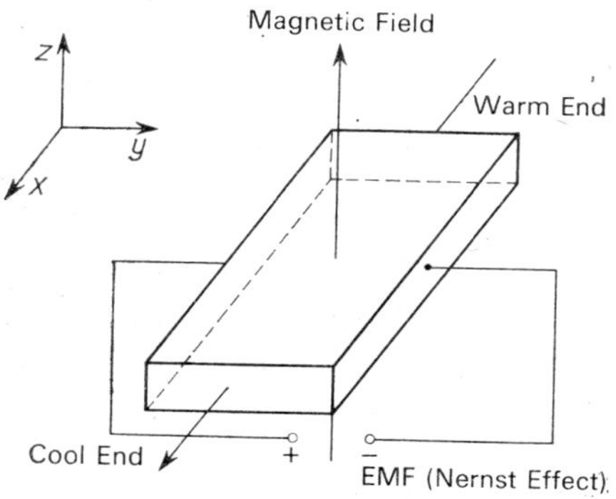

Fig. 4.7 The Nernst effect

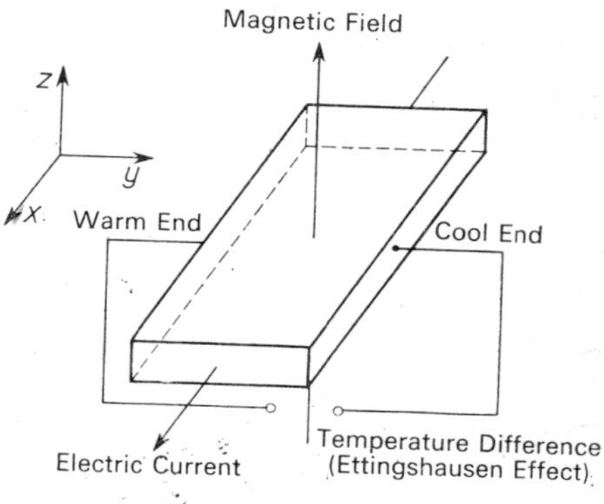

Fig. 4.8 The Ettingshausen effect

The two foregoing effects occur in 'pure' form only if the current circuit or the thermal circuit is open. That is, no current can flow in the case of the Nernst effect (in conformity with the definition of this effect as given in Section 4.8), whereas there is no flow of heat in the case of the Ettingshausen effect (in accordance with the definition in Section 4.8).

In practical applications of the Nernst effect current must flow and hence both effects occur concurrently, whereby there is a coupling of the two effects, the Nernst effect being treated as the principal one and the Ettingshausen effect as secondary. It should be added that it is possible to make a Nernst generator out of two elements, one of which produces a positive Nernst effect and the other, a negative Nernst effect.

The discussion will now concern the efficiency of a Nernst generator made out of one element with a constant cross-sectional area and immersed in a constant, uniform field of magnetic displacement H. Moreover, it is assumed that the material parameters, such as thermal conductivity, electrical resistance, etc. are constant, that the temperature distribution in the element depends on a single coordinate, and that the Righi–Leduc effect is negligible.

With these assumptions, the heat flow supplied to the warmer end of the element is described by

$$\dot{Q}=\dot{Q}_\lambda-\dot{Q}_E-\tfrac{1}{2}\dot{Q}_j\,. \tag{4.169}$$

In this relation

$$\dot{Q}_\lambda=\frac{\lambda bL}{d}\,\Delta T=k\Delta T \tag{4.170}$$

is the conducted heat flow. The following notation has furthermore been adopted in equation (4.170): λ is the thermal conductivity of the material, and b, L, and d are the width, length, and thickness (in the direction of heat conduction) of the generator element, and ΔT is the temperature difference along the x-axis.

The next term in relation (4.169) is

$$\dot{Q}_E = \frac{\lambda b L}{d} \Delta T_E \tag{4.171}$$

which is the Ettingshausen heat flow. Here

$$\Delta T_E = P^t BI/b \tag{4.172}$$

is the temperature difference due to the Ettingshausen effect, and P^t is the Ettingshausen coefficient, B is the external magnetic field strength, and I is the current.

It should be added that the Ettingshausen effect reduces the flow $\dot{Q}$ and counteracts the effect of heat conduction by the generator element.

The last term in relation (4.169),

$$\dot{Q}_j = I^2 R, \tag{4.173}$$

is the Joule heat flow, R here being the resistance of the generator element.

Inserting expression (4.172) into (4.171) gives the formula

$$\dot{Q}_E = \frac{\lambda P^t BL}{d} I = \varepsilon I \tag{4.174}$$

or, making use of the Bridgman equation (4.168) yields $\dot{Q}_E = \dfrac{Q}{d} LTBI = \varepsilon I$, where ε is the analogue of the Peltier coefficient. When relations (4.170), (4.173), and (4.174) are taken into account, expression (4.169) can be rewritten as

$$\dot{Q} = k\Delta T - \varepsilon I - \tfrac{1}{2} I^2 R. \tag{4.175}$$

The electrical power delivered by the generator is

$$P = I^2 R_e, \tag{4.176}$$

where R_e is the electrical resistance of the external lead of the generator. The ratio of the power P to the flow $\dot{Q}$ is the generator efficiency:

$$\eta = \frac{P}{\dot{Q}} = \frac{I^2 R_e}{k\Delta T - \varepsilon I - \frac{1}{2} I^2 R}. \tag{4.177}$$

Just as in the thermoelectric generator the semiconducting material used is characterized by a figure of merit Z, so the material employed in the Nernst generator can be characterized by the parameter

$$\theta = \frac{(Q_i^t)^2 B^2}{\rho \lambda}. \tag{4.178}$$

Here, in addition to the notation used earlier, Q_i^t is the Nernst coefficient, and ρ is the resistivity.

Like Z, the parameter θ does not depend on the geometrical dimensions of the generator, but unlike Z is not a purely material parameter since it contains the value of B which furthermore affects the value of ρ. The electrical resistance of a conductor immersed in a magnetic field is higher than that in the absence of a magnetic field.

The efficiency is optimizable, reaching a maximum when the ratio of resistances, $m = R_e/R$, attains the value

$$m_0 = \sqrt{1 - \theta \bar{T}},$$

the maximum efficiency corresponding to this case is equal to

$$\eta = \eta_C \frac{1 - m_0}{1 + \dfrac{T_0}{T} m_0}, \tag{4.179}$$

where η_C is the efficiency of a Carnot cycle operated between the temperatures T and T_0 of the upper and lower sources. As calculations show, materials now available permit an efficiency of about 2 to 3 per cent to be attained.

The efficiency η can thus be written in a form analogous to the thermoelectric generator efficiency:

$$\eta = \eta_c \alpha,$$

where α plays the part of the efficiency of the generator (exergetic efficiency). As in the case of thermoelectric generators, it is possible to make multi-stage cascade systems permitting efficiencies higher than that of a single-stage system. The relevant relations are similar to those for thermoelectric generators.

The potential use of the Ettingshausen effect for refrigeration emerges distinctly from the diagram in Fig. 4.8. A thermodynamic analysis similar to that for the generator can be carried out for a refrigerator based on the Ettingshausen effect. Such an analysis leads to an expression for the coefficient of performance ε_r which resembles that for the thermoelectric refrigerator, that is,

$$\varepsilon_r = \varepsilon_c \alpha_r,$$

where ε_c is the coefficient of performance for the Carnot cycle, whereas α_r is a function of the temperature difference and the parameter θ described by equation (4.178). Namely, α_r is given by

$$\alpha_r = \frac{1 + 2\left(1 - \dfrac{\Delta T}{\Delta T_E}\right)\theta \bar{T} - \dfrac{T_0}{T}}{2 + 2\left(1 - \dfrac{\Delta T}{\Delta T_E}\right)\theta \bar{T}}, \tag{4.180}$$

whereas ΔT denotes the actual temperature difference $(T - T_0)$; ΔT_E is the temperature difference due to the Ettingshausen effect, that is, when there is no flow of heat between the warm wall and the cool wall; $\bar{T}=(T_0+T)/2$; and T_0 and T are the absolute temperature of the cool and warm walls, respectively.

At present the Ettingshausen effect has little chance of being put to practical use owing to the small values of the parameter θ of materials now available and the very low values of the efficiency α_r of the refrigerating element.

CHAPTER V

THERMIONIC EFFECTS[1]

1. The Saha Equation

During ionization gas atoms A decompose into positive ions A^+ and negative electrons E^- in accordance with the reaction

$$A \rightleftarrows A^+ + E^- .$$

In the equilibrium state the sum of all molar electrochemical potentials μ_i', each multiplied by the relevant stoichiometric coefficient, is zero:

$$\sum_i \nu_i \mu_i' = 0 ,$$

$$\mu_a' - \mu_k' - \mu_e' = 0 ; \tag{5.1}$$

the subscript a has been used here for un-ionized atoms, k for positive ions, and e for electrons.

The electrochemical potentials of ions and electrons can be expressed in terms of the chemical potentials μ_i, the electrostatic potentials ψ in vacuum, and the charge e_e per kilomole of electrons. The charge per kilomole of positive ions will be $-e_e$ in the case under consideration. Hence, the electrochemical potential per kilomole of positive ions is

$$\mu_k' = -e_e \psi + \mu_k . \tag{5.2}$$

The electrochemical potential per kilomole of electrons is

$$\mu_e' = e_e \psi + \mu_e . \tag{5.3}$$

Combination of equations (5.1), (5.2), and (5.3), with $\mu_a' = \mu_a$, leads to

$$\mu_a - \mu_k - \mu_e = 0 . \tag{5.4}$$

As is seen, the equilibrium equation for ionized gas has two similar forms which differ only in that electrochemical potentials appear in equation (5.1) whereas chemical potentials do in equation (5.4).

Although ions and electrons interact through the intermediary of Coulomb forces, in the case of low pressures and high temperatures the gas under consideration may be assumed to obey Dalton's law.

[1] In this chapter all specific quantities are per kilomole as the unit of amount of substance. For simplicity of notation specific quantities per kilomole are designated in the same way as specific quantities per kilogram were earlier.

The chemical potential of component i forming part of a solution subject to Dalton's law is calculated as for the pure component at the solution temperature T and component pressure p_i:

$$\mu_i = \tilde{h}_i - T\tilde{s}_i . \tag{5.5}$$

For this hypothetical case of a solution of ideal gases

$$\mu_i = c_{pi}T - c_{pi}T\ln T + RT\ln p_i + h_{0i} - c_{pi}T_0 + (c_{pi}\ln T_0' - R\ln p_0 - s_{0i})\,T , \tag{5.6}$$

where h_{0i} and s_{0i} denote the molar enthalpy and entropy of component i in the reference state.

The reference temperature T_0 for enthalpy may differ from the reference temperature T_0' for entropy. When $T_0=0$ and when a reference constant is introduced for entropy,

$$s_{00i} = s_{0i} - c_{pi}\ln T_0' + R\ln p_0 , \tag{5.7}$$

equation (5.6) simplifies to

$$\mu_i = c_{pi}T - c_{pi}T\ln T + RT\ln p_i + h_{0i} - s_{00i}T . \tag{5.8}$$

If equation (5.8) is used for the given gas consisting of atoms, ions, and electrons, equation (5.4) yields

$$\Big(\sum_i \nu_i c_{pi}\Big)T - \Big(\sum_i \nu_i c_{pi}\Big)T\ln T + RT\ln\frac{p_{\mathrm{k}}\,p_{\mathrm{e}}}{p_{\mathrm{a}}} + \sum_i \nu_i h_{0i} - \Big(\sum_i \nu_i s_{00i}\Big)T = 0 . \tag{5.9}$$

Here, the expression

$$\sum_i \nu_i c_{pi} = c_{p\mathrm{k}} + c_{p\mathrm{e}} - c_{p\mathrm{a}} \tag{5.10}$$

TABLE 5.1

Ionization Potentials

Substance	Ionization Potential V_i in volts	Substance	Ionization Potential V_i in volts
Ar	15.75	K	4.34
Ba	5.19	Li	5.39
Ca	6.09	Na	5.14
Cs	3.89	Ne	21.56
H	13.59	Rb	4.18
He	24.58	Sr	5.67

determines the change in heat capacity during the ionization reaction at constant pressure, whereas

$$\sum_i \nu_i h_{0i} = h_{0\mathrm{k}} + h_{0\mathrm{e}} - h_{0\mathrm{a}} = V_i \tag{5.11}$$

specifies the enthalpy change during the ionization reaction at 0 K, called the ***ionization potential*** of the monatomic substance under study (Table 5.1), and the expression

$$\sum_i \nu_i s_{00i} = s_{00\mathrm{k}} + s_{00\mathrm{e}} - s_{00\mathrm{a}} \tag{5.12}$$

gives the change in entropy during the reaction at unit pressure and unit temperature.

Equation (5.9) can now be written in the form of the *Saha equation*

$$\frac{p_\mathrm{k}\, p_\mathrm{e}}{p_\mathrm{a}} = BT^{\frac{1}{R}\sum_i \nu_i c_{pi}} \exp\left(-\frac{V_i}{RT}\right), \tag{5.13}$$

where the constant B is defined as

$$B = \exp\left[\frac{1}{R}\sum_i \nu_i (s_{00i} - c_{pi})\right]. \tag{5.14}$$

All of the particles considered (un-ionized atoms, positive ions, and electrons) are single. For an ideal monatomic gas the specific heat at constant pressure is

$$c_{pi} = \tfrac{5}{2}R, \tag{5.15}$$

whereby

$$\frac{1}{R}\sum_i \nu_i c_{pi} = \tfrac{5}{2}. \tag{5.16}$$

For an ionized monatomic gas under conditions such that the solution of un-ionized gas along with the products of its ionization may be treated as an ideal gas the Saha equation becomes

$$\frac{p_\mathrm{k}\, p_\mathrm{e}}{p_\mathrm{a}} = BT^{\frac{5}{2}} \exp\left(-\frac{V_i}{RT}\right) = K_p(T), \tag{5.17}$$

where $K_p(T)$ is the equilibrium constant for the reaction of ionization of the monatomic gas under consideration and its meaning is similar to that of the equilibrium constant for the dissociation reaction.

As is seen from Table 5.1, the alkali metals, especially caesium, have the lowest ionization potentials, and hence the greatest ionizability.

In accordance with the conditions for thermodynamic equilibrium all components of an ionized gas should, in the state of equilibrium, be at the same temperature and have the same electrochemical potential.

Suppose that the state of the gas of the same temperature but with zero electrostatic potential is taken for the reference state, which is also an equilibrium state, and suppose that it is labelled with a subscript 0. Then, insertion of equation (5.8) in equations (5.2) and (5.3) yields expressions for the electrochemical potential differences in any state (without a subscript) and in the reference state (subscript 0) at the same temperature

$$\mu_\mathrm{a} - \mu_{\mathrm{a}0} = RT \ln \frac{p_\mathrm{a}}{p_{\mathrm{a}0}}, \tag{5.18}$$

$$\mu'_k - \mu'_{k0} = RT \ln \frac{p_k}{p_{k0}} - e_e \psi , \tag{5.19}$$

$$\mu'_e - \mu'_{e0} = RT \ln \frac{p_e}{p_{e0}} + e_e \psi . \tag{5.20}$$

It is evident that an equilibrium pressure equal to the reference pressure

$$p_a = p_{a0} \tag{5.21}$$

is obtained only for uncharged particles. For positive ions and electrons the pressures in the equilibrium state are, respectively,

$$p_k = p_{k0} \exp\left(\frac{e_e \psi}{RT}\right) \tag{5.22}$$

$$p_e = p_{e0} \exp\left(-\frac{e_e \psi}{RT}\right) . \tag{5.23}$$

The charge densities of the ions and electrons, respectively, are

$$\rho_k = -e_e c_k = -\frac{e_e p_k}{RT} > 0 , \tag{5.24}$$

$$\rho_e = e_e c_e = \frac{e_e p_e}{RT} < 0 , \tag{5.25}$$

where c_k and c_e are the molar concentrations of the ions and electrons.

Substituting the values of the component pressures in the equilibrium state from expressions (5.22) and (5.23) gives

$$\rho_k = \rho_{k0} \exp\left(\frac{\rho_e \psi}{RT}\right) > 0 , \tag{5.26}$$

$$\rho_e = \rho_{e0} \exp\left(-\frac{\rho_e \psi}{RT}\right) < 0 . \tag{5.27}$$

The total density of electric charge in the ionized gas is

$$\rho_{el} = \rho_k + \rho_e = \rho_{k0} \left[\exp\left(\frac{e_e \psi}{RT}\right) + \frac{\rho_{e0}}{\rho_{k0}} \exp\left(-\frac{e_e \psi}{RT}\right) \right] . \tag{5.28}$$

By Gauss's law,

$$\nabla \cdot \boldsymbol{E} = \frac{\rho_{el}}{\varepsilon_0} , \tag{5.29}$$

where the electric field strength $\boldsymbol{E} = -\nabla\psi$, whereas ε_0 is the permittivity of free space. These relations yield Poisson's equation

$$\nabla \cdot (\nabla\psi) = \nabla^2 \psi = -\frac{\rho_{el}}{\varepsilon_0} , \tag{5.30}$$

which, together with equation (5.28), gives the equilibrium condition for any ionized ideal gas, written in terms of the electrostatic potentials:

$$\nabla^2\psi = -\frac{\rho_{k0}}{\varepsilon_0}\left[\exp\left(\frac{e_e\psi}{RT}\right) + \frac{\rho_{e0}}{\rho_{k0}}\exp\left(-\frac{e_e\psi}{RT}\right)\right]. \tag{5.31}$$

This equation can be integrated for given boundary conditions in order to obtain the distribution of the electrostatic potential.

Example 5.1. Find the distribution of the electrostatic potential in an ionized ideal gas if a potential change occurs only in the x-direction. Adapt the solution to a layer of gas bounded on one side.

Solution. On introduction of the dimensionless parameters

$$\alpha = -\frac{e_e\psi}{RT}, \tag{1}$$

$$z^2 = -x^2\frac{\rho_{k0}\,e_e}{\varepsilon_0 RT}, \tag{2}$$

$$\beta = -\frac{\rho_{e0}}{\rho_{k0}}, \tag{3}$$

equation (5.31) takes on the form

$$\frac{d^2\alpha}{dz^2} = \beta e^{\alpha} - e^{-\alpha}. \tag{4}$$

A single integration of this equation gives

$$\left(\frac{d\alpha}{dz}\right)^2 = 2(\beta e^{\alpha} + e^{-\alpha}) + C_1. \tag{5}$$

The constant of integration is determined for specified boundary conditions. For a layer of gas bounded only on one side a reference state is taken at an infinite distance from the bounding surface, so that for $x = \infty$

$$\psi = 0, \quad \frac{d\psi}{dx} = 0, \quad \frac{d^2\psi}{dx^2} = 0, \tag{6}$$

and hence also

$$\alpha = 0, \quad \frac{d\alpha}{dz} = 0, \quad \frac{d^2\alpha}{dz^2} = 0. \tag{7}$$

These relations are satisfied if $\beta = 1$ by equation (4), whereas $C_1 = -4$ by equation (5). For the given case equation (5) goes over into

$$\left(\frac{d\alpha}{dz}\right)^2 = 4(\cosh\alpha - 1) \tag{8}$$

or, after transformation, into

$$\frac{d\alpha}{dz} = 2\sqrt{2}\sinh\frac{\alpha}{2}. \tag{9}$$

Another integration yields

$$z = \frac{1}{\sqrt{2}}\int_0^{\infty}\frac{d\frac{\alpha}{2}}{\sinh\frac{\alpha}{2}} + C_2 = \frac{1}{\sqrt{2}}\ln\left(\tanh\frac{\alpha}{4}\right) + C_2. \tag{10}$$

The constant of integration C_2 is found from the boundary conditions for $x=0$, that is, $z=0$, where $\psi=\psi_0$ and $\alpha_0=-\frac{e_e\psi_0}{RT}$. Accordingly,

$$C_2=-\frac{1}{\sqrt{2}}\ln\left(\tanh\frac{\alpha_0}{4}\right), \tag{11}$$

whereas

$$z=\frac{1}{\sqrt{2}}\ln\left(\frac{\tanh\frac{\alpha}{4}}{\tanh\frac{\alpha_0}{4}}\right). \tag{12}$$

The potential distribution in a semi-bounded, ionized ideal gas is given by the equation

$$\tanh\frac{\alpha}{4}=-\tanh\left(\frac{e_e\psi}{4RT}\right)=\exp\left[\sqrt{2}(z-C_2)\right]. \tag{13}$$

2. Equilibrium between Ionized Vapour and the Solid Phase

The partially ionized vapour of monatomic substance A is in equilibrium with the solid phase of the same substance which is a conductor of electricity. The vapour pressure is so small that the vapour may be viewed as an ideal gas. The solid phase will be indicated by a superscript s, and the gas phase by a superscript g.

In addition to equality of temperatures, the condition for equilibrium is that the electrochemical potentials of the atoms, positive ions, and electrons in the gas and solid phases be equal:

$$\mu_a'^{g}=\mu_a'^{s}=\mu_a^{g}=\mu_a^{s}, \quad \mu_k'^{g}=\mu_k'^{s}, \quad \mu_e'^{g}=\mu_e'^{s}. \tag{5.32}$$

For un-ionized atoms, by relation (5.8)

$$\mu_a^{g}=c_{pa}^{g}T-c_{pa}^{g}T\ln T+RT\ln p_a+h_{0a}^{g}-s_{00a}^{g}T, \tag{5.33}$$

where the component pressure of the vapour of atoms is

$$p_a=C_a T^{c_{pa}^{g}/R}\exp\left(\frac{\mu_a^{g}-h_{0a}^{g}}{RT}\right), \tag{5.34}$$

where the value of the constant for the given substance is

$$C_a=\exp\left(\frac{s_{00a}^{g}-c_{pa}^{g}}{R}\right). \tag{5.35}$$

For a monatomic gas, since $c_{pa}^{g}/R=5/2$, the component pressure is

$$p_a=C_a T^{\frac{5}{2}}\exp\left(\frac{\mu_a^{g}-h_{0a}^{g}}{RT}\right). \tag{5.36}$$

By relations (5.2), (5.3), and (5.32), the chemical potentials of ions and electrons of the solid phase can be related to their electrochemical potential in the gas phase:

$$\mu_k'^{s}=\mu_k^{g}-e_e\psi, \quad \mu_e'^{s}=\mu_e^{g}+e_e\psi. \tag{5.37}$$

The electrochemical potential of ions or electrons in the solid phase can also be made dependent on ϕ_k or ϕ_e, the work function of ions or electrons in the solid phase, and on ψ^0, the limiting value of the electrostatic potential external to the solid phase but right at the phase interface:

$$\mu_k'^s = e_e(\phi_k - \psi^0), \quad \mu_e'^s = e_e(\phi_e + \psi^0). \tag{5.38}$$

The physical interpretation of the work function is based on the following reasoning. The specific energy of component i, which possesses electric charge, is

$$e_{ui} = u_i(s, v) + e_i\psi,$$

where $e_i = e_e$ for electrons and $e_i = -e_e$ for positive ions.

In the state with zero specific internal energy ($u=0$) and zero specific entropy ($s=0$) external to the solid phase the specific energy of component i is

$$e_{u0i} = e_i\psi^0.$$

Through the free enthalpy the chemical potential is related to the work of a reversible isobaric-isothermal process by

$$\left(\frac{đW_{p,T}}{dn_i}\right)_{\text{rev}} = -\left(\frac{\partial G}{\partial n_i}\right)_{p,T} = -(\mu_i' - e_{u0i}) = -(\mu_i' - e_i\psi^0),$$

and, when equation (5.38) is taken into account, also by

$$\phi_i = \pm\left(\frac{đW_{p,T}}{d(e_i n_i)}\right)_{\text{rev}}$$

The work function ϕ_i for the surface of a solid phase is equal to the work done per unit charge if an infinitesimal amount of component i is removed reversibly to the solid phase from a point external to the surface and from a state of zero entropy and zero internal energy.

Comparison of relations (5.37) and (5.38) for the solid-phase surface reveals that

$$\mu_i^g = e_i\phi_i \quad \text{(at surface)}. \tag{5.39}$$

By relation (5.8) the component pressure of ions or electrons is obtained in a form analogous to that of equation (5.34) derived above for atoms:

$$p_i = C_i T^{c_{pi}^g/R} \exp\left(\frac{\mu_i^g - h_{0i}^g}{RT}\right), \tag{5.40}$$

where the constant C_i is

$$C_i = \exp\left(\frac{s_{00i}^g - c_{pi}^g}{R}\right). \tag{5.41}$$

The reference level for the molar enthalpy of ions and electrons in the gas phase is chosen so that $h_{0i}^g = 0$ at the reference temperature $T_0 = 0$ K. In this case the molar reference enthalpy of atoms is, by relation (5.11), equal to the ionization potential taken with a minus sign, that is

$$h^{g}_{0a} = -V_i, \quad \text{when} \quad T_0 = 0\,\text{K}. \tag{5.42}$$

If these reference values are used for the enthalpies of atoms, ions, and electrons, then in the case of equilibrium with the solid-phase surface equations (5.34) and (5.42) give the component pressures for un-ionized atoms,

$$p_a = C_a T^{c^g_{pa}/R} \exp\left(\frac{\mu^s_a + V_i}{RT}\right), \tag{5.43}$$

and the component pressures of ions or electrons at the solid-phase surface,

$$p_{is} = C_i T^{c^g_{pi}/R} \exp\left(\pm \frac{e_i \phi_i}{RT}\right), \tag{5.44}$$

where the minus sign in front of the fraction applies to ions, and the plus sign to electrons.

Electrons have a monatomic structure, hence $c^g_{pe}/R = 5/2$ and the component pressure of electrons in equilibrium with the surface of the solid phase is

$$p_{es} = C_e T^{\frac{5}{2}} \exp\left(\frac{e_e \phi_e}{RT}\right). \tag{5.45}$$

3. The Potential Distribution and the Saturation Current Density between Parallel Surfaces

Let us consider two solids, *I* and *II* (Fig. 5.1), with parallel surfaces which are electrical conductors. The two surfaces are at the same temperature *T* and are in equilibrium with the electronic gas between them. This system will be considered under

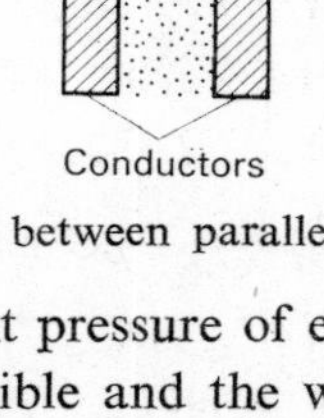

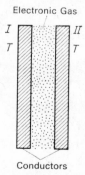

Fig. 5.1 Electronic gas in the gap between parallel surfaces of conductors *I* and *II*

conditions such that the component pressure of each component of the conductors other than electrons will be negligible and the work functions ϕ_e for the removal of electrons from surfaces *I* and *II* are identical.

By equation (5.23), the pressure of electrons of surfaces I and II in the state of equilibrium is

$$p_{es}=p_{e0}\exp\left(-\frac{e_e\psi_s}{RT}\right), \tag{5.46}$$

where ψ_s is the electrostatic potential at the surfaces of conductors I and II.

Comparison of relations (5.45) and (5.46) makes it possible to calculate the reference pressure of electrons for equilibrium with a gas with zero electrostatic potential ψ. This reference pressure is

$$p_{e0}=C_e T^{\frac{5}{2}}\exp\left(\frac{e_e(\phi_e+\psi_s)}{RT}\right)=C_e T^{\frac{5}{2}}\exp\left(\frac{\mu_e'^s}{RT}\right), \tag{5.47}$$

where account has been taken of relation (5.38) defining the electron work function.

Inserting this relation into equation (5.23) yields the component pressure of the electronic gas at any point in the space between the surfaces of bodies I and II,

$$p_e=C_e T^{\frac{5}{2}}\exp\left(\frac{\mu_e'^s-e_e\psi}{RT}\right)=C_e T^{\frac{5}{2}}\exp\left(\frac{\mu_e^s}{RT}\right). \tag{5.48}$$

By equation (5.25) the electronic charge density is

$$\rho_e=\frac{e_e p_e}{RT}=\frac{e_e}{R}C_e T^{\frac{3}{2}}\exp\left(\frac{\mu_e^s}{RT}\right). \tag{5.49}$$

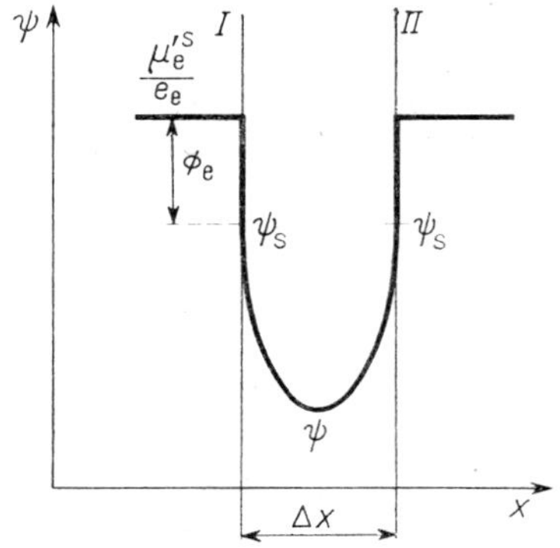

Fig. 5.2 Distribution of the electrostatic potential in the gap between parallel surfaces of conductors I and II with the same temperature, potential and work function

Since the density of positive ions is zero ($\rho_k=0$), for the given case equation (5.31) takes the form

$$\nabla^2\psi=-\frac{e_e C_e}{\varepsilon_0 R}T^{\frac{3}{2}}\exp\left(\frac{\mu_e^s}{RT}\right). \tag{5.50}$$

The foregoing equation can be used to determine the symmetric distribution of the electrostatic potential (Fig. 5.2) in a planoparallel gap of width Δx between surfaces I and II which have the same temperature, the same potentials, and the same work functions for removal of electrons from the surface. This is the case of equilibrium between the surfaces of conductors when the electrochemical potential $\mu_e'^s$ of electrons has the same value at both surfaces.

The equilibrium conditions considered above are perturbed if a difference between the electrochemical potentials of electrons at surfaces *I* and *II* is produced by switching

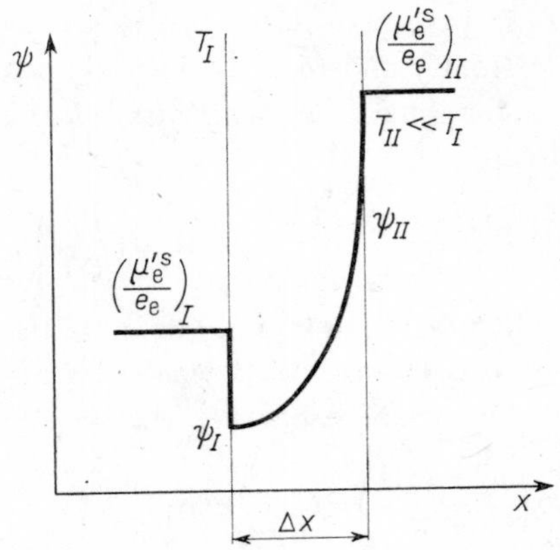

Fig. 5.3 Distribution of the electrostatic potential in the gap between parallel surfaces of conductors *I* and *II* at different temperatures and potentials

on an external current source between these conductors and producing a temperature difference. Figure 5.3 depicts the case when the electrochemical potential of electrons at surface *II* is greater than that at surface *I*, that is, $(\mu_e'^s)_{II} > (\mu_e'^s)_I$, so that electrons flow from *I* to *II*. The difference of electrochemical potential of electrons between the surfaces under consideration is so great that the gradient of electrostatic potential at surface *I* is zero. The temperature of surface *II* is so low that the emission of electrons from it may be neglected.

The density of the current flowing from surface *I* to *II* under the conditions in question, that is, with zero electrostatic potential gradient at surface *I* and zero re-emission from surface *II*, is called the *saturation current density*. By this definition, the saturation current density corresponds to the phenomenon taking place under conditions deviating considerably from the equilibrium state. Closer consideration of this phenomenon makes it possible to employ reasoning based on the assumption of local thermodynamic equilibrium for the conditions discussed below.

The charge per kilomole electrons, ρ_e, is very large. Thus, as implied by the Poisson equation the curvature of the electrostatic potential distribution is very great in the proximity of the surface of zero potential gradient. An electron emitted from surface *I* may find itself in a region in which it experiences large acceleration before it collides with another electron. The greater the slope of the potential distribution curve, the greater the electron acceleration. There is a considerable probability that all electrons emitted by surface *I* reach surface *II* and then the saturation current density becomes equal to the rate of electron charge emission from that surface.

At temperatures below the melting point of the most high-melting solids conducting an electric current (e.g. tungsten) the electronic gas in equilibrium with the surface of the solid conductor has a very low density in comparison with the electron density in the solid conductor. In practice, the pressure on the solid phase does not affect the electron pressure in the gas phase. The emission of electrons from the surface of a solid is a function only of the temperature of the solid and, under the given conditions, does not depend on the electron pressure in the space around the solid. The saturation current thus is equal to the rate of electron transport from the solid phase to the gas (or conversely) under equilibrium conditions, multiplied by the electron charge.

For metals it may be assumed with good approximation that all electrons striking the surface of the solid penetrate from the gas phase into the solid phase. The electronic gas may be treated as an ideal monatomic gas. In the state of local thermodynamic equilibrium the molecules of this gas are subject to the Maxwell distribution. In this case the numerical electron flux reaching the surface perpendicular to the x-axis is

$$j_{-}=\frac{1}{v_e}\left(\frac{m_e}{2\pi kT}\right)^{\frac{3}{2}}\int_{-\infty}^{\infty}\int_{-\infty}^{\infty}\int_{0}^{\infty}\exp\left(-\frac{m_e}{2kT}(v_x^2+v_y^2+v_z^2)\right)v_x\,dv_x\,dv_y\,dv_z\,,$$

where the electron mass is

$$m_e=\frac{M_e}{N_A}\,,$$

the volume of electronic gas per electron is

$$v_e=\frac{kT}{p_e}$$

and $k=R/N_A$ is Boltzmann's constant, N_A being Avogadro's number.

Integration yields

$$j_{-}=\frac{p_e}{\sqrt{2\pi m_e kT}}\,.$$

Dividing both sides of this equation by Avogadro's number gives the number of kilomoles of electrons reaching a surface of unit area per unit time, that is, it gives the molar electron flux

$$j_{es}=\frac{p_{es}}{\sqrt{2\pi M_e RT}}\,, \tag{5.51}$$

where M_e is the electron molar mass, R is the universal gas constant, and p_{es} is the electron pressure on the surface of the solid.

When equation (5.45) is taken into account the saturation current density is

$$J_{es}=e_e j_{es}=\frac{e_e C_e}{\sqrt{2\pi M_e R}}\,T^2\exp\left(\frac{e_e\phi_e}{RT}\right). \tag{5.52}$$

This equation, called the *Richardson–Dushman equation* can also be written in terms of constants relating to individual electrons, bearing in mind that the electron charge

$\epsilon = e_e/N_A$, and then

$$J_{es} = AT^2 \exp\left(\frac{\epsilon\phi_e}{kT}\right), \tag{5.53}$$

where

$$A = \frac{\epsilon C_e}{\sqrt{2\pi m_e k}}. \tag{5.54}$$

Since the quantity C_e, defined by equation (5.41), is a constant which depends solely on the properties of the electronic gas, so does the quantity A, called the *Richardson–Dushman constant.* As follows from theoretical considerations, the Richardson–Dushman constant for pure metals is

$$A = 120.4 \text{ A cm}^{-2}\text{K}^{-1}$$

and does not depend on the material of the conductor on whose surface the saturation current density is determined. In actual fact, the measured magnitude of the Richardson–Dushman constant does depend on the kind of conductor material (Table 5.2).

TABLE 5.2

Work Function[1] ϕ_e and Measured Value of Richardson–Dushman Constant A_m for Some Conductors

Conductor	Work Function ϕ_e in volts	Richardson Constant A_m in A cm^{-2}K^{-1}
Ag	4.08	60.2
Al_2O_3	3.77	1.4
BaO	3.44–1.66	2.5
BaO-SrO	1.0–1.2	0.1–1.0
Cr	4.6	48
Cs	1.81	160
K	1.9–2.46	
LaB_6	2.7	14
Mo	4.41–3.48	60.2
Na	2.5	
Nb	4.0	
Ni	2.77	26.8
Pt	5.32	32
Re	4.74–5.1	720
Ta	4.18–4.07	60.2
ThC_2	3.2	200
ThO_2	2.6	5.0
W	4.52–4.40	60.2
W-Ba	1.6–2.0	1–2
W-Cs	1.4	3
W-Th	2.63	3
ZrC	3.8	134
ZrO_2	3.4	0.35

[1] The term work function is also applied to the quantity $\epsilon\phi_e$ expressed in electron volts with the same numerical values as ϕ_e in volts.

The difference between the measured values of A_m and the theoretical value A is due to the measurement conditions in which it is assumed that the work function of electrons does not depend on the temperature and that the surface of the conductor

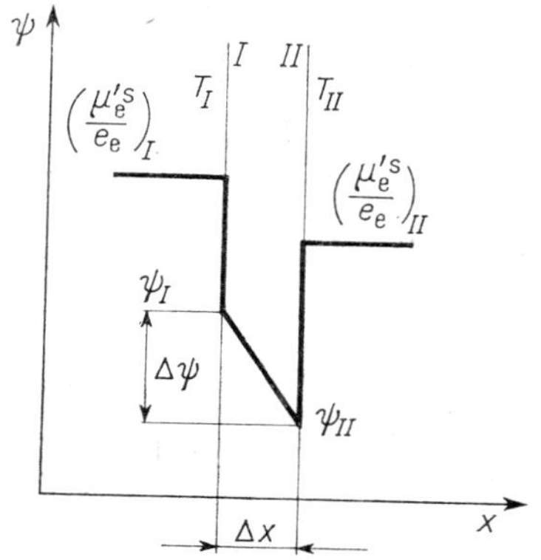

Fig. 5.4 Linear distribution of the electrostatic potential in a very narrow gap between parallel surfaces of conductors *I* and *II* at different temperatures and potentials

is homogeneous. In reality, the electron work function increases with the temperature and the surface of the conductor has an inhomogeneous crystalline structure.

In addition to the electron flux there is an energy flux in the system under discussion. The separation Δx of the surfaces of conductors *I* and *II* (Fig. 5.4) may be made so small that electron collision is negligible. The distribution of the electrostatic potential in such a narrow gap may also be assumed to be linear. All electrons emitted from surface *II* will reach surface *I* if the electrostatic potential gradient ψ is large enough to accelerate them. Conversely, only a fraction f of electrons emitted by *I* possess sufficient energy to overcome the potential difference $\Delta\psi = \psi_{II} - \psi_I$ and do reach *II*.

The electronic gas in the gap and the electronic gas which is emitted by the surface of zero reflectivity and is in equilibrium with the first-named gas are treated as an ideal gas having an electron velocity distribution in accordance with the Maxwell distribution.

The fraction f of electrons emitted from *I* and having a kinetic energy component greater than $\epsilon\Delta\psi$ in the direction perpendicular to the surface is given by the relation

$$f = \frac{\int\limits_{-\infty}^{\infty}\int\limits_{-\infty}^{\infty}\int\limits_{\sqrt{\frac{2\epsilon\Delta\psi}{m_e}}}^{\infty} \exp\left[-\frac{m_e(v_x^2+v_y^2+v_z^2)}{2kT}\right] v_x \, dv_x \, dv_y \, dv_z}{\int\limits_{-\infty}^{\infty}\int\limits_{-\infty}^{\infty}\int\limits_{0}^{\infty} \exp\left[-\frac{m_e(v_x^2+v_y^2+v_z^2)}{2kT}\right] v_x \, dv_x \, dv_y \, dv_z}$$

$$= \exp\left(-\frac{\epsilon\Delta\psi}{kT}\right). \qquad (5.55)$$

The density J_e of the current flowing from I to II can thus be calculated from the saturation current densities J_{esI} and J_{esII} at surfaces I and II as the difference between the density of the current flowing from I to II and that in the opposite direction

$$J_e = fJ_{esI} - J_{esII} = J_{esI} \exp\left(-\frac{\epsilon \Delta\psi}{kT_I}\right) - J_{esII}. \tag{5.56}$$

The isothermal energy of transport of electrons with potential energy ψ and kinetic energy greater that $e_e\Delta\psi$ in the x-direction is

$$U_e^* = e_e \left(\frac{J_u}{J_e}\right)_{\Delta T=0}$$

$$= e_e \frac{\int_{-\infty}^{\infty}\int_{-\infty}^{\infty}\int_{\sqrt{\frac{2\epsilon\Delta\psi}{m_e}}}^{\infty} \left[\frac{m_e(v_x^2+v_y^2+v_z^2)}{2} + \psi\right] \exp\left[-\frac{m_e(v_x^2+v_y^2+v_z^2)}{2kT}\right] v_x \, dv_x \, dv_y \, dv_z}{\int_{-\infty}^{\infty}\int_{-\infty}^{\infty}\int_{\sqrt{\frac{2\epsilon\Delta\psi}{m_e}}}^{\infty} \exp\left[-\frac{m_e(v_x^2+v_y^2+v_z^2)}{2kT}\right] v_x \, dv_x \, dv_y \, dv_z}$$

$$= e_e\left(\psi + \Delta\psi + \frac{2kT}{\epsilon}\right). \tag{5.57}$$

The energy flux transported in an isothermal diode, when $T_I = T_{II} = T$, together with the electrons amounts to

$$J_u = J_{esI}\left[\exp\left(-\frac{\epsilon\Delta\psi}{kT}\right)\right]\left(\psi_I + \Delta\psi + \frac{2kT}{\epsilon}\right) - J_{esII}\left(\psi_{II} + \frac{2kT}{\epsilon}\right)$$

$$= \left[J_{esI}\exp\left(-\frac{\epsilon\Delta\psi}{kT}\right) - J_{esII}\right]\left(\psi_{II} + \frac{2kT}{\epsilon}\right) \tag{5.58}$$

or, when the Richardson–Dushman equation (5.53) is taken into account, is

$$J_u = AT^2\left[\exp\left(\frac{\epsilon\phi_e}{kT}\right)\right]\left[\exp\left(-\frac{\epsilon\Delta\psi}{kT}\right) - 1\right]\left(\psi_{II} + \frac{2kT}{\epsilon}\right)$$

$$= \left\{AT^2\exp\left[\frac{\epsilon(\phi_e - \Delta\psi)}{kT}\right] - AT^2\exp\left(\frac{\epsilon\phi_e}{kT}\right)\right\}\left(\psi_{II} + \frac{2kT}{\epsilon}\right). \tag{5.59}$$

Example 5.2. Calculate the density of the saturation current flowing from the surface of a tungsten electrode heated to a temperature $T = 2500$ K. The electron work function for tungsten is $\phi_e = 4.52$ V.

Solution. The saturation current density is given by equation (5.53):

$$J_{es} = A\,T^2 \exp\left(\frac{\epsilon\phi_e}{kT}\right)$$

This equation contains the following physical constants: Richardson–Dushman constant $A = 120.4$ A cm^{-2} K^{-2}, Boltzmann constant $k = 1.38049 \times 10^{-23}$ J K^{-1}, and electron charge $\epsilon = -1.60209 \times 10^{-19}$ C.

The saturation current density thus is $J_{es} = 120.4(2500)^2 \exp(-21) = 0.845 \times 10^3$ A m^{-2}.

Example 5.3. What fraction of the electrons emitted by a cathode at $T = 2200$ K has an energy in excess of 1 eV in the direction perpendicular to the cathode?

Solution. The fraction of electrons with an energy higher than 1 eV is determined by equation (5.55),

$$f = \exp\left(-\frac{\epsilon \Delta\psi}{kT}\right),$$

where $\epsilon = -1.60209 \times 10^{-19}$ C, $\Delta\psi = -1$ V, and $k = 1.38049 \times 10^{-23}$ J K^{-1}. Hence

$$f = \exp\left(-\frac{1.6 \times 10^{-19}}{1.38 \times 10^{-23} \times 2200}\right) = e^{-5.27} = 0.00517,$$

which is only slightly more than 0.5%.

4. The Nonequilibrium Thermodynamics of the Thermionic Diode

Phenomena occurring in the thermionic diode can be considered in an approximate manner by means of the methods employed by nonequilibrium thermodynamics. This reasoning is based on the assumption of local thermodynamic equilibrium and linear phenomenological equations. Thus, it is strictly valid only for small temperature and potential differences but it also does allow a qualitative appraisal of the phenomena when these conditions are not met. In the case now under discussion (Fig. 5.5)

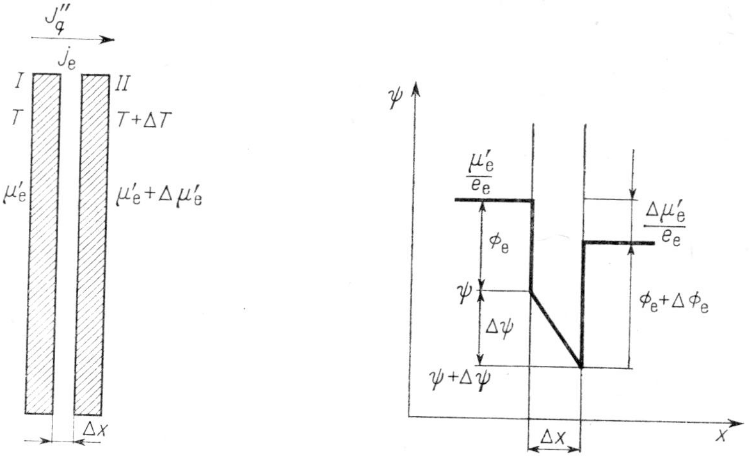

Fig. 5.5 Linear distribution of the electrostatic potential in a very narrow gap between parallel surfaces of conductors *I* and *II* with only small differences between the temperatures, potentials, and work functions

the gap Δx between the surfaces of conductors *I* and *II* is very narrow, the electrochemical potentials for electrons in the conductors are μ_e' and $\mu_e' + \Delta\mu_e'$, respectively, whereas the temperatures are T and $T + \Delta T$. An electric current of electron flux J_e and a heat flux J_q'' flow between the surfaces of conductors *I* and *II*.

Considerations concerning the foregoing effect in the thermionic diode are analogous to the analysis of a flow of electric current and heat through a homogeneous conductor (cf. Section IV.2). By equation (4.31) the dissipation function times the separation between the conductor surfaces gives

$$\Psi_1 \Delta x = T\Phi_{s1} \Delta x = -J_q'' \frac{\Delta T}{T} - j_e (\Delta\mu_e')_{\Delta T=0} . \tag{5.60}$$

On assuming that conductors *I* and *II* are made of the same substance in the chemical respect and have the same pressure acting on them, it can be found by equation (5.39) that the electron work function ϕ_e is a function only of the temperature:

$$(\Delta\phi_e)_{\Delta T=0} = 0 .$$

The definition (5.38) of the electron work function

$$\phi_e = \frac{\mu_e'}{e_e} - \psi$$

implies that the increment in electrostatic potential in the gap is

$$\Delta\psi = \frac{(\Delta\mu_e')_{\Delta T=0}}{e_e} , \tag{5.61}$$

and since the electric current density is

$$J_e = j_e e_e ,$$

the dissipation function can be written as

$$\Psi_1 \Delta x = -J_q'' \frac{\Delta T}{T} - J_e \Delta\psi . \tag{5.62}$$

The differences $-\Delta(\ln T) = -\Delta T/T$ and $-\Delta\psi$ are taken for the thermodynamic forces, whereas the quantities taken for the fluxes are:
the heat flux in the entropy balance equation

$$J_q'' = -L_{qq} \frac{\Delta T}{T} + L_{qe} \Delta\psi , \tag{5.63}$$

the electric current density

$$J_e = -L_{eq} \frac{\Delta T}{T} - L_{ee} \Delta\psi . \tag{5.64}$$

By the Onsager reciprocal relations the matrix of phenomenological coefficients is symmetrical,

$$L_{eq} = L_{qe} . \tag{5.65}$$

The condition that the dissipation function be positive implies that

$$L_{qq}>0, \quad L_{ee}>0, \quad L_{qq}L_{ee}-L_{qe}^2>0. \tag{5.66}$$

In special cases the two conductors may be at the same temperature ($\Delta T=0$), have the same electrostatic potentials ($\Delta\psi=0$), or there may be no flow of electric current ($J_e=0$).

For isothermal conditions equations (5.63) and (5.64) yield the heat of transport of electrons

$$Q_e^{\prime\prime *}=\left(\frac{J_q^{\prime\prime}}{j_e}\right)_{\Delta T=0}=e_e\left(\frac{J_q^{\prime\prime}}{J_e}\right)_{\Delta T=0}=e_e\frac{L_{qe}}{L_{ee}}. \tag{5.67}$$

By equation (5.57) we also have

$$Q_e^{\prime\prime *}=U_e^*-\tilde{h}_e=e_e\psi+e_e\Delta\psi+2RT-\tilde{h}_e. \tag{5.68}$$

Elimination of the electrostatic potential difference from equations (5.63) and (5.64) yields the heat flux in the form

$$J_q^{\prime\prime}=\frac{L_{qe}}{L_{ee}}J_e-\left(L_{qq}-\frac{L_{qe}^2}{L_{ee}}\right)\frac{\Delta T}{T}=\frac{Q_e^{\prime\prime *}}{e_e}J_e-L_k\Delta T, \tag{5.69}$$

where we have introduced a new phenomenological coefficient defined as

$$L_k=\left(L_{qq}-\frac{L_{qe}^2}{L_{ee}}\right)\frac{1}{T}. \tag{5.70}$$

The physical meaning of this coefficient can easily be determined by considering the heat flux when no electric current is flowing:

$$(J_q^{\prime\prime})_{J_e=0}=-L_k\Delta T. \tag{5.71}$$

The phenomenological coefficient L_k thus is the heat flux set up by a unit temperature difference in the absence of an electric current.

The density of electric current flowing through the isothermal diode can be determined from equation (5.64) as

$$(J_e)_{\Delta T=0}=-L_{ee}\Delta\psi \tag{5.72}$$

or on the basis of equation (5.56) as

$$(J_e)_{\Delta T=0}=\left[J_{es}\exp\left(-\frac{e_e\Delta\psi}{RT}\right)-(J_{es}+\Delta J_{es})\right]_{\Delta T=0}. \tag{5.73}$$

The saturation current density is a function of the temperature, pressure, and chemical composition of the conductor, but does not depend on the value of the electrostatic potential at the conductor surface. Hence

$$(\Delta J_{es})_{\Delta T=0}=0$$

and consequently

$$(J_e)_{\Delta T=0}=J_{es}\left[\exp\left(-\frac{e_e\,\Delta\psi}{RT}\right)-1\right]. \tag{5.74}$$

For small potential differences, to which linear nonequilibrium thermodynamics is confined, equation (5.74) takes on the form

$$(J_e)_{\Delta T=0}=-J_s\,\frac{e_e\,\Delta\psi}{RT}\,. \tag{5.75}$$

Comparison of relations (5.75) and (5.72) enables the phenomenological coefficient concerning current conduction to be defined as

$$L_{ee}=J_{es}\,\frac{e_e}{RT}=J_{es}\,\frac{\in}{kT}\,. \tag{5.76}$$

Now that the phenomenological coefficient L_{ee} and the heat of transport (5.67) have been determined, equation (5.64) expressing the electric current density can be rearranged by discarding the second-order term $\Delta\psi\Delta T$ and then

$$J_e=-L_{ee}\left(\frac{L_{eq}}{L_{ee}}\frac{\Delta T}{T}+\Delta\psi\right)=-\frac{J_{es}}{R}\left[(2RT+e_e\psi-\tilde{h}_e)\,\frac{\Delta T}{T^2}+\frac{e_e\,\Delta\psi}{T}\right]. \tag{5.77}$$

For small electrostatic potential differences $\Delta\psi$ equation (5.56) can be rewritten as

$$J_e=J_{es}\left(1-\frac{e_e\,\Delta\psi}{RT}\right)-(J_{es}+\Delta J_{es})=-\Delta J_{es}-J_{es}\,\frac{e_e\,\Delta\psi}{RT}\,. \tag{5.78}$$

When equations (5.77) and (5.78) are combined and account is taken of the fact that

$$\tilde{h}_e=\mu_e'+T\tilde{s}_e-e_e\,\psi\,,\qquad (\Delta\mu_e')_{\Delta T=0}=-e_e\,\Delta\psi\,,$$

the result is

$$\frac{\Delta J_{es}}{J_{es}}=\frac{2RT+e_e\psi-\tilde{h}_e}{RT^2}\,\Delta T=2\,\frac{\Delta T}{T}-\frac{e_e\,\phi_e}{RT^2}\,\Delta T+\frac{e_e\,\Delta\phi_e}{RT}\,.$$

On going over from finite differences to differentials, we obtain

$$\frac{\mathrm{d}J_{es}}{J_{es}}=2\,\frac{\mathrm{d}T}{T}+\mathrm{d}\left(\frac{e_e\,\phi_e}{RT}\right) \tag{5.79}$$

and, finally, on integrating this equation, we find that the saturation current density is

$$J_{es}=AT^2\exp\left(\frac{e_e\,\phi_e}{RT}\right)=AT^2\exp\left(\frac{\in\phi_e}{kT}\right). \tag{5.80}$$

This is the Richardson–Dushman equation for the saturation current density, derived this time by using the Onsager reciprocal relations.

5. The Thermionic Generator

The emission of electrons by a metal under the influence of high temperature was discovered back in 1883 by T. A. Edison, and in 1912 O. W. Richardson busied himself explaining this phenomenon. A complete explanation was not forthcoming until Dushman (1930) provided one on the basis of Fermi's electron theory of solids

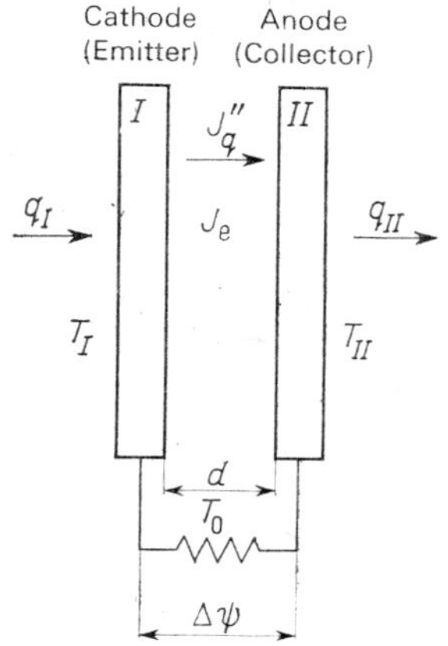

Fig. 5.6 Diagram of a thermionic generator

(1927). Thermionic emission, long familiar in radio engineering, has in recent years finally come to be considered as offering one possible way of converting heat into electricity, although the idea of the thermionic generator is nothing new, being due to Schlichter (1915). The thermionic diode (Fig. 5.6) may be treated as a heat engine when two conductors *I* and *II* with parallel surfaces and different temperatures, T_I and T_{II}, are connected by a resistor as a load. The heat flux supplied to the cathode at the higher temperature T_I will be denoted by q_I. The heat flux extracted from the anode at the lower temperature T_{II} will be denoted by q_{II}. Plate *I* at the higher temperature (cathode) emits electrons and is thus called the *emitter*. These electrons are collected by plate *II* at the lower temperature, this plate being called the *collector*.

The resistive load and the conductors connecting it with the collector and emitter are kept at the ambient temperature T_0. If the emitter–collector distance d is small, the density of the electric current flowing through the diode is given by equation (5.56), that is, by

$$J_e = J_{esI} \exp\left(-\frac{e_e \Delta\psi}{RT_I}\right) - J_{esII}.$$

This formula has been derived with the assumption of a linear distribution of the electrostatic potential between surfaces *I* and *II*.

In an actual thermionic generator the process is complicated by the formation of a space charge. This is an electron cloud in the proximity of the collector, consisting of electrons which do not reach the collector for various reasons. This space charge hinders the flow of electrons to the collector. If the emitter–collector distance is large, the distribution of the electrostatic potential is non-linear (Fig. 5.7) and the density of the current flowing through the diode is a more complicated function of T_I, T_{II}, and $\Delta\psi$. According to the Poisson equation the electrostatic potential falls off first to a minimum value $\psi_{\min}$, and only then attains a value corresponding

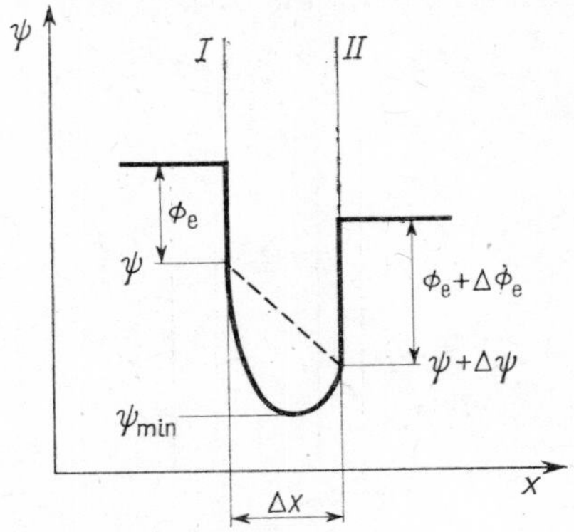

Fig. 5.7 Distribution of the electrostatic potential in a vacuum thermionic generator

to the surface of the other plate. In order to reach the collector *II*, electrons emitted by the emitter *I* must have a kinetic energy greater than $\psi-\psi_{\min}$, that is, as is seen from the drawing, greater than in the case of a linear potential distribution. The density of the current flowing through the diode will thus be correspondingly lower.

In actual practice it is not possible to make a thermionic diode with such a small emitter–collector separation as to obtain a linear potential distribution between them. The influence of the space charge can, however, be reduced by introducing ionized gas into the space between the plates. The positive gas ions which act jointly with the electrons straighten out the curve of the potential distribution. This phenomenon occurs in a very complex manner at considerable distance from the state of equilibrium.

The following analysis of thermodynamic effects in the thermionic diode is confined to the simple case when the emitter–collector separation is so small that the distribution of the electrostatic potential in the gap may be assumed to be linear. To simplify the considerations the calculations will be carried out for a unit surface area of each electrode. The cross-sectional area of the conductor connecting the electrodes with the resistive load will be B_0.

The density of the electric current flowing through the diode can be made dependent on the thermodynamic properties of the emitter and the collector, treated as regions in thermodynamic equilibrium. To this end, we insert into equation (5.56) the saturation current density as expressed by means of the Richardson–Dushman equation (5.80):

$$J_e = AT_I^2 \exp\left[\frac{e_e(\phi_{eI}-\Delta\psi)}{RT_I}\right] - AT_{II}^2 \exp\left(\frac{e_e\,\phi_{eII}}{RT_{II}}\right). \tag{5.81}$$

The potential drop across the interelectrode gap can be written as

$$\Delta\psi = \psi_{II} - \psi_I = \phi_{eI} - \phi_{eII} + \frac{1}{e_e}\,\Delta\mu'_e\,. \tag{5.82}$$

The electric current density can thus be put in the form

$$J_e = AT_I^2 \exp\left(\frac{e_e\,\phi_{eII} - \Delta\mu'_e}{RT_I}\right) - AT_{II}^2 \exp\left(\frac{e_e\,\phi_{eII}}{RT_{II}}\right), \tag{5.83}$$

hence the density J_e of the current flowing through the diode does not depend on the work function ϕ_{eI} for removal of electrons from the emitter.

In the limiting case, when $T_{II} - T_I \to \mathrm{d}T$ and $\mu'_{eII} - \mu'_{eI} \to \mathrm{d}\mu'_e$, we have

$$\delta J_e = -\frac{J_{es}}{RT}\left[(2RT - e_e\,\phi_e)\frac{\mathrm{d}T}{T} + \mathrm{d}\mu'_e\right]. \tag{5.84}$$

As can be seen, equation (5.84) coincides with equation (5.77) derived for the case of local thermodynamic equilibrium. The symbol δJ_e denotes infinitesimal electric current density due to small differences of temperature and electrochemical potential between emitter and collector, whereas the symbol $\mathrm{d}J_e$ will be reserved for changes in the electric current density in the space between two adjoining cross-sections.

Equation (4.54), defining the electric current density in a homogeneous conductor, can be cast into the form

$$J_e\,\mathrm{d}x = -\frac{1}{\rho e_e}\,(S^*_{eB}\,\mathrm{d}T + \mathrm{d}\mu'_e), \tag{5.85}$$

where the subscript B refers to the material of conductor B.

This equation can be used to integrate the differential of the electrochemical potential of electrons in conductor B, of cross-sectional area B_0, connecting the diode with the resistive load; in this operation the electrostatic potential difference $e_e\,\varphi$ between the terminals of the resistive load,

$$\mu'_{eII} - \mu'_{eI} = \Delta\mu'_e = -e_e\int_{T_I}^{T_{II}} \varepsilon_B\,\mathrm{d}T - e_e J_e \int_{I}^{II} \frac{\rho_B}{B_0}\,\mathrm{d}x + e_e\,\varphi\,, \tag{5.86}$$

must be added to the right-hand side. Hence, the potential difference across the electrodes is

$$\varphi = \frac{\Delta\mu'_e}{e_e} - \int_{T_{II}}^{T_I} \varepsilon_B\,\mathrm{d}T + J_e\int_{I}^{II} \frac{\rho_B}{B_0}\,\mathrm{d}x\,. \tag{5.87}$$

where ε_B is the absolute Seebeck coefficient of conductor B, defined by equation (4.65), and ρ_B is the resistivity of conductor B.

The potential difference of an open circuit follows from equations (5.83) and (5.86) on the assumption of zero flow of electric current ($J_e=0$). With $J_e=0$ equation (5.83) gives the electrostatic potential difference of the electrons as

$$\Delta\mu_e' = 2RT_I \ln \frac{T_I}{T_{II}} - e_e \phi_{eII} \left(\frac{T_I}{T_{II}} - 1 \right). \tag{5.88}$$

On inserting this expression into equation (5.86) and noting that $J_e=0$, we get the potential difference of the open circuit

$$\varphi_0 = \int_{T_I}^{T_{II}} \varepsilon_B \, \mathrm{d}T + \frac{2RT_I}{e_e} \ln \frac{T_I}{T_{II}} - \phi_{eII} \left(\frac{T_I}{T_{II}} - 1 \right). \tag{5.89}$$

With $T_{II}=T=\text{const}$, differentiation of this equation with respect to T_I yields

$$\varepsilon_{BT} = \frac{\mathrm{d}\varphi_0}{dT} = -\frac{\phi_{eII}}{T} - \varepsilon_B + \frac{2R}{e_e}. \tag{5.90}$$

By analogy with the thermoelectric generator, ε_{BT} may be regarded as the Seebeck coefficient of the thermionic diode with conductor of metal B connecting the electrodes with the resistive load.

By the relations (4.40) derived for effects in a homogeneous conductor, the heat of transport of electrons is

$$Q_e''^* = T(S_e^* - \tilde{s}_e) = TS_e^* + \mu_e' - e_e \psi - \tilde{h}_e, \tag{5.91}$$

and when the electron work function

$$\phi_e = \frac{\mu_e'}{e_e} - \psi$$

and the specific heat of electrons at constant pressure

$$c_{pe} = \tfrac{5}{2} R$$

are introduced, the result is

$$Q_e''^* = TS_e^* + e_e \phi_e - \tfrac{5}{2} RT. \tag{5.9)}$$

The Seebeck coefficient of the thermionic diode may thus be expressed in terms of the heat of transport as

$$\varepsilon_{BT} = -\frac{1}{e_e T} (Q_{eB}''^* + \tfrac{1}{2} RT). \tag{5.93}$$

The energy flux $J_{uI,II}$ flowing through the diode is equal to the difference between the energy flux associated with the electrons emitted by the emitter I and the energy flux associated with the electrons emitted by the collector II, with due account for the energy transport by radiation between the emitter and collector surfaces:

$$J_{uI,\,II}=J_{uI}-J_{uII}+F(T_I^4-T_{II}^4)\,. \tag{5.94}$$

The energy flux can be determined by taking account of the electric current density, as given by expression (5.83):

$$\begin{aligned} J_{uI,II}=&AT_I^2\left(\psi_{II}+\frac{2RT_I}{e_e}\right)\exp\left(\frac{e_e\,\phi_{eII}-\Delta\mu_e'}{RT_I}\right)\\ &-AT_{II}^2\left(\psi_{II}+\frac{2RT_{II}}{e_e}\right)\exp\left(\frac{e_e\,\phi_{eII}}{RT_{II}}\right)+F(T_I^4-T_{II}^4) \end{aligned} \tag{5.95}$$

or

$$\begin{aligned} J_{uI,II}=&J_e\psi_{II}+\frac{2RA}{e_e}\left[T_I^3\exp\left(\frac{e_e\,\phi_{eII}-\Delta\mu_e'}{RT_I}\right)-T_{II}^3\exp\left(\frac{e_e\,\phi_{eII}}{RT_{II}}\right)\right]\\ &+F(T_I^4-T_{II}^4)\,. \end{aligned} \tag{5.96}$$

For the limiting case $T_{II}-T_I\to\mathrm{d}T$ and $\mu_{eII}'-\mu_{eI}'\to\mathrm{d}\mu_e'$, taking relation (5.84) into consideration, equation (5.96) yields the infinitesimal energy flux between the electrodes:

$$\begin{aligned} \delta J_{uI,II}&=\psi_{II}\,\delta J_e-\frac{2J_{es}}{e_e}\left[(3RT-e_e\,\phi_{eII})\frac{\mathrm{d}T}{T}+\mathrm{d}\mu_e'\right]-4FT^3\mathrm{d}T\\ &=\left(\psi_{II}+\frac{2RT}{e_e}\right)\delta J_e-\frac{2J_{es}}{e_e}R\,\mathrm{d}T-4FT^3\mathrm{d}T\,. \end{aligned} \tag{5.97}$$

In the absence of any flow of electric current ($J_e=0$, $\delta J_e=0$), we have

$$(\delta J_{uI,II})_{J_e=0}=-\left(\frac{2J_{es}R}{e_e}+4FT^3\right)\mathrm{d}T=-K\,\mathrm{d}T\,, \tag{5.98}$$

where the heat transfer coefficient in the thermionic diode has been defined as

$$K=\frac{2J_{es}R}{e_e}+4FT^3\,. \tag{5.99}$$

The transfer of energy by radiation which has been taken into account in the equations above by means of the term $F(T_I^4-T_{II}^4)$ is one of the main losses reducing the efficiency of the thermionic generator. The evaluation of the radiation energy flux is based on the empirical observations of Hottel who proposed a simple formula similar to that of Christiansen:

$$J_r=\frac{\sigma}{\dfrac{1}{\varepsilon_I(T_I)}+\dfrac{1}{\varepsilon_{II}(\bar{T})}-1}\,(T_I^4-T_{II}^4)\,. \tag{5.100}$$

In this formula the relative emissivity ε_{II} of the collector is calculated for the geometric mean of the emitter and collector temperatures,

$$\bar{T}=\sqrt{T_I T_{II}}\,.$$

The symbol $\sigma=5.67\times10^{-8}\ \mathrm{W\,m^{-2}\,K^{-4}}$ is the radiation constant for a blackbody surface.

The heat flux q_I required to maintain the emitter at the constant temperature T_I is the sum of energy fluxes flowing between emitter and collector ($J_{uI,\,II}$), given by equation (5.96), and extracted from the emitter by conductor B (J_{uIB}):

$$q_I=J_{uI,II}+J_{uIB}\,. \tag{5.101}$$

The energy flux carried off by the conductor can be determined by means of the equation

$$\begin{aligned} J_{uIB}&=-\frac{J_\mathrm{e}}{e_\mathrm{e}}\,U_\mathrm{e}^*-B_0\,\lambda_{\infty B}\left(\frac{\mathrm{d}T_B}{\mathrm{d}x}\right)_{x=0}\\ &=-(TS_{\mathrm{e}B}^*+\mu_\mathrm{e})\,\frac{J_\mathrm{e}}{e_\mathrm{e}}-B_0\,\lambda_{\infty B}\left(\frac{\mathrm{d}T_B}{\mathrm{d}x}\right)_{x=0}\,. \end{aligned} \tag{5.102}$$

A minus sign has been put in front of the first term on the right-hand side of the equation since J_e is the electric current density flowing from emitter to conductor B, that is, opposite to the direction taken for the x-axis. If the conductor B of length l is adiabatically insulated and if its resistivity ρ_B, thermal conductivity $\lambda_{\infty B}$, and Seebeck coefficient $\varepsilon_{\infty B}$ do not depend on the temperature, then the gradient temperature at the junction of conductor B with the emitter is

$$\left(\frac{\mathrm{d}T_B}{\mathrm{d}x}\right)_{x=0}=\frac{T_{II}-T_I}{l}+\frac{J_\mathrm{e}^2\,\rho_B\,l}{2B_0^2\,\lambda_{\infty B}}\,. \tag{5.103}$$

Finally, when the values of the fluxes as given by equations (5.97) and (5.102) are inserted into equation (5.101) and when relation (5.103) is taken into account, the heat flux q_I carried off from the emitter is obtained for the limiting case $T_{II}-T_I \to \mathrm{d}T$:

$$\begin{aligned} q_I&=\frac{\delta J_\mathrm{e}}{e_\mathrm{e}}\,(e_\mathrm{e}\,\psi-\mu_\mathrm{e}-TS_{\mathrm{e}B}^*+2RT)-\left(\frac{2J_\mathrm{es}\,R}{e_\mathrm{e}}+4FT^3+\frac{B_0\,\lambda_{\infty B}}{l}\right)\mathrm{d}T\\ &=\delta J_\mathrm{e}\left(\frac{2RT}{e_\mathrm{e}}-\phi_\mathrm{e}-T\varepsilon_B\right)-\left(\frac{2J_\mathrm{es}\,R}{e_\mathrm{e}}+4FT^3+\frac{B_0\,\lambda_{\infty B}}{l}\right)\mathrm{d}T. \end{aligned} \tag{5.104}$$

The thermal efficiency of the thermionic generator treated as a heat engine is the ratio of the power $P=\varphi J_\mathrm{e}$ to the heat flux delivered to the emitter:

$$\eta=\frac{P}{q_I}=\frac{\varphi J_\mathrm{e}}{q_I}\,. \tag{5.105}$$

This efficiency depends on the properties of the conducting materials used to make the diode electrodes and the conductors connecting them with the resistive load and also on the emitter and collector temperatures.

The thermal efficiency of the thermionic generator is optimizable with respect to the difference $\Delta\mu_e'$ of the electrochemical potentials of electrons. Such optimization can be carried out analytically only in the case of small temperature differences $\Delta T \to 0$ and small differences $\Delta\mu_e' \to 0$ of electrochemical potentials of electrons, so that linear relations are valid. For this case H. N. Hatsopoulos established the formula

$$\eta_{\max} = \frac{\mathrm{d}T}{T} \frac{\sqrt{ZT+1}-1}{\sqrt{ZT+1}+1}, \tag{5.106}$$

where

$$Z = \frac{\varepsilon_{BT}^2}{\left(\sqrt{\lambda_{\infty B}\rho_B} + \sqrt{\dfrac{2R^2T}{e_e^2} + \dfrac{4RFT^4}{J_{es}e_e}}\right)^2}. \tag{5.107}$$

The expression

$$\frac{2R^2T}{e_e^2} + \frac{4RFT^4}{J_{es}e_e},$$

appearing in formula (5.107) can be transformed, on noting that

$$\lambda_{\infty T}\rho_T = -K\left(\frac{\mathrm{d}\varphi}{J_e}\right)_{\mathrm{d}T=0},$$

$$(J_e)_{\mathrm{d}T=0} = -\left(\frac{J_{es}\,\mathrm{d}\mu_e'}{RT}\right)_{\mathrm{d}T=0},$$

$$(\mathrm{d}\mu_e')_{\mathrm{d}T=0} = e_e(\mathrm{d}\varphi)_{\mathrm{d}T=0}.$$

Then, for the thermionic diode (subscript T) we get

$$\lambda_{\infty T}\rho_T = \frac{2R^2T}{e_e^2} + \frac{4RFT^4}{J_{es}e_e} \tag{5.108}$$

and, just as for the thermoelectric generator, we also have

$$Z = \frac{\varepsilon_{BT}^2}{(\sqrt{\lambda_{\infty B}\rho_B} + \sqrt{\lambda_{\infty T}\rho_T})^2}. \tag{5.109}$$

This result is similar to that obtained when calculating the efficiency of the thermoelectric generator. It confirms the feasibility of considering the thermionic generator as a thermocouple consisting of a thermionic diode and conductors B.

Suppose that the thermionic generator consists of a large number of diodes in series between two reservoirs at temperatures T_I and T_{II} so that the collector of one diode transfers heat directly to the emitter of another. Then, by the definition of efficiency, for each stage we can write the relation

$$\frac{đq}{q} = \alpha_0 \frac{dT}{T}, \tag{5.110}$$

where

$$\alpha_0 = \frac{\sqrt{ZT+1}-1}{\sqrt{ZT+1}+1}. \tag{5.111}$$

Integration of expression (5.110) from the first stage to the last yields

$$\frac{|q_{II}|}{q_I} = \exp\left[\int_{T_{II}}^{T_I} \alpha_0 \, d(\ln T)\right] \tag{5.112}$$

and, finally, the efficiency is

$$\eta = 1 - \frac{|q_{II}|}{q_I} = 1 - \exp\left[\int_{T_{II}}^{T_I} \alpha_0 \, d(\ln T)\right]. \tag{5.113}$$

This efficiency has been derived for the case of optimal choice of difference $\Delta\mu_e'$ of electrochemical potentials of electrons and for linear relations following from small deviations from the equilibrium state. If losses due to radiation are neglected in equation (5.107), this is the highest efficiency attainable by a thermionic generator made of given materials (electrodes and conductors), operating between the temperatures T_I and T_{II}, and having an infinite number of stages. Every real generator with a finite number of stages which operates under these same conditions will have a lower efficiency.

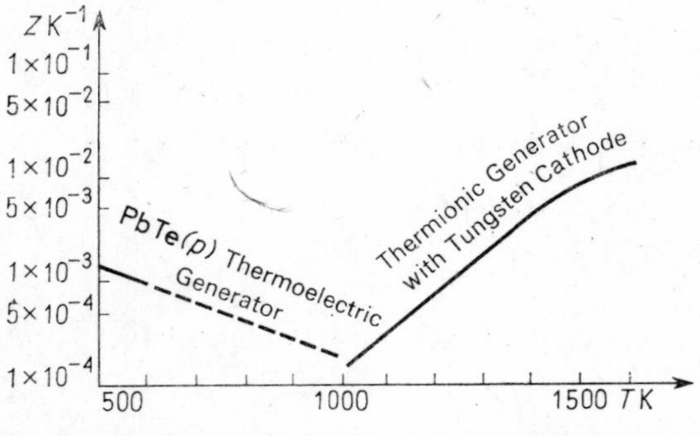

Fig. 5.8 The figure of merit Z vs. the temperature for a PbTe(p) thermoelectric generator and a thermionic generator

The fact that formulae (4.102) and (5.106) for the efficiencies of thermoelectric and thermionic generators, respectively, are similar allows the two types of generators to be compared by considering the figure of merit Z, which is dependent only on the material properties.

Figure 5.8 gives Z as a function of the temperature for PbTe(p) used in thermoelectric generators and for tungsten electrodes used in thermionic generators in the absence of a space-charge barrier. As can be seen from the plot, at high temperatures thermionic generators are superior to thermoelectric generators with respect to efficiency. At lower temperatures, the converse is true.

The best thermoelectric material at present available has a figure of merit of $Z = 2\times 10^{-3}\ \mathrm{K}^{-1}$ and can be used at temperatures of up to 1000 K, that is, $ZT=2$, whereas the maximum efficiency of the thermoelectric generator is

$$\eta = 0.267\,\frac{\Delta T}{T}.$$

For a thermionic generator with tungsten electrodes, operating at 1500 K the product ZT has a value of 15, and hence the maximum efficiency is

$$\eta = 0.6\,\frac{\Delta T}{T}.$$

Similar conclusions are arrived at by analysing the efficiencies of actual generators operating with finite temperature differences.

6. Problems in the Construction of Thermionic Generators

In the theoretical discussion above we have considered the general scheme of a thermionic generator. It would seem that it is a simple engine, formed by two plates, one of which must be kept at a high temperature by the addition of heat and the other kept at a low temperature by the extraction of heat.

The simplified description of a thermionic generator would be incomplete if there were no discussion, at least qualitatively, of the principal problems which come up in the operation of an actual generator.

One of the main goals of development work on thermionic generators is to come up with generators boasting high efficiencies and large power ratings. For this purpose the emitter (cathode) should be kept at the highest possible temperature and should be made of materials with suitably high electron work functions. Conversely, the collector (anode) should be kept at the lowest possible temperature and made of material with the smallest electron work functions. Careful consideration shows that the efficiency of the thermionic generator rises with an increase in the work function for the emitter only to a certain maximum and then falls off. The work function for the collector, on the other hand, should be as small as possible so as to counteract the re-emission of electrons from collector to emitter, which reduces the diode power.

Attempts are being made to fashion emitters out of materials which are resistant to temperatures of up to 2500 K, without suffering damage due to evaporation, melting, or mechanical effects. The consensus is that a material is suitable for emitters if the layer vaporized off the emitter in the course of 1000 hours of operation is less than 0.127 mm. The vaporization pressures and the melting points for various materials are given in Fig. 5.9.

Typical materials used for emitters are: boron-saturated tungsten, for the temperature range 870 to 1200°C and with a work function of 1.7 V; thorium on a tung-

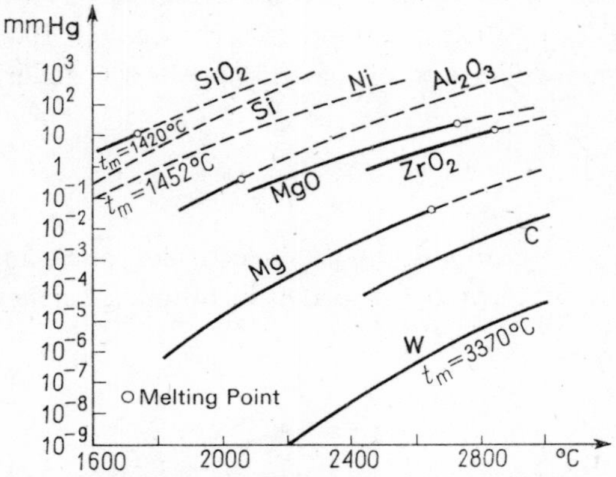

Fig. 5.9 Vaporization pressures and melting points for various gases

sten matrix, for the temperature range 1760 to 1980°C and with a work function of 2.55 V; and caesium on a tungsten matrix, for the temperature range 1370 to 1600°C and with a work function of 1.7 V.

Apart from a low work function, materials used for collectors should have a low relative emissivity, that is, a high capability for reflecting thermal radiation:

$$r = 1 - \varepsilon .$$

This is necessary in order to reduce the loss due to the collector being heated by the thermal radiation of the emitter. The collector usually works at a temperature that is 550 to 1100 K below the emitter temperature.

Typical collector materials are: barium oxide or strontium oxide on a nickel matrix, with a work function of 1 V; caesium on a silver oxide matrix, with a work function of 0.75 V; and caesium on a tungsten oxide matrix, with a work function of 0.71 V. In view of the small differences between the work functions of the emitter and collector, thermionic generators operate at low voltages and high currents. Obviously, they generate direct current.

The figures given above for the electron work functions are approximate since the work function depends on the space charge in the diode. A space charge is set up as a result of the mutual repulsion of electrons with charges of the same sign. Hence, a fraction of the electrons which escape from the emitter do not reach the collector but return to the emitter.

To mitigate the undesirable influence the space charge has on the generator power, the electrode spacing is decreased, thus eliminating the nonlinearity in the potential distribution, or the space charge is neutralized by the introduction of positive ions.

In the first case the time of flight of electrons between emitter and collector is shortened, thereby diminishing the possibility of groups of electrons being formed and thus setting up a space charge. To remove the influence of the space charge,

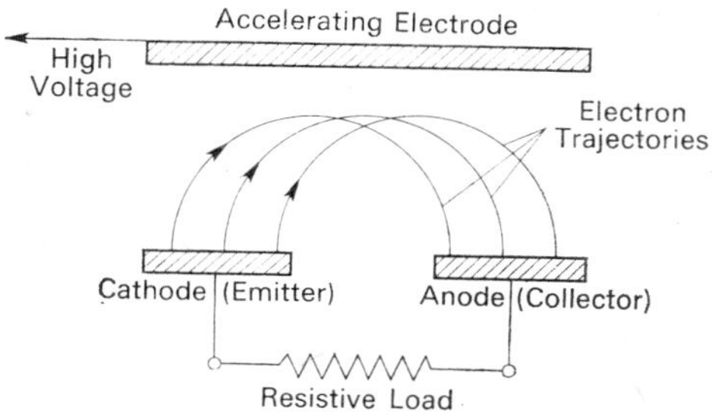

Fig. 5.10 Neutralization of space charge by means of an accelerating electrode

the electrodes must be brought closer together, so that they are no more than 0.0127 mm apart. Although at first it seemed that technological considerations would not allow such small gaps to be produced, the difficulties were soon overcome. On the other hand, with such a small interelectrode separation material vaporized from the emitter condenses on the anode. During the operation of the generator the difference between the electron work functions decreases and the generator power is reduced.

The second method of eliminating space charge is illustrated in Fig. 5.10. This is the method that was first used in the magnetrons of radar equipment and done by a triode. The cathode and anode are in one plane and opposite them is the accelerating electrode connected with a high-voltage source. The magnetic field thus set up deflects the electron paths towards the anode, thus preventing the electrons from congregating.

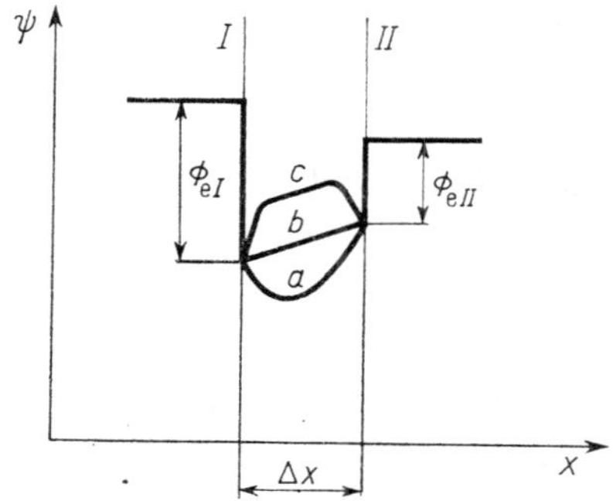

Fig. 5.11 The distribution of the electrostatic potential in the gap of a thermionic generator at (a) low, (b) intermediate, and (c) high ceasium vapour pressure

The third way is to neutralize the space charge of electrons by introducing positive ions between the electrodes. Such ions can be generated in the interelectrode space by introducing an easily ionized gas at low pressure. An energy called the *ionization potential* is required to remove an electron from a gas atom. If the ionization potential is smaller than the work function for removal of electrons from the cathode, the gas atoms will undergo ionization through collision with electrons emitted by the cathode. The electrons freed in the ionization of the atoms travel quickly to the anode whereas the much heavier positive ions remain almost motionless. The complete neutralization of the space charge, that is a state where there is an equal number of positive ions and electrons, is attained by a suitable choice of pressure for the ionized gas for the given temperature and electron flux. In general, for this purpose use is made of caesium with an ionization potential of 3.89 V while the cathode is of tungsten with a work function of 4.52 V. The gas pressures are of the order of 10^{-3} to 10^{-6} atmosphere. The pressures of caesium atoms and positive ions and of electrons are determined by the Saha equation (5.13). The distribution of the electrostatic potential in the gap of a thermionic generator after the introduction of caesium is given in Fig. 5.11.

As stated earlier, the thermionic generator is superior to the thermoelectric generator in that it has a much higher efficiency at high temperatures. Combination of ordinary chemical fuels with air does not make it possible to attain the temperatures of the order of 1650 to 2200°C necessary for economic operation of the thermionic generator. Such temperatures can be reached by making use of the heat generated in nuclear reactions.

CHAPTER VI

NONEQUILIBRIUM THERMOELECTROMAGNETIC EFFECTS IN GASES

1. The Formation of Plasma

With a rise in temperature gas molecules dissociate into simpler molecules or even individual atoms. This phenomenon, known as *thermal dissociation*, is discussed in textbooks on classical thermodynamics. At sufficiently high temperatures a gas may be assumed, with great accuracy, to consist only of atoms. At even higher temperatures, of the order of 5000 K, there occurs a process called *ionization* in which electrons are detached from atoms, forming an electronic gas. In some cases negative ions also appear.

Ionized gas is often characterized by the degree of ionization defined as the ratio of the number of kilomoles of ionized atoms, n_i, to the number of kilomoles of atoms before ionization begins, n_b, that is,

$$\beta = \frac{n_i}{n_b} . \tag{6.1}$$

The number of electrons leaving the electronic shell of an atom determines the ionization multiplicity. A solution of gases containing a considerable number of singly and multiply ionized atoms is called a *plasma*.

The energy required to remove an outer-shell electron from its normal position in the un-ionized atom to outside the nuclear interaction range determines the ionization potential of a gas (Table 5.1). Large amounts of energy are required to ionize gases and this energy is in turn evolved during recombination in which electrons combine with positive ions. If all electrons removed from atoms remain in the gas, the plasma is electrically neutral.

The energy needed to change the energy level or to detach an outer-shell electron from an atom may come from the collision of the atom with an electron or positive ion, from absorption of a quantum of radiation energy, and from the collision of atoms which remain un-ionized at high temperatures. All the molecules of a gas at high temperatures have high velocities and, hence, also have sufficient kinetic energy to detach electrons in collisions with atoms. Hydrogen plasma and subsequently argon plasma will be considered below as examples of the ionization of gases.

The dissociation of diatomic hydrogen gas H_2 begins at a temperature of about 2000 K, and above 5000 K hydrogen has been almost completely dissociated, that is to say, consists of individual atoms H. At temperatures in the interval from 2000 to 5000 K hydrogen is partially dissociated and, knowing its dissociation constant, one can determine its composition in the equilibrium state on the basis of relations given in the classical thermodynamics.

Ionization of hydrogen begins above 8000 K. Hydrogen atoms lose negative electrons so that only positive protons remain. Since the hydrogen atom consists of only one proton and one electron, hydrogen can be only singly ionized.

Hydrogen ionization proceeds according to the scheme

$$H \rightleftarrows H^+ + E^-,$$

that is, monatomic hydrogen H (subscript H) gives rise to protons H^+ (subscript p) and electrons E^- (subscript e), forming, respectively, proton gas and electron gas which are treated as ideal gases. The molar fractions of these components will be labelled, respectively, as z_H – the molar fraction of monatomic hydrogen, z_p – the molar fraction of protons, and z_e – the molar fraction of electrons, whereby

$$z_H + z_p + z_e = 1. \tag{6.2}$$

In electrically neutral hydrogen plasma the number of protons and electrons is the same, and hence

$$z_p = z_e, \tag{6.3}$$

$$z_H = 1 - 2z_e = 1 - 2z_p. \tag{6.4}$$

The total pressure of hydrogen plasma is

$$p = p_H + p_p + p_e, \tag{6.5}$$

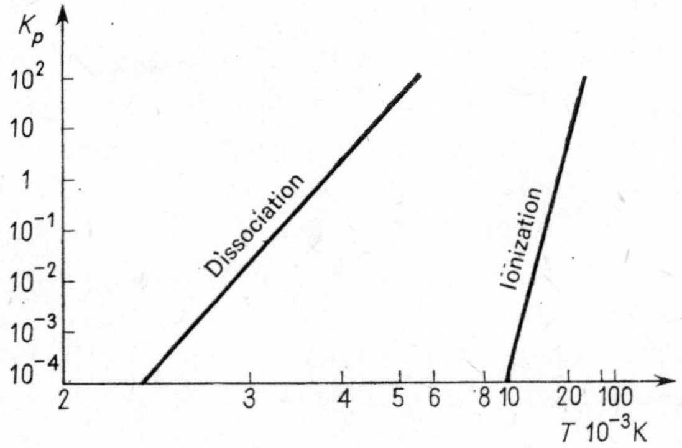

Fig. 6.1 Equilibrium constants for hydrogen dissociation and ionization

and, since

$$p_e = p_p, \tag{6.6}$$

for electrically neutral plasma it may also be written as

$$p = p_H + 2p_e = p_H + 2p_p. \tag{6.7}$$

The component pressures of monatomic hydrogen, protons, and electrons, respectively, are

$$p_{\mathrm{H}}=z_{\mathrm{H}}\,p\,,\qquad p_{\mathrm{p}}=z_{\mathrm{p}}\,p\,,\qquad p_{\mathrm{e}}=z_{\mathrm{e}}\,p\,. \tag{6.8}$$

These component pressures of hydrogen plasma are related by the ionization equilibrium constant which is a function of only the temperature (Fig. 6.1), this constant being determined by means of the Saha equation (5.17):

$$K_p=\frac{p_{\mathrm{p}}\,p_{\mathrm{e}}}{p_{\mathrm{H}}}=\frac{z_{\mathrm{p}}\,z_{\mathrm{e}}}{z_{\mathrm{H}}}\,p=BT^{5/2}\exp\left(-\frac{V_i}{RT}\right). \tag{6.9}$$

The results of computations of the volumetric composition of hydrogen plasma, on the basis of equations (6.3), (6.4), and (6.9), are given in Fig. 6.2.

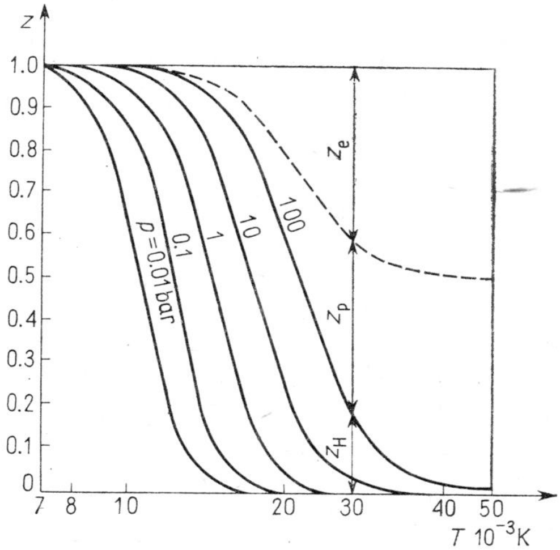

Fig. 6.2 The composition of hydrogen plasma

The equilibrium constant for hydrogen ionization can be associated with the degree of ionization β by means of the relations below. By the time equilibrium is reached, $n_{\mathrm{b}}\,\beta$ kilomoles out of n_{b} kilomoles of hydrogen will have been ionized; the amount of un-ionized hydrogen remaining, in kilomoles, is

$$n_{\mathrm{H}}=n_{\mathrm{b}}(1-\beta)\,.$$

At the same time, ionization products will have evolved in the form of protons and electrons with the same number of kilomoles:

$$n_{\mathrm{p}}=n_{\mathrm{e}}=n_{\mathrm{b}}\,\beta\,.$$

The total number of kilomoles of plasma components in the equilibrium state is

$$n=n_{\mathrm{H}}+n_{\mathrm{p}}+n_{\mathrm{e}}=n_{\mathrm{b}}(1+\beta)\,.$$

If the total pressure of the plasma is p, the respective component pressures are:

$$p_{H}=\frac{1-\beta}{1+\beta}\,p\,, \tag{6.10}$$

$$p_{p}=p_{e}=\frac{\beta}{1+\beta}\,p\,. \tag{6.11}$$

The equilibrium constant for hydrogen ionization may be written in terms of the degree of ionization as

$$K_p=\frac{p_p\,p_e}{p_H}=\frac{\beta^2}{1-\beta^2}\,p\,. \tag{6.12}$$

The degree of ionization of hydrogen can be calculated from the equilibrium constant by the relation

$$\beta=\sqrt{\frac{K_p}{p+K_p}}=\left(1+\frac{p}{K_p}\right)^{-0.5} \tag{6.13}$$

and for low degrees of ionization by the relation

$$\beta\approx\sqrt{\frac{K_p}{p}}=\left(\frac{B}{p}\right)^{0.5}T^{5/4}\exp\left(-\frac{V_i}{2RT}\right). \tag{6.14}$$

The degree of hydrogen ionization is plotted in Fig. 6.3 against the temperature and the pressure.

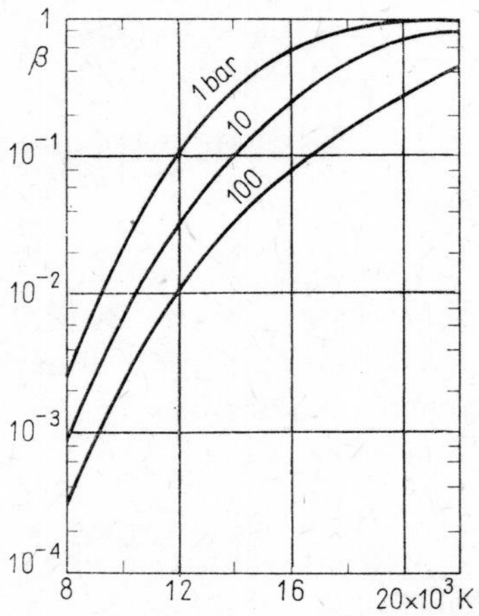

Fig. 6.3 The degree of ionization of hydrogen plasma vs. the temperature and the pressure

The ionization of argon is a more complicated process than is the ionization of hydrogen. Argon atoms have 18 electrons in three shells surrounding the nucleus. In the temperature interval from 10,000 to 15,000 K argon undergoes only single ionization. As the temperature rises further, the number of electrons detached from the outermost shell of the argon atom increases sharply, reaching seven at 100,000 K. Thus, multiple ionization can occur in argon.

The ionization of argon atoms proceeds according to the scheme

$$A_{(i)} \rightleftarrows A_{(i+1)} + E^{-},$$

where i is the ionization multiplicity, that is the number of electrons detached from the atom.

In the case of k-fold ionization, the sum total of molar fractions of the individual plasma components is

$$\sum_{i=0}^{k} z_{(i)} + z_e = 1. \tag{6.15}$$

For electrically neutral plasma the molar fraction of electrons is

$$z_e = \sum_{i=0}^{k} i z_{(i)}, \tag{6.16}$$

and thus combination of these two equations yields

$$\sum_{i=0}^{k} (i+1) z_{(i)} = 1. \tag{6.17}$$

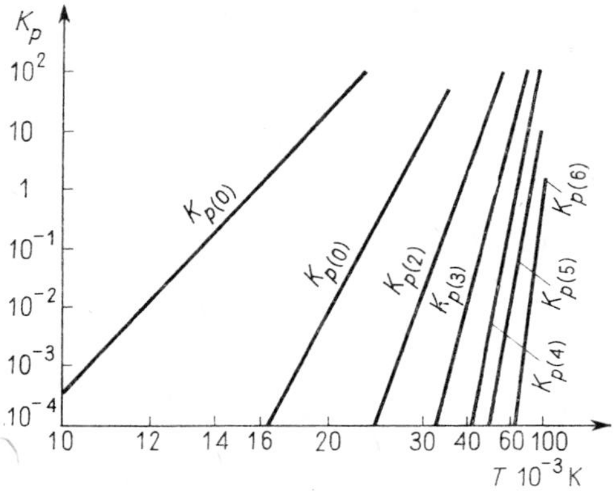

Fig. 6.4 The equilibrium constants for the ionization of argon plasma

The component pressures of argon plasma are

$$p_{(i)} = z_{(i)} p, \qquad p_e = z_e p = p \sum_{i=0}^{k} i z_{(i)} = \sum_{i=0}^{k} i p_{(i)}. \tag{6.18}$$

The component pressures of argon plasma are related by equilibrium constants (Fig. 6.4):

$$K_{p(i)} = \frac{p_{(i+1)} p_e}{p_{(i)}} = \frac{p_{(i+1)} \sum_{i=0}^{k} i p_{(i)}}{p_{(i)}} = \frac{z_{(i+1)} \sum_{i=0}^{k} i z_{(i)}}{z_{(i)}} p = f(T). \tag{6.19}$$

For k-fold ionization there are k such equilibrium constants, labelled consecutively by subscripts from 0 to $k-1$.

Although argon plasma may be 18-fold ionized, in practice it consists of at most ions with three successive ionization multiplicities (Fig. 6.5) and calculation of its

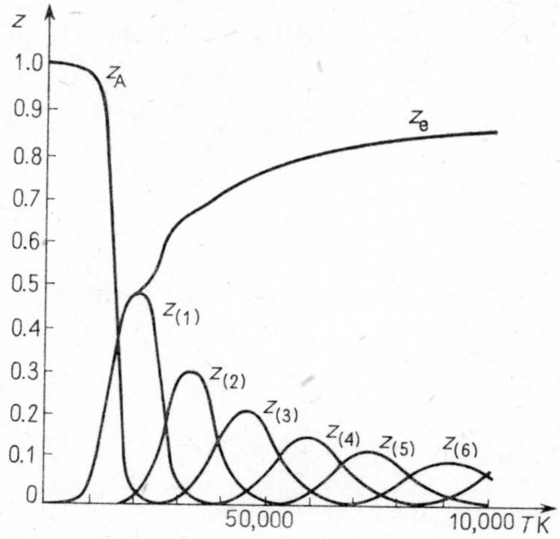

Fig. 6.5 Volumetric composition of argon plasma at pressure of 1 bar

composition reduces to determination of the three sought-after molar fractions which occur in the system of equations

$$\sum_{i=k-2}^{k} (i+1)\, z_{(i)} = (k-1)\, z_{(k-2)} + k z_{(k-1)} + (k+1)\, z_k = 1\,, \tag{6.20}$$

$$\frac{z_{(k)} \sum\limits_{i=k-2}^{k} i z_{(i)}}{z_{(k-1)}} = \frac{1}{p} K_{p(k-1)}\,, \tag{6.21}$$

$$\frac{z_{(k-1)} \sum\limits_{i=k-2}^{k} i z_{(i)}}{z_{(k-2)}} = \frac{1}{p} K_{p(k-2)}\,. \tag{6.22}$$

The degree of k-fold ionization is

$$\beta_{(k)} = \frac{n_k}{n_{(k-1)b}}\,. \tag{6.23}$$

The procedure involved in the calculations is simplified considerably for some substances, e.g. lithium, since only two adjacent degrees of ionization then appear. The equilibrium constant for single ionization of lithium can be calculated, depending on the degree of single ionization, by analogy with equation (6.12) as

$$K_{p(1)}=\frac{\beta_{(1)}^2}{1-\beta_{(1)}^2}\,p\,, \tag{6.24}$$

and the degree of single ionization, by analogy with equation (6.13), is

$$\beta_{(1)}=\sqrt{\frac{K_{p(1)}}{p+K_{p(1)}}}\,. \tag{6.25}$$

For double ionization of lithium or some other gas subject to multiple ionization (e.g. helium) with only two adjacent degrees of ionization the reasoning proceeds as follows. By the time equilibrium is attained $n_{(1)\mathrm{b}}\,\beta_{(2)}$ out of $n_{(1)\mathrm{b}}$ kilomoles singly-ionized lithium has become doubly-ionized, leaving

$$n_{(1)}=n_{(1)\mathrm{b}}(1-\beta_{(2)})$$

kilomoles not doubly-ionized. At the same time ionization products have appeared in the form of doubly-ionized lithium, the amount being

$$n_{(2)}=n_{(1)\mathrm{b}}\,\beta_{(2)}$$

and $n_{(1)\mathrm{b}}$ and $n_{(1)\mathrm{b}}\,\beta_{(2)}$ kilomoles of electrons from single and double ionization, respectively, the total being

$$n_{\mathrm{e}}=n_{(1)\mathrm{b}}(1+\beta_{(2)})\,.$$

The overall number of kilomoles of components of lithium plasma in which there is double ionization is

$$n=n_{(1)}+n_{(2)}+n_{\mathrm{e}}=n_{(1)\mathrm{b}}(2+\beta_{(2)})\,.$$

When the total pressure of the plasma is p, the component pressures are

$$p_{(1)}=\frac{1-\beta_{(2)}}{2+\beta_{(2)}}\,p\,,\qquad p_{(2)}=\frac{\beta_{(2)}}{2+\beta_{(2)}}\,p\,,\qquad p_{\mathrm{e}}=\frac{1+\beta_{(2)}}{2+\beta_{(2)}}\,p\,. \tag{6.26}$$

The equilibrium constant for the double ionization of lithium, calculated on the basis of the degree of ionization, is

$$K_{p(2)}=\frac{p_{(2)}\,p_{\mathrm{e}}}{p_{(1)}}=\frac{(1+\beta_{(2)})\,\beta_{(2)}}{(2+\beta_{(2)})(1-\beta_{(2)})}\,p\,. \tag{6.27}$$

This equation allows the degree of double ionization to be calculated:

$$\beta_{(2)}=-\frac{1}{2}+\sqrt{\frac{1}{4}+\frac{2}{1+\dfrac{p}{K_{p(2)}}}}\,. \tag{6.28}$$

Analogous reasoning for triple ionization shows that the component pressures are

$$p_{(2)}=\frac{1-\beta_{(3)}}{3+\beta_{(3)}}\,p\,,\qquad p_{(3)}=\frac{\beta_{(3)}}{3+\beta_{(3)}}\,p\,,\qquad p_{e}=\frac{2+\beta_{(3)}}{3+\beta_{(3)}}\,p\,, \tag{6.29}$$

and the equilibrium constant is

$$K_{p(3)}=\frac{p_{(3)}\,p_{e}}{p_{(2)}}=\frac{(2+\beta_{(3)})\,\beta_{(3)}}{(3+\beta_{(3)})(1-\beta_{(3)})}\,p\,, \tag{6.30}$$

whereas the degree of triple ionization is

$$\beta_{(3)}=-1+\sqrt{1+\frac{3}{1+\dfrac{p}{K_{p(3)}}}}\,. \tag{6.31}$$

Plasma can have a high degree of ionization only at low pressures, since the recombination of ions and electrons is directly proportional to the number of molecular collisions and thus to the pressure as well. At low pressures, and hence large interionic distances, the potential energy of the van der Waals forces which is inversely proportional to the sixth power of the interionic distances is negligible in comparison to the other components of the internal energy. The potential energy of Coulomb electrostatic forces which occur between electrons and oppositely charged ions is another component of the integral energy of the plasma. This energy is directly proportional to the first power of the distance between ions and electrons. At high temperatures and low pressures, that is, for low plasma densities, the energy of the van der Waals attractions and of the Coulomb forces may be neglected and the plasma treated as an ideal gas.

The limit from which the ideal gas laws and equations are applicable to plasma is determined by the condition [13]

$$\frac{n_i \epsilon^6}{2^{1.5}(kT)^3}\left\{\ln\left[1+\left(\frac{kT}{\epsilon^2 n_i^{1/3}}\right)^2\right]\right\}^{1.5} \ll 1\,, \tag{6.32}$$

where n_i is the number of ions per cm^3 plasma, ϵ is the electron charge, k is the Boltzmann constant, and T is the absolute temperature.

Total ionization of the plasma of ordinary gases is achieved at very high temperatures, not applied in technical equipment because available structural materials do not have the necessary resistance to such high temperatures. Hence the high electrical conductivity which is required of plasma and which corresponds to a high degree of ionization is attained not by raising the temperature but by seeding the gas with easily ionizing vapours of alkali metals which have a low ionization potential. Alkali metals have a single easily detachable electron in their outermost shell and hence attain a considerable degree of ionization even at moderate temperatures of the order of 3000 K which can be achieved by the combustion of hydrocarbon fuels with oxygen or in suitably preheated air. Higher temperatures occur in stars, shock waves set up when objects enter dense layers of the atmosphere at high speeds, and in electric arcs. At temperatures of the order of 10^5 K the density of the radiation energy approaches that of the internal energy of the plasma.

In considering a multicomponent solution, one cannot use the equations given above for a plasma formed out of atoms of only one species. The equilibrium of the plasma electrons is affected by the conditions under which are all the other plasma components.

The degree of ionization β_r of component r is defined by the ratio of the number n_{ir} of kilomoles of ions of this component to the number n_{br} of kilomoles of atoms of the component prior to ionization, that is,

$$\beta_r = \frac{n_{ir}}{n_{br}}. \tag{6.33}$$

The average degree of ionization for the plasma as a whole is determined by taking the sums for all n plasma components in the numerator and denominator:

$$\bar{\beta} = \frac{\sum_{r=1}^{n} n_{ir}}{\sum_{r=1}^{n} n_{br}}. \tag{6.34}$$

In this case, the Saha equation reduces to

$$\frac{\beta_r}{1-\beta_r} \frac{\bar{\beta}}{1+\bar{\beta}} p = BT^{5/2} \exp\left(-\frac{V_{ir}}{RT}\right). \tag{6.35}$$

For n species of atoms it is necessary to solve a system of $n+1$ equations consisting of n equations of the form of equations (6.34) and (6.35). Atoms with a lower ionization potential will be ionized to a greater degree than those having a higher ionization potential.

Example 6.1. Calculate the degree of ionization and the composition of hydrogen plasma at a pressure of $p=1$ bar and a temperature of $T=16{,}000$ K.

Solution. At $T=16{,}000$ K the equilibrium constant for the ionization of hydrogen is $K_p=0.3$. By equation (6.13) the degree of hydrogen ionization is

$$\beta = \sqrt{\frac{K_p}{p+K_p}} = \sqrt{\frac{0.3}{1+0.3}} = 0.48. \tag{1}$$

According to equations (6.8) and (6.11) the molar fractions of the individual plasma components are:
the molar fraction of atomic hydrogen

$$z_H = \frac{p_H}{p} = \frac{1-\beta}{1+\beta} = \frac{0.52}{14.8} = 0.35, \tag{2}$$

the molar fractions of protons and electrons

$$z_p = z_e = \frac{\beta}{1+\beta} = \frac{0.48}{1.48} = 0.325. \tag{3}$$

Example 6.2. Calculate the composition of lithium plasma at a pressure of $p=1$ bar and a temperature of $T=55{,}000$ K.

Solution. For the parameters given only the value $K_{p(2)}=0.1365$ should be taken into account for the equilibrium constant for double ionization.

By equation (6.28) the degree of double ionization of lithium plasma is

$$\beta_2=-\frac{1}{2}+\sqrt{\frac{1}{4}+\frac{2}{1+\dfrac{p}{K_{p(2)}}}}=0.2. \qquad (1)$$

In view of equations (6.26) the molar fractions of the various components are: singly-ionized lithium

$$z_{(1)}=\frac{1-\beta_{(2)}}{2+\beta_{(2)}}=0.364, \qquad (2)$$

doubly-ionized lithium

$$z_{(2)}=\frac{\beta_{(2)}}{2+\beta_{(2)}}=0.091, \qquad (3)$$

electrons

$$z_e=\frac{1+\beta_{(2)}}{2+\beta_{(2)}}=0.545. \qquad (4)$$

2. The Fundamental Equations for Plasma in an Electromagnetic Field

Plasma can be considered as a k-component fluid consisting of atoms, positive ions, and electrons. Each of the components is characterized by temperature, component pressure, mass (or molar) fraction, and three velocity components. The state of the plasma under study also depends on the electric and magnetic field strengths $\boldsymbol{E}$ and $\boldsymbol{B}$ which, being vectors, have three components each. In all, in order to determine the local state of such a plasma one must know the values of $6n+6$ variables related to each other by $6n+6$ equations of a thermodynamic ($6n$ equations) and electromagnetic (6 equations) nature. The number of unknowns is reduced considerably if global quantities concerning the whole of the plasma can be introduced. In particular, one temperature is taken for all components.

Low-density plasma can in general be treated as an ideal gas and can have equations of state like those for ideal gases applied to it, that is

$$p=\rho RT, \qquad (6.36)$$

$$du=c_v\,dT. \qquad (6.37)$$

The fundamental equations for an unpolarized medium, as derived in Section IV.1, are also applicable to plasma. These are: the principle of conservation of charge (4.1), the momentum balance equation (4.5), the internal energy balance equation (4.12), and the entropy balance equation (4.13).

The entropy source strength for plasma is of the form (4.14)

$$\Phi_s = \boldsymbol{J}_u \cdot \nabla\left(\frac{1}{T}\right) - \frac{1}{T}\sum_{i=1}^{k} \boldsymbol{j}_i \cdot \left\{T\nabla\left(\frac{\mu_i}{T}\right) - e_i\left[\boldsymbol{E} + \frac{1}{c}(\boldsymbol{v}\times\boldsymbol{B})\right]\right\}$$

$$+\frac{1}{T}\,\boldsymbol{\tau} : (\nabla \boldsymbol{v}) \geqslant 0. \qquad (6.38)$$

Our considerations henceforth will be confined to the flow of a nonviscous fluid for which

$$\boldsymbol{\tau} : (\nabla \boldsymbol{v}) = 0.$$

These considerations will be analogous to those in Chapter III concerning vectorial effects.

Going over from the conduction energy flux $\boldsymbol{J}_u$ to the heat flux in the entropy balance equation (3.133),

$$\boldsymbol{J}_q'' = \boldsymbol{J}_u - \sum_{i=1}^{k} \tilde{h}_i \boldsymbol{j}_i, \qquad (6.39)$$

gives the entropy source strength in the form

$$\Phi_s = \boldsymbol{J}_q'' \cdot \nabla\left(\frac{1}{T}\right) - \frac{1}{T}\sum_{i=1}^{k} \boldsymbol{j}_i \cdot \left\{\nabla_T \mu_i - e_i\left[\boldsymbol{E} + \frac{1}{c}(\boldsymbol{v}\times\boldsymbol{B})\right]\right\} \geqslant 0 \qquad (6.40)$$

or, when it is noted that by equation (3.26)

$$\sum_{i=1}^{k} \boldsymbol{j}_i = 0,$$

also in the form

$$\Phi_s = \boldsymbol{J}_q'' \cdot \nabla\left(\frac{1}{T}\right) - \frac{1}{T}\sum_{i=1}^{k-1} \boldsymbol{j}_i \cdot \left\{\nabla_T(\mu_i - \mu_k) - (e_i - e_k)\left[\boldsymbol{E} + \frac{1}{c}(\boldsymbol{v}\times\boldsymbol{B})\right]\right\} \geqslant 0. \qquad (6.41)$$

The dissipation function is now equal to

$$\Psi = T\Phi_s = -\boldsymbol{J}_q'' \cdot \nabla \ln T$$

$$-\sum_{i=1}^{k-1} \boldsymbol{j}_i \cdot \left\{\nabla_T(\mu_i - \mu_k) - (e_i - e_k)\left[\boldsymbol{E} + \frac{1}{c}(\boldsymbol{v}\times\boldsymbol{B})\right]\right\} \geqslant 0 \qquad (6.42)$$

or by equation (4.16)

$$\Psi = -\boldsymbol{J}_s \cdot \nabla T - \sum_{i=1}^{k-1} \boldsymbol{j}_i \cdot \left\{\nabla(\mu_i - \mu_k) - (e_i - e_k)\left[\boldsymbol{E} + \frac{1}{c}(\boldsymbol{v}\times\boldsymbol{B})\right]\right\} \geqslant 0. \qquad (6.43)$$

If the fluid is in mechanical equilibrium, then by equation (3.206)

$$\sum_{i=1}^{k} \rho_i(\nabla_T \mu_i - \boldsymbol{F}_i) = \sum_{i=1}^{k} \rho_i \left\{\nabla_T \mu_i - e_i\left[\boldsymbol{E} + \frac{1}{c}(\boldsymbol{v}_i\times\boldsymbol{B})\right]\right\} = 0 \qquad (6.44)$$

or

$$\sum_{i=1}^{k} \rho_i \left\{ \nabla \mu_i - e_i \left[\boldsymbol{E} + \frac{1}{c} (\boldsymbol{v}_i \times \boldsymbol{B}) \right] \right\} = -\rho s \nabla T. \tag{6.45}$$

For ideal gases the isothermal gradient of the chemical potential can be written as

$$\nabla_T \mu_i = -T \nabla_T \tilde{s}_i = R_i T \nabla \ln p_i = \frac{1}{\rho_i} \nabla p_i \tag{6.46}$$

or

$$\nabla_T \mu_i = R_i T \nabla \ln (z_i p) = \tilde{v}_i \nabla p + R_i T \frac{\nabla z_i}{z_i} = \tilde{v}_i \nabla p + \frac{1}{\rho_i} p \nabla z_i. \tag{6.47}$$

For a binary system, the change from molar to mass fractions can be made by taking account of the relations

$$z_1 = x_1 \frac{M}{M_1}, \tag{6.48}$$

$$\nabla z_1 = \frac{M^2}{M_1 M_2} \nabla x_1, \tag{6.49}$$

and then

$$\nabla_T \mu_1 = \tilde{v}_1 \nabla p + \frac{R_1 R_2}{R} T \nabla \ln x_1. \tag{6.50}$$

For ideal gases, the expressions

$$\boldsymbol{X}_q = -\nabla \ln T, \tag{6.51}$$

$$\boldsymbol{X}_i = -\frac{1}{\rho_i} \nabla p_i + \frac{1}{\rho_k} \nabla p_k + (e_i - e_k) \left[\boldsymbol{E} + \frac{1}{c} (\boldsymbol{v} \times \boldsymbol{B}) \right] \tag{6.52}$$

or

$$\boldsymbol{X}_i = -(\tilde{v}_i - \tilde{v}_k) \nabla p - p \left(\frac{1}{\rho_i} \nabla z_i - \frac{1}{\rho_k} \nabla z_k \right) + (e_i - e_k) \left[\boldsymbol{E} + \frac{1}{c} (\boldsymbol{v} \times \boldsymbol{B}) \right] \tag{6.53}$$

can be taken as the thermodynamic forces, conjugated with the generalized flows in the form of the heat flux $\boldsymbol{J}_q''$ in the entropy balance equation and the diffusion flux $\boldsymbol{j}_i$.

Now the phenomenological equations can be written for the heat fluxes which occur in the entropy balance equation,

$$\boldsymbol{J}_q'' = \mathbf{L}_{qq} \cdot \boldsymbol{X}_q + \sum_{i=1}^{k-1} \mathbf{L}_{qi} \cdot \boldsymbol{X}_i, \tag{6.54}$$

and for the diffusion flux of component m,

$$\boldsymbol{j}_m = \mathbf{L}_{mq} \cdot \boldsymbol{X}_q + \sum_{i=1}^{k-1} \mathbf{L}_{mi} \cdot \boldsymbol{X}_i. \tag{6.55}$$

In the presence of a magnetic field, plasma becomes nonisotropic and the phenomenological coefficients are tensors of rank two, associated by Onsager–Casimir relations (2.37) of the form

$$\mathbf{L}_{qi}(\boldsymbol{B})=\tilde{\mathbf{L}}_{iq}(-\boldsymbol{B}), \qquad \mathbf{L}_{mi}(\boldsymbol{B})=\tilde{\mathbf{L}}_{im}(-\boldsymbol{B}),$$
$$L_{qi,\,\alpha\beta}(\boldsymbol{B})=L_{iq,\,\beta\alpha}(-\boldsymbol{B}), \qquad L_{mi,\,\alpha\beta}(\boldsymbol{B})=L_{im,\,\beta\alpha}(-\boldsymbol{B}). \tag{6.56}$$

In the absence of a magnetic field, thermodynamic forces (6.52) and (6.53) simplify to

$$\boldsymbol{X}_i=-\frac{1}{\rho_i}\nabla p_i+\frac{1}{\rho_k}\nabla p_k+(e_i-e_k)\boldsymbol{E} \tag{6.57}$$

or

$$\boldsymbol{X}_i=-(\tilde{v}_i-\tilde{v}_k)\nabla p-p\left(\frac{1}{\rho_i}\nabla z_i-\frac{1}{\rho_k}\nabla z_k\right)+(e_i-e_k)\boldsymbol{E}, \tag{6.58}$$

and the phenomenological coefficients are scalars.

In the particular case of the binary system the thermodynamic forces $\boldsymbol{X}_i$ are limited to a single thermodynamic force

$$\begin{aligned}\boldsymbol{X}_1&=-\nabla_T(\mu_1-\mu_2)+(e_1-e_2)\left[\boldsymbol{E}+\frac{1}{c}(\boldsymbol{v}\times\boldsymbol{B})\right]\\&=-(\tilde{v}_1-\tilde{v}_2)\nabla p-\frac{R_1R_2}{R}\frac{T}{x_1x_2}\nabla x_1+(e_1-e_2)\left[\boldsymbol{E}+\frac{1}{c}(\boldsymbol{v}\times\boldsymbol{B})\right].\end{aligned} \tag{6.59}$$

3. Phenomenological Equations for Plasma

First of all, let us consider totally-ionized plasma under conditions such that it may be treated as an ideal gas. If it was formed from a chemically pure gas, plasma is a binary fluid consisting exclusively of positive ions (subscript i) and electrons (subscript e).

For neutral plasma the electrical charge of the positive ions is equal to minus the charge of the same number of electrons, and hence per kilomole we have

$$M_\mathrm{i}e_\mathrm{i}+M_\mathrm{e}e_\mathrm{e}=0. \tag{6.60}$$

The sum of the diffusion fluxes of all components is zero and hence the ion diffusion flux is thus equal to the electron diffusion flux

$$\boldsymbol{j}_\mathrm{i}=-\boldsymbol{j}_\mathrm{e}. \tag{6.61}$$

The density of the electric current arising as a result of diffusion of electrically charged particles is

$$\boldsymbol{J}_\mathrm{e}=e_\mathrm{i}\boldsymbol{j}_\mathrm{i}+e_\mathrm{e}\boldsymbol{j}_\mathrm{e}=e_\mathrm{e}\left(\frac{M_\mathrm{e}}{M_\mathrm{i}}+1\right)\boldsymbol{j}_\mathrm{e}=e_\mathrm{e}M_\mathrm{e}\left(\frac{1}{M_\mathrm{i}}+\frac{1}{M_\mathrm{e}}\right)\boldsymbol{j}_\mathrm{e}. \tag{6.62}$$

In many cases the molecular mass of ions is much greater than that of electrons, $M_i \gg M_e$, and then equation (6.62) simplifies to

$$J_e = e_e j_e . \tag{6.63}$$

The electron diffusion flux is defined by means of the phenomenological equation

$$j_e = -j_i = -\mathbf{L}_{eq} \cdot \nabla \ln T$$
$$- \mathbf{L}_{ee} \cdot \left\{ (\tilde{v}_e - \tilde{v}_i) \nabla p + \frac{R_e R_i}{R} \frac{T}{x_e x_i} \nabla x_e - e_e \left(1 + \frac{M_e}{M_i}\right) \left[E + \frac{1}{c} (v \times B) \right] \right\} . \tag{6.64}$$

When the heat of transport of electrons $Q_e''^*$, defined by the relation

$$Q_e''^* \mathbf{L}_{ee} = \mathbf{L}_{eq} , \tag{6.65}$$

is introduced, the electron diffusion flux becomes

$$j_e = -\mathbf{L}_{ee} \cdot \left\{ Q_e''^* \nabla \ln T + (\tilde{v}_e - \tilde{v}_i) \nabla p + \frac{R_e R_i}{R} \frac{T}{x_e x_i} \nabla x_e \right.$$
$$\left. - e_e \left(1 + \frac{M_e}{M_i}\right) \left[E + \frac{1}{c} (v \times B) \right] \right\} . \tag{6.66}$$

If only an electromagnetic field acts on the plasma, the electric current density due to the diffusion of charged particles is

$$J_e = e_e^2 \left(\frac{M_e}{M_i} + 1 \right)^2 \mathbf{L}_{ee} \cdot \left[E + \frac{1}{c} (v \times B) \right] = \sigma \cdot \left[E + \frac{1}{c} (v \times B) \right] , \tag{6.67}$$

where electrical conductivity has been defined as

$$\sigma = e_e^2 \left(\frac{M_e}{M_i} + 1 \right)^2 \mathbf{L}_{ee} \approx e_e^2 \mathbf{L}_{ee} . \tag{6.68}$$

In view of the above, the phenomenological coefficient is

$$\mathbf{L}_{ee} \approx \frac{\sigma}{e_e^2} . \tag{6.69}$$

The electron flux can be defined in terms of the electrical conductivity as

$$j_e = -\frac{\sigma}{e_e^2} \cdot \left\{ Q_e''^* \nabla \ln T + (\tilde{v}_e - \tilde{v}_i) \nabla p + \frac{R_e R_i}{R} \frac{T}{x_e x_i} \nabla x_e \right.$$
$$\left. - e_e \left(1 + \frac{M_e}{M_i}\right) \left[E + \frac{1}{c} (v \times B) \right] \right\} . \tag{6.70}$$

Elimination of the thermodynamic force X_e from the system of phenomenological equations

$$J_q''=\mathbf{L}_{qq}\cdot X_q+\mathbf{L}_{qe}\cdot X_e, \tag{6.71}$$

$$j_e=\mathbf{L}_{eq}\cdot X_q+\mathbf{L}_{ee}\cdot X_e, \tag{6.72}$$

leads to

$$J_q''=\mathbf{L}_{qe}\cdot\mathbf{L}_{ee}^{-1}\cdot j_e+(\mathbf{L}_{qq}-\mathbf{L}_{qe}\cdot\mathbf{L}_{eq}\cdot\mathbf{L}_{ee}^{-1})\cdot X_q=\mathbf{L}_{qe}\cdot\mathbf{L}_{ee}^{-1}\cdot j_e-\boldsymbol{\lambda}_\infty\cdot\nabla T, \tag{6.73}$$

where

$$\boldsymbol{\lambda}_\infty=\frac{1}{T}(\mathbf{L}_{qq}-\mathbf{L}_{qe}\cdot\mathbf{L}_{eq}\cdot\mathbf{L}_{ee}^{-1}) \tag{6.74}$$

is the thermal conductivity in the absence of any flow of electric current.

In the case of incompletely ionized plasma, three components are considered: un-ionized atoms (subscript a), positive ions (subscript i), and electrons (subscript e). The phenomenological equations (6.54) and (6.55) in this case take the form: heat flux in the entropy balance equation

$$J_q''=\mathbf{L}_{qq}\cdot X_q+\mathbf{L}_{qi}\cdot X_i+\mathbf{L}_{qe}\cdot X_e, \tag{6.75}$$

diffusion flux of positive ions

$$j_i=\mathbf{L}_{iq}\cdot X_q+\mathbf{L}_{ii}\cdot X_i+\mathbf{L}_{ie}\cdot X_e, \tag{6.76}$$

diffusion flux of electrons

$$j_e=\mathbf{L}_{eq}\cdot X_q+\mathbf{L}_{ei}\cdot X_i+\mathbf{L}_{ee}\cdot X_e. \tag{6.77}$$

The diffusion flux of un-ionized atoms is

$$j_a=-j_i-j_e, \tag{6.78}$$

and hence the diffusion flux of charged particles is

$$j_n=j_e+j_i=-j_a. \tag{6.79}$$

The electric current density for neutral plasma (6.60) is

$$J_e=e_ij_i+e_ej_e=e_e\left(j_e-\frac{M_e}{M_i}j_i\right)=e_eM_e\left(\frac{j_e}{M_e}-\frac{j_i}{M_i}\right). \tag{6.80}$$

In the case of incomplete ionization of plasma the thermodynamic forces are

$$X_q=-\nabla\ln T, \tag{6.81}$$

$$X_i=-(\tilde{v}_i-\tilde{v}_a)\nabla p-p\left(\frac{1}{\rho_i}\nabla z_i-\frac{1}{\rho_a}\nabla z_a\right)+e_i\left[E+\frac{1}{c}(v\times B)\right], \tag{6.82}$$

$$X_e=-(\tilde{v}_e-\tilde{v}_a)\nabla p-p\left(\frac{1}{\rho_e}\nabla z_e-\frac{1}{\rho_a}\nabla z_a\right)+e_e\left[E+\frac{1}{c}(v\times B)\right]. \tag{6.83}$$

The phenomenological equations (6.75) to (6.77) can be transformed by introducing the heats of electron and ion transport, $Q_e''^*$ and $Q_i''^*$, respectively, which for the given ternary system are related by

$$\mathbf{L}_{qe}=\tilde{\mathbf{L}}_{eq}=\mathbf{L}_{ee}\,Q_e''^*+\mathbf{L}_{ie}\,Q_i''^*, \tag{6.84}$$

$$\mathbf{L}_{qi}=\tilde{\mathbf{L}}_{iq}=\mathbf{L}_{ei}\,Q_e''^*+\mathbf{L}_{ii}\,Q_i''^*. \tag{6.85}$$

Consequently, the heat flux in the entropy balance equation is

$$J_q''=Q_e''^*\,j_e+Q_i''^*\,j_i+(\mathbf{L}_{qq}-\mathbf{L}_{eq}\,Q_e''^*-\mathbf{L}_{iq}\,Q_i''^*)\cdot X_q, \tag{6.86}$$

the diffusion flux of positive ions is

$$j_i=\mathbf{L}_{ie}\cdot(X_e+Q_e''^*\,X_q)+\mathbf{L}_{ii}\cdot(X_i+Q_i''^*\,X_q), \tag{6.87}$$

and the diffusion flux of electrons is

$$j_e=\mathbf{L}_{ee}\cdot(X_e+Q_e''^*\,X_q)+\mathbf{L}_{ei}\cdot(X_i+Q_i''^*\,X_q). \tag{6.88}$$

In the stationary state, that is, in the absence of diffusion flows, equation (6.86) yields a relation of the form of Fourier's law

$$J_q''=-\boldsymbol{\lambda}_\infty\cdot\nabla T, \quad j_a=j_i=j_e=0, \tag{6.89}$$

with the thermal conductivity in the stationary state

$$\boldsymbol{\lambda}_\infty=\frac{1}{T}(\mathbf{L}_{qq}-\mathbf{L}_{eq}\,Q_e''^*-\mathbf{L}_{iq}\,Q_i''^*). \tag{6.90}$$

Example 6.3. Find the electric current density in completely ionized, neutral plasma as a function of the gradients of the component pressures.

Solution. The density of the electric current set up by the diffusion of charged particles is determined by equation (6.62)

$$J_e=e_e\,M_e\left(\frac{1}{M_i}+\frac{1}{M_e}\right)j_e. \tag{1}$$

The electron diffusion flux [cf. equation (6.72)] is

$$j_e=\mathbf{L}_{eq}\cdot X_q+\mathbf{L}_{ee}\cdot X_e, \tag{2}$$

and, when the heat of electron transport (6.65) is introduced,

$$j_e=\mathbf{L}_{ee}\cdot(Q_e''^*X_q+X_e). \tag{3}$$

In accordance with equations (6.51) and (6.52), the following may be taken as the thermodynamic forces:

$$X_q=-\nabla\ln T \tag{4}$$

$$X_e=-\frac{1}{\rho_e}\nabla p_e+\frac{1}{\rho_i}\nabla p_i+(e_e-e_i)\left[E+\frac{1}{c}(v\times B)\right]$$

$$=-\frac{1}{\rho_e}\nabla p_e+\frac{1}{\rho_i}\nabla p_i+e_e\,M_e\left(\frac{1}{M_e}+\frac{1}{M_i}\right)\left[E+\frac{1}{c}(v\times B)\right]. \tag{5}$$

Combination of the foregoing equations yields electric current density in the form

$$J_e = e_e M_e \left(\frac{1}{M_i}+\frac{1}{M_e}\right) \mathbf{L}_{ee} \cdot \left\{ -\frac{1}{\rho_e} \nabla p_e + \frac{1}{\rho_i} \nabla p_i + e_e M_e \left(\frac{1}{M_i}+\frac{1}{M_e}\right) \left[E + \frac{1}{c}(\boldsymbol{v} \times \boldsymbol{B})\right] - Q_e''^* \nabla \ln T \right\} \tag{6}$$

or, on introduction of the empirical coefficient (6.69)

$$\sigma \approx e_e^2 \mathbf{L}_{ee} \tag{7}$$

in place of the phenomenological coefficient $\mathbf{L}_{ee}$, also in the form

$$J_e = \frac{\sigma}{e_e} \cdot \left\{ \frac{1}{\rho_i} \nabla p_i - \frac{1}{\rho_e} \nabla p_e + e_e \left[E + \frac{1}{c}(\boldsymbol{v} \times \boldsymbol{B})\right] - Q_e''^* \nabla \ln T \right\} . \tag{8}$$

Example 6.4. Determine the diffusion coefficient D and the thermal diffusion ratio k_T for completely ionized plasma in the absence of an electromagnetic field.

Solution. By equations (6.72), (6.65), and (6.69), the electron flux may be written as

$$j_e = \frac{\sigma}{e_e^2} (Q_e''^* X_q + X_e) \tag{1}$$

with

$$X_q = -\nabla \ln T, \tag{2}$$

$$X_e = -\nabla_T (\mu_e - \mu_i) \tag{3}$$

as the thermodynamic forces.

In the state of mechanical equilibrium

$$x_e \nabla_T \mu_e + x_i \nabla_T \mu_i = 0 \tag{4}$$

by equation (3.206). Since $x_i \gg x_e$, the thermodynamic force given by relation (3) can be written as

$$X_e = -\nabla_T \mu_e = -\left(\frac{\partial \mu_e}{\partial x_e}\right)_{T,p} \nabla x_e \tag{5}$$

and, when relation (3.257)

$$\left(\frac{\partial \mu_e}{\partial x_e}\right)_{T,p} = \frac{1}{x_e} \frac{M^2 R T}{M_e M_i} \tag{6}$$

is taken into account, also as

$$X_e = -\frac{1}{x_e} \frac{M^2 R T}{M_e M_i} \nabla x_e . \tag{7}$$

The equation for the electron diffusion flux now is of the form

$$-j_e = \frac{\sigma}{e_e^2} \left(Q_e''^* \nabla \ln T + \frac{1}{x_e} \frac{M^2 R T}{M_e M_i} \nabla x_e \right) \tag{8}$$

which, upon comparison with relation (3.242),

$$-j_e = j_i = \rho D (k_T \nabla \ln T + \nabla x_e), \tag{9}$$

yields the diffusion coefficient

$$D = \frac{\sigma M^2 RT}{p e_e^2 x_e M_e M_i} \tag{10}$$

and the thermal diffusion ratio

$$k_T = \frac{Q_e^{\prime\prime *} x_e M_e M_i}{M^2 RT} . \tag{11}$$

4. The Hall Effect in Gases

An electromagnetic field acting on plasma causes the Hall effect to appear. As in the case of solids, this is a cross-effect, the essence of which is that a flow of electric current perpendicular to the magnetic displacement sets up a potential difference in the direction perpendicular to both previous vectors (Fig. 6.6). The magnetic field strength $\boldsymbol{B}$ is perpendicular to the paper and directed out of it. Plasma flowing in the direction of the x_α-axis with barycentric velocity $\boldsymbol{v}$ is subjected to an induced force determined by the vector product $\frac{1}{c}(\boldsymbol{v} \times \boldsymbol{B})$. Accordingly, the strength $\boldsymbol{E}$ of a motionless external electric field can be replaced by an electric field strength in

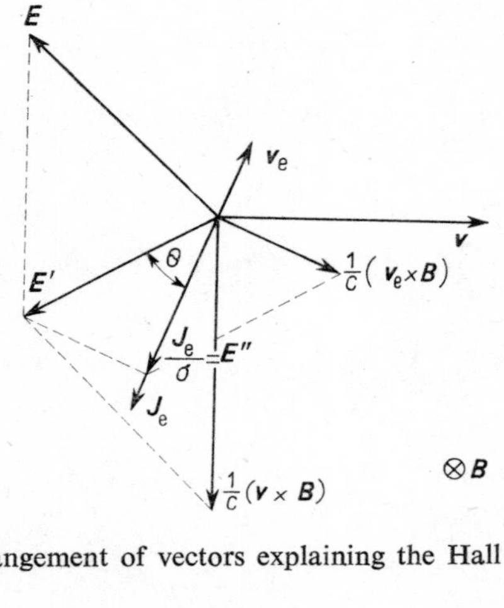

Fig. 6.6 Arrangement of vectors explaining the Hall effect in gases

a coordinate system moving together with the fluid and specified by the resultant of the vectors $\boldsymbol{E}$ and $\frac{1}{c}(\boldsymbol{v} \times \boldsymbol{B})$, that is,

$$\boldsymbol{E}' = \boldsymbol{E} + \frac{1}{c}(\boldsymbol{v} \times \boldsymbol{B}) . \tag{6.91}$$

The current flowing in the direction of vector $\boldsymbol{E}'$ suffers a change under the influence of the current produced by the Hall effect. In actual fact the electrons move with velocity $\boldsymbol{v}_e$ and the electric field strength relative to the moving electrons is given by

$$E'' = E' + \frac{1}{c}(\boldsymbol{v}_e \times \boldsymbol{B}). \tag{6.92}$$

If the scalar electrical conductivity σ of the plasma is introduced and the current transported by the ions is neglected this yields for the electric current density the expression

$$J_e = -\boldsymbol{v}_e \rho_e = \sigma \left[E' + \frac{1}{c}(\boldsymbol{v}_e \times \boldsymbol{B}) \right], \tag{6.93}$$

where ρ_e is the electronic charge density. The last term in this equation, $\frac{\sigma}{c}(\boldsymbol{v}_e \times \boldsymbol{B})$, specifies the Hall current density.

The electrical conductivity σ of the plasma can be expressed in terms of the electron mobility ν_e and electronic charge density ρ_e as

$$\sigma = \nu_e \rho_e, \tag{6.94}$$

whence

$$\sigma \boldsymbol{v}_e = \nu_e \rho_e \boldsymbol{v}_e = -\nu_e J_e, \tag{6.95}$$

and hence the electric current density may also be written as

$$J_e = \sigma E' - \frac{\nu_e}{c}(J_e \times \boldsymbol{B}), \tag{6.96}$$

where the Hall current density is $-\frac{\nu_e}{c}(J_e \times \boldsymbol{B})$.

The Hall angle θ between vectors E' and $E'' = J_e/\sigma$ is determined by

$$\tan\theta = \frac{v_e B \sigma}{J_e c} = \frac{\nu_e}{c} B. \tag{6.97}$$

With the choice of coordinate system as in Fig. 6.6, the respective components of the electric current density are:

$$J_{e\alpha} = \sigma E'_\alpha - \frac{\nu_e}{c} J_{e\beta} B, \tag{6.98}$$

$$J_{e\beta} = \sigma E'_\beta + \frac{\nu_e}{c} J_{e\alpha} B, \tag{6.99}$$

$$J_{e\gamma} = \sigma E'_\gamma. \tag{6.100}$$

If the coordinate system is chosen so that the electric vector is in the $x_\alpha x_\beta$-plane and if the Hall effect coefficient

$$\beta_e = \frac{\nu_e}{c} B \tag{6.101}$$

is introduced, equations (6.98) and (6.99) transform into

$$J_{e\alpha}=\frac{\sigma(E'_\alpha-\beta_e E'_\beta)}{1+\beta_e^2}, \tag{6.102}$$

$$J_{e\beta}=\frac{\sigma(E'_\beta+\beta_e E'_\alpha)}{1+\beta_e^2}. \tag{6.103}$$

The conductivity tensor $\boldsymbol{\sigma}$ of ionized gas in a magnetic field then is of the form

$$\boldsymbol{\sigma}=\begin{vmatrix} \frac{\sigma}{1+\beta_e^2} & -\frac{\sigma\beta_e}{1+\beta_e^2} & 0 \\ \frac{\sigma\beta_e}{1+\beta_e^2} & \frac{\sigma}{1+\beta_e^2} & 0 \\ 0 & 0 & 0 \end{vmatrix} \tag{6.104}$$

The Hall effect is negligible at sufficiently low electron and ion mobilities. The electron mobility is inversely proportional to the pressure and hence the Hall effect may be diminished by increasing the gas pressure.

In the microscopic approach the Hall effect may be explained as follows. Suppose that an electron moves with velocity $\boldsymbol{v}_e$ perpendicular to the magnetic field strength $\boldsymbol{B}$ in the absence of an electric field. Between two consecutive collisions the electron moves along a path which is an arc of a circle. The radius r of this circle can be determined by comparing the magnetic force and the centrifugal force which balance it,

$$\epsilon v_e\frac{B}{c}=\frac{m_e v_e^2}{r}, \qquad r=\frac{m_e v_e c}{\epsilon B},$$

where m_e is the electron mass, and ϵ is the electron charge.

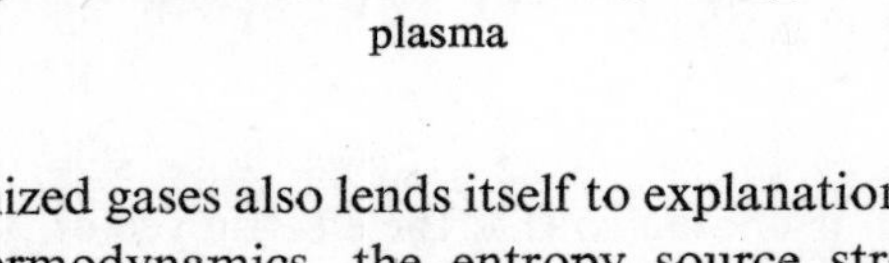

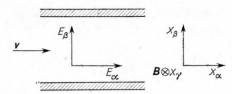

Fig. 6.7 Electric and magnetic fields assumed in deriving entropy source strength formula for plasma

The Hall effect in ionized gases also lends itself to explanation by means of methods of nonequilibrium thermodynamics, the entropy source strength being taken as the point of departure.

Consider plasma (Fig. 6.7) flowing with velocity $\boldsymbol{v}$ along the x_α-axis into a space in which there is a constant magnetic field of induction $\boldsymbol{B}$ directed along the x_γ-axis.

The electric field acts only in the $x_\alpha x_\beta$-plane, that is, the electric field strength has only the components E_α and E_β. The effect will be considered in a coordinate system which is moving along with the fluid; this enables the fluid to be treated as being motionless and the magnetic field as moving with velocity $\boldsymbol{v}$.

Plasma consists of neutral molecules, ions, and electrons. To simplify the reasoning, the ion and electron fluxes will be assumed to depend mainly on the field forces and not on the gradients of the intensive parameters of the plasma. Thus, the flows of charged particles will be considered as being not connected with the flows of neutral particles.

By our earlier considerations, the entropy source strength for the case in question takes on the form

$$\Phi_s = j_{i\alpha}\frac{\partial\left(\frac{\mu_i'}{T}\right)}{\partial x_\alpha} + j_{i\beta}\frac{\partial\left(\frac{\mu_i'}{T}\right)}{\partial x_\beta} + j_{e\alpha}\frac{\partial\left(\frac{\mu_e'}{T}\right)}{\partial x_\alpha} + j_{e\beta}\frac{\partial\left(\frac{\mu_e'}{T}\right)}{\partial x_\beta}, \tag{6.105}$$

where j_i and j_e are the ion and electron fluxes, respectively, and μ_i' and μ_e' are the electrochemical potentials of ions and electrons.

In view of the assumption that the fluid element under consideration is isothermal, it is possible to go over directly to the dissipation function

$$\Psi = T\Phi_s = j_{i\alpha}\frac{\partial\mu_i'}{\partial x_\alpha} + j_{i\beta}\frac{\partial\mu_i'}{\partial x_\beta} + j_{e\alpha}\frac{\partial\mu_e'}{\partial x_\alpha} + j_{e\beta}\frac{\partial\mu_e'}{\partial x_\beta}. \tag{6.106}$$

If the magnetic field is absent or motionless, the change in the electrochemical potential of an isothermal medium is equal to the change in electrical potential. On the other hand, if the magnetic field moves with velocity $\boldsymbol{v}$ in the x_α-direction, a charged particle is acted on by a force of $\frac{1}{c}vB$ in the direction of the x_β-axis, the result being that the effective electric field strength in the x_β-direction is $E_\beta + \frac{1}{c}vB$ but does not change in the x_α-direction, remaining equal to E_α. For this reason the electric potential gradients of the ions and electrons, respectively, are

$$\frac{\partial\mu_i'}{\partial x_\alpha} = e_e E_\alpha, \tag{6.107}$$

$$\frac{\partial\mu_i'}{\partial x_\beta} = e_e\left(E_\beta + \frac{1}{c}vB\right), \tag{6.108}$$

$$\frac{\partial\mu_e'}{\partial x_\alpha} = -e_e E_\alpha, \tag{6.109}$$

$$\frac{\partial\mu_e'}{\partial x_\beta} = -e_e\left(E_\beta + \frac{1}{c}vB\right). \tag{6.110}$$

It is also convenient to introduce an electric current density with the components

$$J_{e\alpha}=e_e(j_{i\alpha}-j_{e\alpha}), \tag{6.111}$$

$$J_{e\beta}=e_e(j_{i\beta}-j_{e\beta}). \tag{6.112}$$

The foregoing relations can be used to reduce the dissipation function to the form

$$\Psi=J_{e\alpha}E_\alpha+J_{e\beta}\left(E_\beta+\frac{1}{c}vB\right), \tag{6.113}$$

to which the phenomenological equations

$$J_{e\alpha}=L_{11}E_\alpha+L_{12}\left(E_\beta+\frac{1}{c}vB\right), \tag{6.114}$$

$$J_{e\beta}=L_{21}E_\alpha+L_{22}\left(E_\beta+\frac{1}{c}vB\right) \tag{6.115}$$

correspond.

The phenomenological coefficients satisfy the Onsager–Casimir relations

$$L_{12}(B)=L_{21}(-B). \tag{6.116}$$

If no electric current flows in the direction of the plasma flow, that is, if $J_{e\alpha}=0$, then equation (6.114) yields

$$E_\alpha=-\frac{L_{12}}{L_{11}}\left(E_\beta+\frac{1}{c}vB\right) \tag{6.117}$$

which, on insertion into equation (6.115), gives the electric current density in the direction perpendicular to the plasma flux

$$J_{e\beta}=\left(L_{22}-\frac{L_{12}L_{21}}{L_{11}}\right)\left(E_\beta+\frac{1}{c}vB\right). \tag{6.118}$$

This current vanishes at an electric field strength of

$$E_\beta=-\frac{vB}{c}. \tag{6.119}$$

5. Magnetohydrodynamic Generators

The operating principle of the magnetohydrodynamic (MHD) generator is very simple, consisting as it does in obtaining an electric current directly from the kinetic energy of a partially ionized gas. This current arises from the motion of charge in an external magnetic field. The MHD generator differs from the dynamo in that the flow of current through a moving metal conductor has been replaced by the flow of plasma through a motionless channel. Owing to the action of Lorentz forces, particles with opposite charges move towards different electrodes connected into the electrical circuit. Electric current flows in the MHD generator perpendicular to the direction of both the plasma flow and the magnetic induction (Fig. 6.8). Under the influence of the interaction between the external magnetic field and the

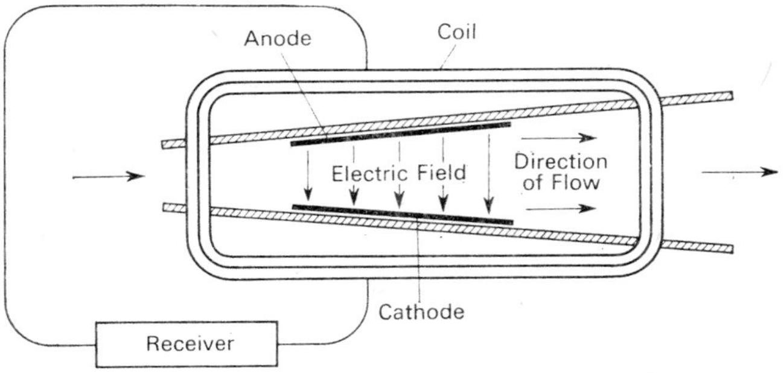

Fig. 6.8 Diagram of straight-channel magnetohydrodynamic generator

ionized particles of the plasma, the plasma velocity decreases as the electric current is drawn from the electrodes. The inverse effect can also be achieved. Application of a current to the electrodes causes the gas flow to be accelerated. The three main types of MHD generators are considered below, with the Hall effect being taken into account.

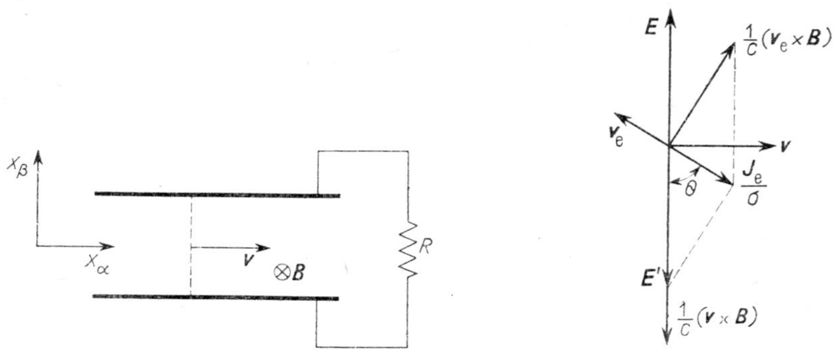

Fig. 6.9 Continuous-electrode MHD generator

Continuous-electrode MHD Generator (Fig. 6.9). As a result of the Hall effect the flow of electrons is deflected from the x_β-direction perpendicular to the electrode surface. In this arrangement the components of the electric field are $E_\alpha=0$, $E'_\alpha=0$, and $E'_\beta=E_\beta-\dfrac{1}{c}vB$. If

$$\alpha=\frac{cE_\beta}{vB} \tag{6.120}$$

denotes the ratio of the strength of the acting electric field to the magnetic induction for the open circuit, the components of the electric current density can be written as [equations (6.102) and (6.103)]

$$J_{e\alpha}=\frac{\sigma vB\beta_e(1-\alpha)}{c(1+\beta_e^2)}, \tag{6.121}$$

$$J_{e\beta}=-\frac{\sigma vB(1-\alpha)}{c(1+\beta_e^2)}. \tag{6.122}$$

The power generated per unit volume of plasma in this case amounts to

$$P=-E_\beta J_{e\beta}=\frac{\sigma v^2B^2\alpha(1-\alpha)}{c^2(1+\beta_e^2)}. \tag{6.123}$$

The ratio of the power conveyed to the electrodes per unit volume of plasma to the power used to ensure plasma flow is the electrical efficiency of the generator. In the case in question the power supplied is

$$J_{e\beta}\,v\,\frac{B}{c}=\frac{\sigma v^2B^2(1-\alpha)}{c^2(1+\beta_e^2)}. \tag{6.124}$$

Hence the electrical efficiency of the continuous-electrode generator is

$$\eta_e=\frac{E_\beta J_{e\beta}c}{J_{e\beta}vB}=\alpha. \tag{6.125}$$

Segmented-electrode MHD Generator (Fig. 6.10). The undesirable influence the Hall effect has on the generation of electricity can be eliminated by segmenting the

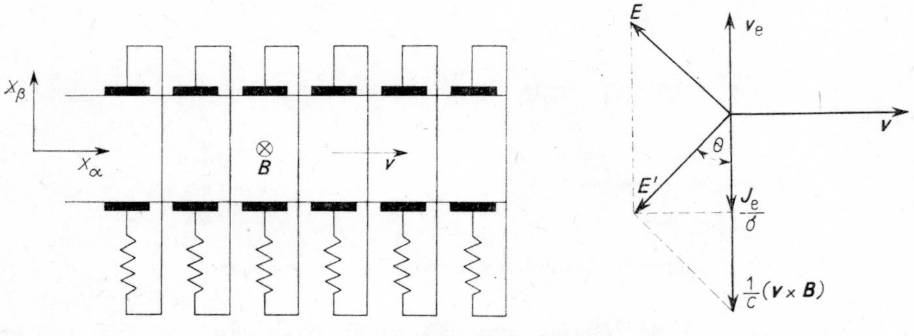

Fig. 6.10 Segmented-electrode MHD generator

electrodes, in which case the components of the electric field strength are

$$E'_\alpha=\beta_e E'_\beta, \quad E_\alpha=\beta_e\left(E_\beta-\frac{1}{c}\,vB\right), \tag{6.126}$$

and the components of the electric current density are

$$J_{e\alpha}=0, \quad J_{e\beta}=-\sigma\left(E_\beta-\frac{1}{c}\,vB\right)=\frac{\sigma vB}{c}(1-\alpha). \tag{6.127}$$

The power generated per unit volume of plasma in this case amounts to

$$P=\frac{\sigma v^2 B^2 \alpha}{c^2}(1-\alpha), \tag{6.128}$$

which corresponds fully to the previous case when the Hall effect is neglected [equation (6.124) with Hall coefficient $\beta_e=0$]. The electrical efficiency of the generator is the same as in the previous case, that is to say, is given by equation (6.125).

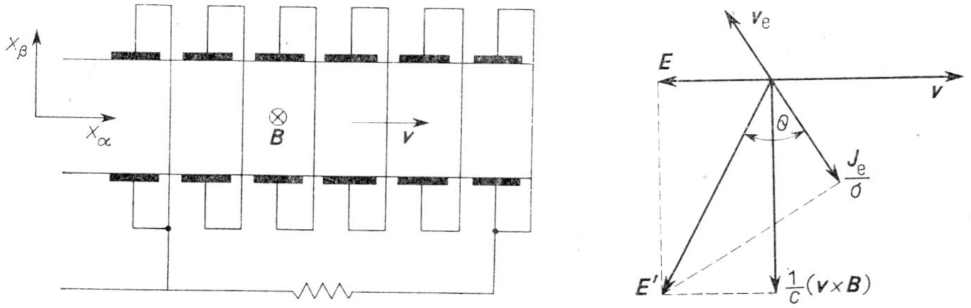

Fig. 6.11 Hall generator

Hall Generator (Fig. 6.11). Instead of eliminating its influence on the generation of electrocity, one can use the Hall effect to perform work in an electrical circuit in the manner shown in the drawing. In this case the respective components of the electric field strength are

$$E_\beta=0, \quad E'_\alpha=E_\alpha, \quad E'_\beta=-\frac{1}{c}vB, \tag{6.129}$$

whereas the components of the electric current density are

$$J_{e\alpha}=\frac{\sigma\left(E_\alpha+\frac{1}{c}\beta_e vB\right)}{1+\beta_e^2}, \tag{6.130}$$

$$J_{e\beta}=\frac{\sigma\left(\beta_e E_\alpha-\frac{1}{c}vB\right)}{1+\beta_e^2}. \tag{6.131}$$

For an open electrical circuit, since $J_{e\alpha}=0$,

$$E_{\alpha 0}=-\frac{1}{c}\beta_e vB, \quad J_{e\beta 0}=-\frac{1}{c}\sigma vB, \tag{6.132}$$

and hence the parameter α specifying the ratio of the electric field strength in the x_α-direction with the circuit closed to that with the circuit open is

$$\alpha=\frac{E_\alpha}{E_{\alpha 0}}=-\frac{cE_\alpha}{\beta_e vB}. \tag{6.133}$$

The power generated per unit volume of plasma is now equal to

$$P = \frac{\sigma v^2 B^2 \alpha \beta_e^2 (1-\alpha)}{c^2(1+\beta_e^2)}. \tag{6.134}$$

Now, the power consumed is

$$-J_{e\beta}\frac{vB}{c} = \frac{\sigma v^2 B^2 (1+\alpha\beta_e^2)}{c^2(1+\beta_e^2)}, \tag{6.135}$$

whereby the electrical efficiency of the Hall generator is

$$\eta_e = \frac{\alpha\beta_e^2(1-\alpha)}{1+\alpha\beta_e^2}. \tag{6.136}$$

In order to compare the properties of the three types of magnetohydrodynamic generators discussed above, the dimensionless power

$$P^* = \frac{Pc^2}{\sigma v^2 B^2} \tag{6.137}$$

has been plotted against the electrical efficiency η_e for each type in Fig. 6.12.

In summary, it may be said that these three principal types of MHD generators with straight channels have the following advantages and disadvantages.

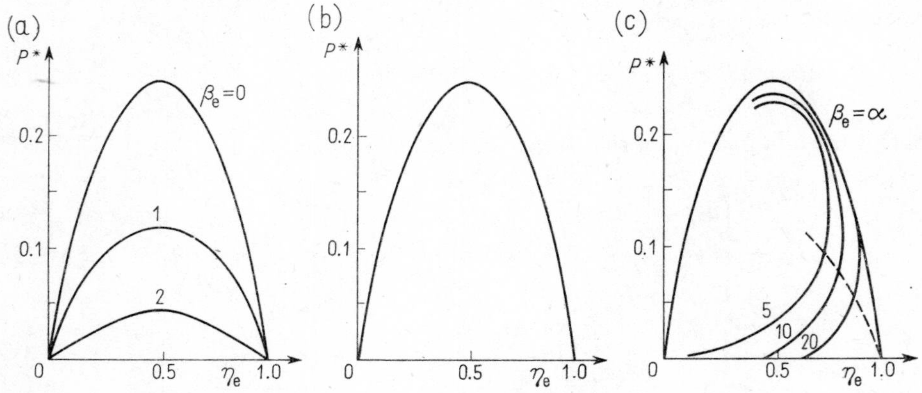

Fig. 6.12 Dimensionless power P^* vs. the electrical efficiency η_e of a MHD generator with (a) continuous and (b) segmented electrodes, and of a Hall generator (c)

The continuous-electrode generator is simple in construction and insulating its electrodes presents no major difficulty. A disadvantage is that it has a small power output P per unit volume of plasma for a large Hall coefficient β_e.

The segmented-electrode generator is characterized by higher unit power output P, by the occurrence of only axial forces, and by the fact that it can be employed at moderate Hall coefficients β_e. However, it has the drawbacks of a complicated electrical circuit and a large potential gradient in the axial direction.

The advantages of the Hall generator include a single electrical circuit, applicability for large Hall coefficients, and large currents at relatively low voltages. Its disadvantages are low unit power output *P*, a limited low maximum efficiency, and a large potential gradient in the axial direction.

6. Design Problems of MHD Generators

Under ordinary conditions gases are poor conductors of electricity. The gas flowing through the channel of an MHD generator should have a high electrical conductivity, that is, should be ionized. Even with an ionization of only 1% the electrical conductivity of a gas (Fig. 6.13) exceeds 90% of the maximum conductivity corresponding to the complete ionization of the gas. This is due to the fact that electrons have a mass many times smaller than that of positive ions, and have a greater mobility than the ions, and consequently the flow of current through the gas depends mainly upon the motion of free electrons. As the temperature of the gas rises, its conductivity increases at first rather substantially owing to the growth in the degree of ionization and hence also in the density of free electrons. With the degree of ionization a mere 0.1% the electrical conductivity of the gas is equal to almost half the maximum, this being enough for using this gas in the channel of an MHD generator. At higher degrees of ionization the increase in electrical conductivity begins to taper off as a result of the greater number of collisions between electrons and positive ions and the consequent association of charged particles into neutral particles.

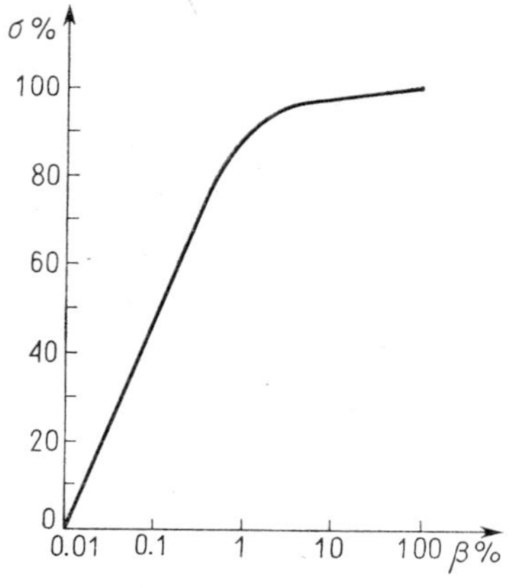

Fig. 6.13 Dependence of electrical conductivity of a gas on the degree of ionization

Materials used to construct MHD generators exhibit a limited resistance to high temperatures and the electrical resistance of insulating materials deteriorates at high temperatures. As a consequence of this, the temperature of the walls and electrodes of the channel must not significantly exceed 2500 K. Ordinary gases are very poorly ionized at this temperature. As is known, a very small addition of alkali metals (sodium, potassium, caesium) to the gas (argon, helium, air, combustion gases) increase the conductivity of the gas considerably. For instance, it is sufficient to add 1% potassium to argon at 2700 K, 1% caesium to helium at 2500 K and 1% potassium to combustion gases at 2900 K. Alkali metals have a single valence electron (in the outermost orbit) which is readily detached from the atom under a relatively small ionization energy. For this reason gases with traces of alkali metals as additives are used to obtain plasma in magnetohydrodynamic generators. Besides forming charge carriers, an easily ionizable additive makes it possible for collisions to occur between free electrons and electrons bound with the atoms of the principal gas and the subsequent detachment of the bound electrons results in a higher degree of ionization.

To diminish the impact of the Hall effect, one should reduce the mean free path of particles between collisions. This is done by raising the plasma pressure and choosing a high enough electron pressure. The Hall effect prevents the pressure of the working gas from being decreased to such a value that significant ionization of the gas could be attained through collisions. Hence, there are only two ways of increasing the degree of ionization: employing high temperatures, and seeding with easily ionizable additives.

The most difficult problem to be solved in order to develop truly useful MHD generators is that of obtaining suitable materials that would be resistant to high temperatures and abrupt changes in temperature, and resistant to the corrosive action of the alkali metal additives used to seed the gases. If the generator is to be used in conjunction with a nuclear reactor, these materials must also be resistant to the action of gamma rays and neutrons. As for electrodes, there is the additional requirement of good electrical conductivity and a high electron emissivity (which is a characteristic of tungsten, for instance).

The temperature of 2700 K or thereabouts to which the combustion gases must be heated if the MHD generator is to operate properly requires the use of oxygen or suitably preheated air for the combustion of hydrocarbon fuels. The first method of raising the temperature of combustion gases is uneconomical. In the second case air must be preheated to a temperature in the vicinity of 1400 K and this entails overcoming difficulties encountered in building heat exchangers for use at such high temperatures.

CHAPTER VII

NONEQUILIBRIUM EFFECTS IN DISCONTINUOUS SYSTEMS WITHOUT CHEMICAL REACTIONS

1. The Entropy Balance Equation

Instances of nonequilibrium transport effects in discontinuous (heterogeneous) systems are phenomena involving transport of substance, heat, and electrical energy between two parts of the same system, henceforth called phases (Fig. 7.1). In the case under consideration here the phases will be not only uniform parts of the system but also homogeneous, that is to say, the values of the intensive parameters will be the same throughout the volume of each phase. The two phases constitute a closed system which does not exchange substance with the environment, whereas every phase is an open system, that is, substance may be transported from one phase to another. No chemical reactions will occur, either within a single phase or between the phases. Each phase may contain k components which are labelled by the running index i. Phases in different states of matter have a natural interface, whereas liquid or gas phases must be separated by means of a porous diaphragm, capillary, small aperture, or semipermeable membrane.

The postulate of local thermodynamic equilibrium is in this case replaced by the requirement that the variations in the intensive parameters be slow enough for each of the phases to be in thermodynamic equilibrium at every instant. The intensive parameters of each phase are only functions of time, whereas at the interface they may change by jumps.

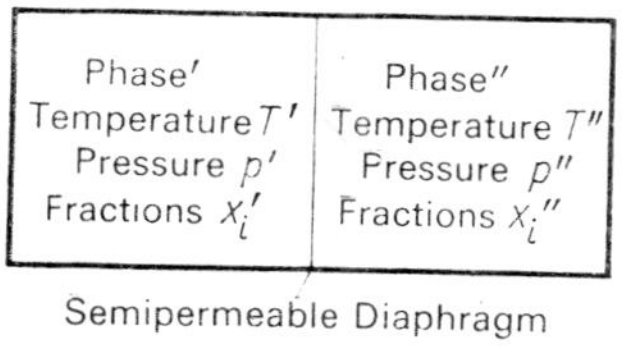

Fig. 7.1 Multicomponent fluids separated by a semipermeable diaphragm

Quantities concerning the first phase will be indicated by a prime whereas those pertaining to the second phase will be indicated by a double prime.

The conservation law for the amount of substance of component i of the given discontinuous system without chemical reactions implies that

$$\mathrm{d}m_i' + \mathrm{d}m_i'' = 0. \tag{7.1}$$

The amount of substance transported per unit time between the two subsystems is characterized by the flow of component i from the second phase to the first:

$$J_i = \frac{\mathrm{d}m_i'}{\mathrm{d}t} = -\frac{\mathrm{d}m_i''}{\mathrm{d}t}. \tag{7.2}$$

The equation expressing the first law of thermodynamics for an open system can be written per unit mass in the form

$$đq = \mathrm{d}u + p\,\mathrm{d}v. \tag{7.3}$$

Since

$$đQ = m\,đq, \quad \mathrm{d}U = m\,\mathrm{d}u + u\,\mathrm{d}m, \quad \mathrm{d}V = m\,\mathrm{d}v + v\,\mathrm{d}m, \quad h = u + pv,$$

for the entire volume of the subsystem

$$đQ' = \mathrm{d}U' + p'\mathrm{d}V' - h'\mathrm{d}m'. \tag{7.4}$$

The elementary heat $đQ'$ consists of the heat of interaction $đQ_e'$ of the system with the environment and the heat $đQ_i'$ resulting from the actions between the two subsystems, so that

$$đQ' = đQ_e' + đQ_i'. \tag{7.5}$$

At the same time it should also be taken into account that since there are many components in the system

$$h'\mathrm{d}m' = \sum_{i=1}^{k} \tilde{h}_i'\,\mathrm{d}m_i'.$$

Finally, the equation expressing the first law of thermodynamics for the first phase becomes

$$đQ_e' + đQ_i' = \mathrm{d}U' + p'\mathrm{d}V' - \sum_{i=1}^{k} \tilde{h}_i'\,\mathrm{d}m_i', \tag{7.6}$$

and for the second phase

$$đQ_e'' + đQ_i'' = \mathrm{d}U'' + p''\mathrm{d}V'' - \sum_{i=1}^{k} \tilde{h}_i''\mathrm{d}m_i'', \tag{7.7}$$

whereas for the whole system comprising both phases it takes the form

$$đQ_e = đQ_e' + đQ_e'' = \mathrm{d}U' + \mathrm{d}U'' + p'\mathrm{d}V' + p''\mathrm{d}V''. \tag{7.8}$$

Comparison of expressions (7.6), (7.7), and (7.8) yields

$$đQ_i' + đQ_i'' + \sum_{i=1}^{k} \tilde{h}_i'\,\mathrm{d}m_i' + \sum_{i=1}^{k} \tilde{h}_i''\mathrm{d}m_i'' = 0. \tag{7.9}$$

The heat flow from the second phase to the first is

$$J_q'' = \frac{đQ_i'}{\mathrm{d}t} = -\frac{đQ_i''}{\mathrm{d}t} + \sum_{i=1}^{k} (\tilde{h}_i'' - \tilde{h}_i')\,J_i. \tag{7.10}$$

Since each phase is in thermodynamic equilibrium, the Gibbs relation can be applied to them in the form

$$T'\mathrm{d}S' = \mathrm{d}U' + p'\mathrm{d}V' - \sum_{i=1}^{k} \mu_i' \,\mathrm{d}m_i', \tag{7.11}$$

$$T''\mathrm{d}S'' = \mathrm{d}U'' + p''\mathrm{d}V'' - \sum_{i=1}^{k} \mu_i'' \mathrm{d}m_i''. \tag{7.12}$$

The equations above contain the expressions

$$\mathrm{d}U' + p'\mathrm{d}V' = \text{đ}Q_\mathrm{e}' + \text{đ}Q_\mathrm{i}' + \sum_{i=1}^{k} \tilde{h}_i' \,\mathrm{d}m_i', \tag{7.13}$$

$$\begin{aligned} \mathrm{d}U'' + p''\mathrm{d}V'' &= \text{đ}Q_\mathrm{e}'' + \text{đ}Q_\mathrm{i}'' + \sum_{i=1}^{k} \tilde{h}_i''\mathrm{d}m_i'' \\ &= \text{đ}Q_\mathrm{e}'' - (\text{đ}Q_\mathrm{i}' + \sum_{i=1}^{k} \tilde{h}_i' \,\mathrm{d}m_i'). \end{aligned} \tag{7.14}$$

The change in the entropy of the system is equal to the sum of the changes in the entropies of the two phases which constitute subsystems, and by equations (7.11) to (7.14) the result is

$$\begin{aligned} \mathrm{d}S &= \mathrm{d}S' + \mathrm{d}S'' \\ &= \frac{\text{đ}Q_\mathrm{e}'}{T'} + \frac{\text{đ}Q_\mathrm{e}''}{T''} + \left(\frac{1}{T'} - \frac{1}{T''}\right)(\text{đ}Q_\mathrm{i}' + \sum_{i=1}^{k} \tilde{h}_i' \,\mathrm{d}m_i') \\ &\quad - \sum_{i=1}^{k} \left(\frac{\mu_i'}{T'} - \frac{\mu_i''}{T''}\right) \mathrm{d}m_i'. \end{aligned} \tag{7.15}$$

The time variation of the entropy of the system,

$$\frac{\mathrm{d}S}{\mathrm{d}t} = \frac{\mathrm{d}S_\mathrm{e}}{\mathrm{d}t} + \frac{\mathrm{d}S_\mathrm{i}}{\mathrm{d}t}, \tag{7.16}$$

is due to the entropy flow transported from the environment to the system,

$$\frac{\mathrm{d}S_\mathrm{e}}{\mathrm{d}t} = \frac{1}{T'}\frac{\text{đ}Q_\mathrm{e}'}{\mathrm{d}t} + \frac{1}{T''}\frac{\text{đ}Q_\mathrm{e}''}{\mathrm{d}t}, \tag{7.17}$$

and also to the entropy production inside the system per unit time,

$$\begin{aligned} \Phi_s &= \frac{\mathrm{d}S_\mathrm{i}}{\mathrm{d}t} \\ &= (J_q'' + \sum_{i=1}^{k} \tilde{h}_i' J_i)\left(\frac{1}{T'} - \frac{1}{T''}\right) - \sum_{i=1}^{k} J_i \left(\frac{\mu_i'}{T'} - \frac{\mu_i''}{T''}\right) \geqslant 0. \end{aligned} \tag{7.18}$$

By the principle of entropy increase this is a positive quantity for all nonequilibrium processes and zero for reversible processes.

Henceforth the discussion will be confined to small differences in the intensive parameters in both phases:

$$|\Delta T| = |T'' - T'| \ll T = T',$$
$$|\Delta\mu_i| = |\mu_i'' - \mu_i'| \ll \mu_i = \mu_i'\,; \tag{7.19}$$

this will permit a dissipation function in the form

$$\Psi = T\Phi_s = (J_q'' + \sum_{i=1}^{k} \tilde{h}_i J_i)\frac{\Delta T}{T} + \sum_{i=1}^{k} J_i T \Delta\left(\frac{\mu_i}{T}\right) \tag{7.20}$$

to be introduced. If use is made of the relations

$$T\Delta\left(\frac{\mu_i}{T}\right) = -\frac{\tilde{h}_i}{T}\Delta T + \Delta_T \mu_i, \tag{7.21}$$

$$\Delta_T \mu_i = \tilde{v}_i \Delta p + \sum_{j=1}^{k-1}\left(\frac{\partial \mu_i}{\partial x_j}\right)_{T,\,p,\,x_{m\neq j}} \Delta x_j, \tag{7.22}$$

the result is

$$\Psi = J_q'' \frac{\Delta T}{T} + \sum_{i=1}^{k} J_i \Delta_T \mu_i$$
$$= J_q'' \frac{\Delta T}{T} + \sum_{i=1}^{k} J_i [\tilde{v}_i \Delta p + \sum_{j=1}^{k-1}\left(\frac{\partial \mu_i}{\partial x_j}\right)_{T,\,p,\,x_{m\neq j}} \Delta x_j]. \tag{7.23}$$

2. Heat and Mass Transport in Discontinuous Multicomponent Systems

The dissipation function (7.23) introduced above makes it possible to choose thermodynamic forces of the form

$$X_q = \frac{\Delta T}{T}, \tag{7.24}$$

$$X_i = \tilde{v}_i \Delta p + \sum_{j=1}^{k-1}\left(\frac{\partial \mu_i}{\partial x_j}\right)_{T,\,p,\,x_{m\neq j}} \Delta x_j \quad (i = 1, 2, 3, \ldots, k), \tag{7.25}$$

conjugated with, respectively, the heat and substance flows which can be determined by means of the phenomenological equations

$$J_q'' = L_{qq} X_q + \sum_{i=1}^{k} L_{qi} X_i$$
$$= L_{qq}\frac{\Delta T}{T} + \sum_{i=1}^{k} L_{qi}\left[\tilde{v}_i \Delta p + \sum_{j=1}^{k-1}\left(\frac{\partial \mu_i}{\partial x_j}\right)_{T,\,p,\,x_{m\neq j}} \Delta x_j\right], \tag{7.26}$$

$$J_m = L_{mq} X_q + \sum_{i=1}^{k} L_{mi} X_i$$

$$= L_{mq} \frac{\Delta T}{T} + \sum_{i=1}^{k} L_{mi} \left[\tilde{v}_i \Delta p + \sum_{j=1}^{k-1} \left(\frac{\partial \mu_i}{\partial x_j} \right)_{T,\, p,\, x_{m \neq j}} \Delta x_j \right] \tag{7.27}$$

$$(m = 1, 2, 3, \ldots, k).$$

By the Onsager reciprocal relations the matrix of phenomenological coefficients is symmetric

$$L_{qi} = L_{iq}, \qquad L_{mi} = L_{im} \qquad (i,\ m = 1, 2, 3, \ldots, k). \tag{7.28}$$

The dissipation function is a positive quantity

$$\Psi = J_q X_q + \sum_{i=1}^{k} J_i X_i$$

$$= L_{qq} X_q^2 + \sum_{i=1}^{k} (L_{iq} + L_{qi}) X_i X_q + \sum_{i,\, m=1}^{k} L_{mi} X_m X_i > 0, \tag{7.29}$$

and hence the phenomenological coefficients with the same subscripts must be positive,

$$L_{qq} > 0, \qquad L_{ii} > 0 \qquad (i = 1, 2, 3, \ldots, k) \tag{7.30}$$

and coefficients with different subscripts must satisfy the inequality

$$L_{qq} L_{ii} - L_{qi}^2 > 0, \qquad L_{mm} L_{ii} - L_{mi}^2 > 0 \qquad (i,\ m = 1, 2, 3, \ldots, k). \tag{7.31}$$

The isothermal heats of transport $Q_i''^*$ of the components can be introduced into the phenomenological equations, these heats being defined by the relations

$$L_{mq} = \sum_{i=1}^{k} L_{mi} Q_i''^* \qquad (\Delta T = 0) \qquad (m = 1, 2, 3, \ldots, k). \tag{7.32}$$

In this case equation (7.27) yields the substance flow of component m in the form

$$J_m = \sum_{i=1}^{k} L_{mi} (X_i + Q_i''^* X_q) \qquad (m = 1, 2, 3, \ldots, k). \tag{7.33}$$

By equation (7.26) the heat flow is

$$J_q'' = L_{qq} X_q + \sum_{i,\, m=1}^{k} L_{im} Q_m''^* X_i$$

$$= (L_{qq} - \sum_{m=1}^{k} L_{mq} Q_m''^*) X_q + \sum_{m=1}^{k} Q_m''^* J_m. \tag{7.34}$$

In isothermal flow ($\Delta T = 0$) the thermodynamic force $X_q = \Delta T / T$ vanishes and the heat of transport of component m then is

$$Q_m''^* = \left(\frac{J_q''}{J_m} \right)_{J_{i \neq m} = 0,\ \Delta T = 0}, \tag{7.35}$$

that is, it is the heat transported per unit flow of component m in the absence of the flow of other components ($J_{i\neq m}=0$) and in the absence of any temperature difference ($\Delta T=0$).

The explicit form of the thermodynamic forces (7.24) and (7.25) can be used to rewrite equation (7.33) as

$$J_m=\sum_{i=1}^{k} L_{mi}\left[\tilde{v}_i\,\Delta p+\sum_{n=1}^{k-1}\left(\frac{\partial\mu_i}{\partial x_n}\right)_{T,\,p,\,x_{j\neq n}}\Delta x_n+\frac{Q_i''^*}{T}\,\Delta T\right]$$

$$(m=1,2,3,\ldots,k)\,. \tag{7.36}$$

The substantial flows obtained for a binary fluid are

$$J_1=L_{11}\left[\tilde{v}_1\,\Delta p+\left(\frac{\partial\mu_1}{\partial x_1}\right)_{T,\,p}\Delta x_1+\frac{Q_1''^*}{T}\,\Delta T\right]$$

$$+L_{12}\left[\tilde{v}_2\,\Delta p+\left(\frac{\partial\mu_2}{\partial x_1}\right)_{T,\,p}\Delta x_1+\frac{Q_2''^*}{T}\,\Delta T\right] \tag{7.37}$$

$$J_2=L_{21}\left[\tilde{v}_1\,\Delta p+\left(\frac{\partial\mu_1}{\partial x_1}\right)_{T,\,p}\Delta x_1+\frac{Q_1''^*}{T}\,\Delta T\right]$$

$$+L_{22}\left[\tilde{v}_2\,\Delta p+\left(\frac{\partial\mu_2}{\partial x_1}\right)_{T,\,p}\Delta x_1+\frac{Q_2''^*}{T}\,\Delta T\right]. \tag{7.38}$$

In a binary fluid the heat of transport (7.32) is related to the phenomenological coefficients by

$$L_{1q}=L_{11}\,Q_1''^*+L_{12}\,Q_2''^*\,, \tag{7.39}$$

$$L_{2q}=L_{21}\,Q_1''^*+L_{22}\,Q_2''^*=L_{12}\,Q_1''^*+L_{22}\,Q_2''^*, \tag{7.40}$$

where, in the last equation, account has been taken of the Onsager reciprocal relation

$$L_{21}=L_{12}\,. \tag{7.41}$$

The unknown heats of transport $Q_1''^*$ and $Q_2''^*$ are calculated from the system of two linear equations (7.39) and (7.40) as

$$Q_1''^*=\frac{\begin{vmatrix}L_{1q} & L_{12}\\ L_{2q} & L_{22}\end{vmatrix}}{\begin{vmatrix}L_{11} & L_{12}\\ L_{12} & L_{22}\end{vmatrix}},\qquad Q_2''^*=\frac{\begin{vmatrix}L_{11} & L_{1q}\\ L_{12} & L_{2q}\end{vmatrix}}{\begin{vmatrix}L_{11} & L_{12}\\ L_{12} & L_{22}\end{vmatrix}}\,. \tag{7.42}$$

The two phases may be imagined to be homogeneously mixed at the beginning of the process so that $\Delta p=0$, $\Delta x_j=0$, the thermodynamic force $X_i=0$ [equation (7.25)], and the two phases differ only as to temperature, $\Delta T\neq 0$. In this case the heat flow (7.26) is

$$J_q''=L_{qq}\,\frac{\Delta T}{T}=\lambda_0\,\frac{A}{\delta}\,\Delta T\,, \tag{7.43}$$

where A is the area of the flow cross section, and δ is the thickness of the porous membrane or the length of the conduit between the phases.

With the components homogeneously mixed, the thermal conductivity is

$$\lambda_0 = \frac{\delta}{A} \frac{L_{qq}}{T}. \tag{7.44}$$

In the stationary state, when the flows of substance vanish ($J_i = 0$), by equation (7.34) the heat flow is limited to

$$J_q'' = (L_{qq} - \sum_{i=1}^{k} L_{iq} Q_i''^*) \frac{\Delta T}{T} = \lambda_\infty \frac{A}{\delta} \Delta T. \tag{7.45}$$

The thermal conductivity λ_∞ in the stationary state is related to the phenomenological coefficients by

$$\lambda_\infty = \frac{\delta}{AT} (L_{qq} - \sum_{i=1}^{k} L_{iq} Q_i''^*). \tag{7.46}$$

The difference between the thermal conductivities at the beginning and the end of the process, λ_0 and λ_∞, can be written in terms of the phenomenological coefficients and the heat of transport as

$$\lambda_0 - \lambda_\infty = \frac{\delta}{AT} \sum_{i=1}^{k} L_{iq} Q_i''^* = \frac{\delta}{AT} \sum_{i,m=1}^{k} L_{im} Q_i''^* Q_m''^*. \tag{7.47}$$

As follows from inequality (7.30) and equation (7.44) the thermal conductivity is a positive quantity when the components have been homogeneously mixed:

$$\lambda_0 > 0. \tag{7.48}$$

In conformity with inequalities (7.31) it can be said that the difference $\lambda_0 - \lambda_\infty$ specified by equation (7.47) is a positive quadratic form, that is,

$$\lambda_0 > \lambda_\infty. \tag{7.49}$$

The thermal conductivity when the components are mixed homogeneously is always greater than in the stationary state. It may also be shown that

$$L_{qq} - \sum_{i=1}^{k} L_{iq} Q_i''^* > 0, \tag{7.50}$$

that is, in view of relation (7.46),

$$\lambda_\infty > 0. \tag{7.51}$$

Since thermal conductivity is always positive, the heat flow is directed from the warmer region to the cooler region. This statement may not hold if the heat flow is defined otherwise; this may be regarded as an advantage of using the heat flow which appears in the entropy balance equation for open multicomponent systems.

The flows of substance vanish ($J_i=0$) in the stationary state and thus equation (7.36) implies that

$$\tilde{v}_i \Delta p + \sum_{m=1}^{k-1} \left(\frac{\partial \mu_i}{\partial x_m}\right)_{T,p,x_{j\neq m}} \Delta x_m + \frac{Q_i''^*}{T} \Delta T = 0 \qquad (i=1,2,3,\ldots,k). \tag{7.52}$$

The pressure difference and $k-1$ differences of mass fractions, Δx_m, can be calculated from the k equations above. The value Δx_k is found from the relation

$$\sum_{i=1}^{k} x_i = 1, \qquad \sum_{i=1}^{k} \Delta x_i = 0. \tag{7.53}$$

For a binary fluid

$$\tilde{v}_1 \Delta p + \left(\frac{\partial \mu_1}{\partial x_1}\right)_{T,p} \Delta x_1 + \frac{Q_1''^*}{T} \Delta T = 0, \tag{7.54}$$

$$\tilde{v}_2 \Delta p + \left(\frac{\partial \mu_2}{\partial x_1}\right)_{T,p} \Delta x_1 + \frac{Q_2''^*}{T} \Delta T = 0, \tag{7.55}$$

and, when the Gibbs–Duhem relation

$$x_1 \left(\frac{\partial \mu_1}{\partial x_1}\right)_{T,p} + x_2 \left(\frac{\partial \mu_2}{\partial x_1}\right)_{T,p} = 0$$

has been taken into account, this becomes

$$\tilde{v}_2 \Delta p - \frac{x_1}{x_2} \left(\frac{\partial \mu_1}{\partial x_1}\right)_{T,p} \Delta x_1 + \frac{Q_2''^*}{T} \Delta T = 0. \tag{7.56}$$

When the pressure difference is eliminated from equations (7.54) and (7.56) and the specific volume of the solution

$$v = x_1 \tilde{v}_1 + x_2 \tilde{v}_2 \tag{7.57}$$

is taken into account, the result is

$$\frac{\Delta x_1}{\Delta T} = \frac{x_2(\tilde{v}_1 Q_2''^* - \tilde{v}_2 Q_1''^*)}{vT \left(\frac{\partial \mu_1}{\partial x_1}\right)_{T,p}}. \tag{7.58}$$

This equation specifies the difference which a temperature difference ΔT causes in the mass fraction (or concentration) of component 1 in the stationary state. This effect is called *thermal effusion*.

Elimination of the mass-fraction difference Δx_1 from equations (7.54) and (7.56) yields

$$\frac{\Delta p}{\Delta T} = -\frac{x_1 Q_1''^* + x_2 Q_2''^*}{vT}. \tag{7.59}$$

This equation defines the pressure difference Δp set up in the stationary state by a temperature difference ΔT. This pressure difference is called the *thermomolecular pressure*.

In the special case two vessels may be separated by a membrane which is permeable only to the first component such that J_2 is always zero, whereas in the stationary state J_1 is also zero. If the membrane is movable or deformable, then there can be no pressure difference ($\Delta p=0$); by equation (7.54) the mass-fraction difference is

$$\Delta x_1 = -\frac{Q_1''^*}{T\left(\frac{\partial \mu_1}{\partial x_1}\right)_{T,p}} \Delta T. \tag{7.60}$$

A difference of mass fractions (concentrations) arising under the influence of a temperature difference is called *thermo-osmosis,* whereas the temperature difference arising from a difference of mass fractions (concentrations) is called the *osmotic temperature.*

If the membrane is motionless and if there is no temperature difference ($\Delta T=0$), then by equation (7.54) the pressure difference is

$$\Delta p = -\frac{\left(\frac{\partial \mu_1}{\partial x_1}\right)_{T,p}}{\tilde{v}_1} \Delta x_1 . \tag{7.61}$$

The pressure difference due to the difference of mass fractions (concentrations) in the absence of a temperature difference is the *osmotic pressure*. This latter effect is reversible since if $\Delta T=0$, $J_2=0$ and in the stationary state also $J_1=0$, equation (7.34) yields $J_q''=0$ and the entropy production rate is zero. The stationary state in this case is an equilibrium state. Equation (7.61) does not contain the heat of transport $Q_1''^*$, a quantity that is characteristic of the description of nonequilibrium transport phenomena.

The foregoing relations concerning phenomena of heat and mass transport can easily be supplemented with terms pertaining to electrical effects [22], [25].

Example 7.1. Transform the thermodynamic forces and generalized flows for a discontinuous system by introducing the flow

$$J_u' = J_q'' + T \sum_{i=1}^{k} \tilde{s}_i J_i \tag{1}$$

in the place of J_q''.

Solution. As the starting point for the considerations here take the dissipation function defined by equation (7.20), that is

$$\begin{aligned} \Psi = T\Phi_s &= (J_q'' + \sum_{i=1}^{k} \tilde{h}_i J_i)\frac{\Delta T}{T} + \sum_{i=1}^{k} J_i T\Delta\left(\frac{\mu_i}{T}\right) \\ &= (J_u' + \sum_{i=1}^{k} \mu_i J_i)\frac{\Delta T}{T} + \sum_{i=1}^{k} J_i T\Delta\left(\frac{\mu_i}{T}\right) \\ &= J_u' \frac{\Delta T}{T} + \sum_{i=1}^{k} J_i \Delta\mu_i . \end{aligned} \tag{2}$$

The thermodynamic forces now are given by the expressions

$$X_u = X_q = \frac{\Delta T}{T}, \tag{3}$$

$$X_i' = \Delta\mu_i \tag{4}$$

conjugated with the flows J_u' and J_i, for which the phenomenological equations take the form

$$J_u' = L_{uu}\frac{\Delta T}{T} + \sum_{i=1}^{k} L_{ui}\Delta\mu_i, \tag{5}$$

$$J_m = L_{mu}\frac{\Delta T}{T} + \sum_{i=1}^{k} L_{mi}\Delta\mu_i \tag{6}$$

$$(m = 1, 2, 3, \ldots, k).$$

It is easily seen that the expression

$$\frac{J_u'}{T} = \frac{J_q''}{T} + \sum_{i=1}^{k} \tilde{s}_i J_i = J_s \tag{7}$$

defines the conduction entropy flow and hence also

$$J_s = L_{uu}\frac{\Delta T}{T^2} + \sum_{i=1}^{k} L_{ui}\frac{\Delta\mu_i}{T}. \tag{8}$$

The temperatures which appear in the terms on the left-hand sides of equations (5), (6), and (8) can be combined with the phenomenological coefficients.

The phenomenological coefficients under study are associated by the Onsager reciprocal relations

$$L_{ui} = L_{iu}, \quad L_{mi} = L_{im} \quad (i, m = 1, 2, 3, \ldots, k). \tag{9}$$

They satisfy the following inequalities which emerge from the positive value of the dissipation function:

$$L_{uu} > 0, \quad L_{ii} > 0,$$

$$L_{uu}L_{ii} - L_{ui}^2 > 0, \quad L_{mm}L_{ii} - L_{mi}^2 > 0 \tag{10}$$

$$(i, m = 1, 2, 3, \ldots, k).$$

Finally, the entropy of transport defined by the relation

$$J_s = \frac{J_u'}{T} = \sum_{i=1}^{k} S_i^* J_i \tag{11}$$

or

$$TS_i^* = U_i^* - \mu_i \tag{12}$$

can be introduced into the phenomenological equations. The phenomenological coefficient then is

$$L_{mu} = \sum_{i=1}^{k} L_{mi}TS_i^* \quad (m = 1, 2, 3, \ldots, k) \tag{13}$$

whereas the phenomenological equations are of the form

$$J_u' = L_{uu}\frac{\Delta T}{T} + \sum_{i,m=1}^{k} L_{im}TS_m^*\Delta\mu_i, \tag{14}$$

$$J_m = \sum_{i=1}^{k} L_{mi}(S_i^*\Delta T + \Delta\mu_i) \tag{15}$$

$$(m = 1, 2, 3, \ldots, k).$$

3. Thermomechanical Effects in a One-component Fluid

The system to be considered now is a system containing a one-component fluid. It consists of two parts, called phases as a matter of convention, which are separated by a diaphragm with a small aperture, by a capillary, porous wall, or permeable membrane (Fig. 7.2). The two parts of the system may differ as to temperature and

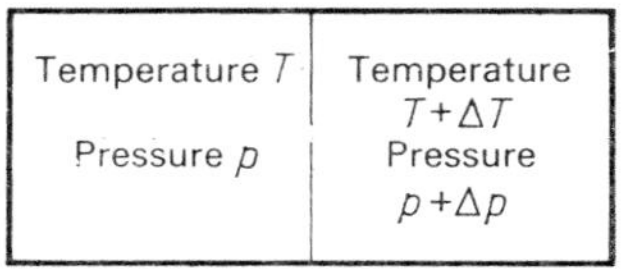

Fig. 7.2 One-component fluid separated by a permeable membrane

pressure owing to the transport of heat and substance between them. Since there are two thermodynamic forces and two generalized flows in this case, two cross effects ensue: transport of mass under the influence of a temperature difference, and the reverse effect of heat transport owing to a pressure difference.

If the temperature difference ΔT and pressure difference Δp are small, the heat flow J_q and substance flow J_1 can be written as linear phenomenological equations emerging directly from equations (7.26) and (7.27),

$$J_q = L_{qq}\frac{\Delta T}{T} + L_{q1}\,v\Delta p, \tag{7.62}$$

$$J_1 = L_{1q}\frac{\Delta T}{T} + L_{11}\,v\Delta p. \tag{7.63}$$

On introduction of the cross-sectional area A of the flow and the thickness δ of the porous diaphragm, they may also be written as

$$J_q = \frac{A}{\delta}\,C\Delta p + \frac{A}{\delta}\,\lambda_0 \Delta T, \tag{7.64}$$

$$J_1 = \frac{A}{\delta}\,E\Delta p + \frac{A}{\delta}\,B\Delta T. \tag{7.65}$$

The thermal conductivity λ_0 corresponds to the case of no pressure difference, that is, since

$$J_q = L_{qq}\frac{\Delta T}{T} = \frac{A}{\delta}\,\lambda_0 \Delta T \qquad (\Delta p = 0),$$

it is equal to

$$\lambda_0 = \frac{L_{qq}}{T}\frac{\delta}{A}. \tag{7.66}$$

The filtration coefficient E may be defined by considering a system in which there is no temperature difference:

$$J_1 = L_{11}\, v\Delta p = \frac{A}{\delta}\, E\Delta p \qquad (\Delta T = 0),$$

that is,

$$E = L_{11}\, v\, \frac{\delta}{A}. \tag{7.67}$$

In the mechanocaloric effect heat is transported only as a result of a substance flow due to pressure difference, and not as a result of a temperature difference. This effect is described by an empirical coefficient C, called the *osmotic heat conductivity*; hence, equation

$$J_q = L_{q1}\, v\Delta p = \frac{A}{\delta}\, C\Delta p \qquad (\Delta T = 0)$$

yields

$$C = L_{q1}\, v\, \frac{\delta}{A}. \tag{7.68}$$

A thermomechanical effect consists in a flow of substance being caused only by a temperature difference; examples of such effects are the Knudsen effect, thermo-osmosis, and the fountain effect in liquid He II. It is described by the empirical coefficient B, called the *coefficient of thermo-osmotic filtration*. This coefficient is related to the phenomenological coefficient by

$$J_1 = L_{1q}\, \frac{\Delta T}{T} = \frac{A}{\delta}\, B\Delta T \qquad (\Delta p = 0),$$

that is

$$B = \frac{L_{1q}}{T}\, \frac{\delta}{A}. \tag{7.69}$$

The empirical coefficients B, C, E, and λ_0 depend on the kind of fluid, the geometrical dimensions of the flow channels connecting the two reservoirs, the mean temperature and, to a lesser extent, the mean pressure in the system.

The pressure difference Δp may be eliminated from equations (7.64) and (7.65) and then the heat flux is

$$J_q = \frac{C}{E} J_1 + \lambda_\infty \frac{A}{\delta} \Delta T, \tag{7.70}$$

where the thermal conductivity, defined as

$$\lambda_\infty = \lambda_0 - \frac{BC}{E}, \tag{7.71}$$

has been introduced.

Since the substance flow vanishes ($J_1=0$) in the stationary state, λ_∞ is the thermal conductivity in the stationary state. This is seen distinctly from the form of the heat flow in the stationary state:

$$J_q=\lambda_\infty \frac{A}{\delta} \Delta T. \tag{7.72}$$

The Onsager reciprocal relations for a one-component, discontinuous system yield the relation between the coefficients,

$$L_{q1}=L_{1q}, \quad C=BvT. \tag{7.73}$$

For an isothermal system $\Delta T=0$ and hence equations (7.64) and (7.65) give rise to the following expression containing the heat of transport:

$$\left(\frac{J_q}{J_1}\right)_{\Delta T=0}=\frac{L_{q1}}{L_{11}}=\frac{B}{E}vT=\frac{C}{E}=Q^*. \tag{7.74}$$

The difference between the thermal conductivities given earlier can now be written as

$$\lambda_0-\lambda_\infty=\frac{B^2}{E}vT=BQ^*. \tag{7.75}$$

The inequalities concerning the phenomenological coefficients

$$L_{qq}>0, \quad L_{11}>0, \quad L_{qq}L_{11}-L_{q1}^2>0 \tag{7.76}$$

lead to inequalities for the empirical coefficients

$$E>0, \quad \lambda_0>\lambda_\infty>0, \quad B^2<\frac{E\lambda_0}{vT}. \tag{7.77}$$

In the stationary state, when the flow of substance vanishes ($J_1=0$), the pressure difference resulting from the existence of a temperature difference, that is, the thermomolecular pressure,

$$\Delta p=-\frac{L_{1q}}{L_{11}vT}\Delta T=-\frac{B}{E}\Delta T=-\frac{Q^*}{vT}\Delta T \tag{7.78}$$

is obtained from equations (7.63), (7.65), and (7.74).

The heat of transport for the Kundsen effect in rarefied ideal gases can be determined by theoretical consideration. The Knudsen effect occurs when the gas is contained in two reservoirs connected by means of an aperture much smaller in diameter than the mean free path of the gas molecules. In this case each molecule reaching the aperture passes through it without collision with other molecules.

The coordinate system is chosen so that the z-axis of polar coordinates is directed perpendicular to the surface of the aperture. In conformity with the Maxwellian velocity distribution law, out of a total of n molecules contained in a unit volume the fraction of molecules whose velocities lie in the interval between v and $v+\mathrm{d}v$ and whose directions lie in the solid angle $\sin\theta\,\mathrm{d}\theta\,\mathrm{d}\varphi$ is

$$n\left(\frac{m}{2\pi kT}\right)^{\frac{3}{2}}\left[\exp\left(-\frac{mv^2}{2kT}\right)\right]v^2\mathrm{d}v\sin\theta\,\mathrm{d}\theta\,\mathrm{d}\varphi\,,$$

where m is the mass of the molecule and k is Boltzmann's constant.

Of this number, only those molecules which are contained in the volume $v\cos\theta$ arrive in a unit area of the aperture in a unit time; hence, the number of molecules per unit area of aperture per unit time is

$$n\left(\frac{m}{2\pi kT}\right)^{\frac{3}{2}}\left[\exp\left(-\frac{mv^2}{2kT}\right)\right]v^3\mathrm{d}v\sin\theta\cos\theta\,\mathrm{d}\theta\,\mathrm{d}\varphi\,.$$

The number of particles n_v passing through an aperture of unit area per unit time with velocities between v and $v+\mathrm{d}v$ is calculated by integrating the expression above with respect to θ between 0 and $\pi/2$ and with respect to φ between 0 and 2π. The result is

$$n_v\,\mathrm{d}v=n\pi\left(\frac{m}{2\pi kT}\right)^{\frac{3}{2}}\left[\exp\left(-\frac{mv^2}{2kT}\right)\right]v^3\mathrm{d}v\,.$$

To obtain the mean kinetic energy transported by a molecule the expression above should be multiplied by $mv^2/2$, then integrated for velocities varying from 0 to infinity, and divided by the total number of molecules passing through the aperture per second:

$$\left\langle\frac{mv^2}{2}\right\rangle=\frac{\int_0^\infty \frac{1}{2}mv^2 n_v\,\mathrm{d}v}{\int_0^\infty n_v\,\mathrm{d}v}=\frac{1}{2}m\,\frac{\int_0^\infty\left[\exp\left(-\frac{mv^2}{2kT}\right)\right]v^5\mathrm{d}v}{\int_0^\infty\left[\exp\left(-\frac{mv^2}{2kT}\right)\right]v^3\mathrm{d}v}=2kT\,.$$

The mean kinetic energy transported by 1 kilomole of gas is $2R_{\mathrm{M}}T$. The energy of transport of ideal gases per kg is

$$U^*=2\frac{R_{\mathrm{M}}}{M}T=2RT\,. \tag{7.79}$$

The heat of transport of monatomic ideal gases for the Knudsen effect,

$$Q^*=U^*-h=2RT-\tfrac{5}{2}RT=-\tfrac{1}{2}RT\,, \tag{7.80}$$

is always negative and does not depend on the reference level of the internal energy.

When the equation of state for ideal gases

$$pv=RT \tag{7.81}$$

is taken into account equations (7.78) and (7.80) yield

$$\frac{\Delta p}{\Delta T}=\frac{1}{2}\frac{R}{v}=\frac{1}{2}\frac{p}{T} \tag{7.82}$$

or

$$\frac{\Delta p}{p}=\frac{1}{2}\frac{\Delta T}{T}. \tag{7.83}$$

It was assumed at the beginning of this reasoning that the temperature difference between the two parts of the system is small,

$$\left|\frac{\Delta T}{T}\right| \ll 1$$

and

$$1+\frac{1}{2}\frac{\Delta T}{T}\approx\sqrt{1+\frac{\Delta T}{T}},$$

whereby

$$\frac{p+\Delta p}{p}=\sqrt{\frac{T+\Delta T}{T}}.$$

On introduction of the parameters p', T' and p'', T'' (respectively, of the first and second parts of the system), the result is

$$\frac{p''}{p'}=\sqrt{\frac{T''}{T'}}. \tag{7.84}$$

This is the equation describing the Knudsen effect in the stationary state of ideal gases; this effect occurs when the aperture through which the gas flows is so small

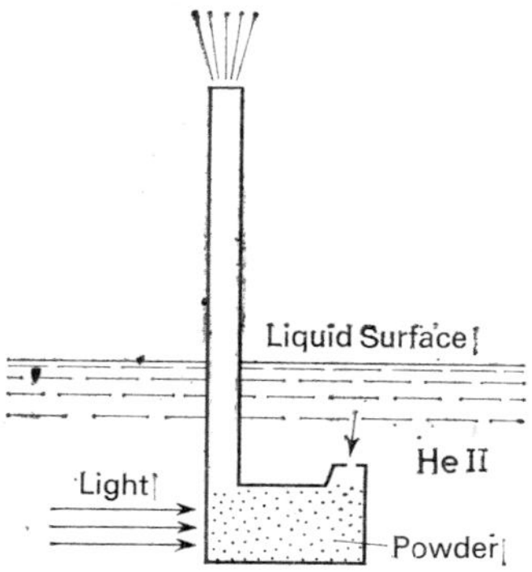

Fig. 7.3 The fountain effect in liquid He II

that the flow is not coherent. The aperture diameter must be comparable with the mean free path of the molecules. If the flow aperture is large enough, the energy of transport should be supplemented with the work of pumping pv, and then by equation (7.80) the heat of transport Q^* is zero. As follows from equation (7.78), the thermomolecular pressure is zero under such conditions. The Knudsen effect does not occur.

The thermomechanical effect in liquid He II may take place as the so-called fountain effect. It occurs in an arrangement as shown in Fig. 7.3. The small vessel ending the fine capillary tube is filled with a fine dark powder for thermal radiation absorption and has an aperture. If this vessel is immersed in liquid He II and the capillary is allowed to protrude above the surface of the liquid and a beam of light is directed on the vessel, liquid helium shoots out of the capillary as a fountain. The liquid in the capillary rises to as much as 30 cm above the free surface of the liquid.

The heat of transport for the fountain effect of liquid He II is always negative and depends only on the mean temperature and the capillary diameter. Experimental investigations by Brewer and Edwards have confirmed the validity of the Onsager reciprocal relation for this case.

The heat of transport for effects of thermo-osmosis through a permeable membrane depends on many factors such as the kind of fluid, the material used for the membrane, and the mean temperature of the fluid. In this case the heat of transport assumes positive as well as negative values.

TENSOR NOTATION IN NONEQUILIBRIUM THERMODYNAMICS

Definition of a Tensor

Tensor calculus is used extensively in physics, but its applications in engineering are much less widespread. Tensor notation is the most suitable 'language' in field theory, in the theory of combined fields, in fluid mechanics, and in nonequilibrium thermodynamics which is linked with the aforementioned fields of science. For an understanding of the fundamentals of nonequilibrium thermodynamics one can limit oneself to the basics of tensor calculus in the rectangular, Cartesian coordinate system.

Physical quantities, called *scalars*, are specified by a single numerical value. Example of scalars are mass, volume, density, and temperature. Direction as well as numerical value is required in order to define other physical quantities; these are *vectors*, e.g. force, displacement, and velocity. Scalars and vectors do not make it possible to determine all physical quantities as some quantities require even more numerical values. All physical quantities are *tensors* of various ranks.

In three-dimensional space a tensor of rank n is determined by 3^n numbers (or functions) called its components. A scalar, which is specified by $3^0=1$ number, is a tensor of rank zero. A vector, which is determined by $3^1=3$ numbers, is a tensor of rank one. To specify a tensor of rank two it is necessary to have $3^2=9$ numbers, etc.

When the notation for tensors is written their rank may be denoted by the number of bars above the symbol, e.g. a – scalar, $\bar{v}$ – tensor, $\bar{\bar{T}}$ – tensor of rank two. Tensors of various ranks are denoted in print by different kinds of type: scalars by italic, e.g. a; vectors by italic boldface, e.g. $\boldsymbol{v}$; and tensors of rank two by roman boldface, e.g. $\mathbf{T}$.

Another method of denoting tensors, in print and in writing, is that of indicial notation. Each letter index appearing once in an expression may take the values 1, 2, 3 in three-dimensional space; for instance, w_α denotes the set w_1, w_2, w_3, and $T_{\alpha\beta}$ stands for the set T_{11}, T_{12}, T_{13}, T_{21}, T_{22}, T_{23}, T_{31}, T_{32}, T_{33}. As can be seen, the number of indices (pertaining to the coordinate axes) which appear once determines the rank of the tensor. By the Einstein summation convention, whenever an index is repeated the expression is to be summed over the index from 1 to 3 (in three-dimensional space), and hence

$$T_{\alpha\alpha}=\sum_{\alpha=1}^{3} T_{\alpha\alpha}=T_{11}+T_{22}+T_{33}=T_{\beta\beta}=a \quad \text{(scalar)}. \tag{I.1}$$

Indices of summation are called *dummy indices* since they can be replaced by any other letter. As follows from the example above, $T_{\alpha\alpha}=T_{\beta\beta}=a$ (scalar), the dummy indices are not counted in establishing the rank of the tensor.

The experimentally verified fact that physical quantities are independent of the choice of coordinates requires satisfaction of a relevant transformation law for the components of the tensor describing the physical quantity in question. Such laws ensure the invariance of physical quantities under a change of coordinate system. Although specified by various values of components in different coordinate systems, a given physical quantity remains the same.

That scalars are independent of the choice of coordinate system is obvious. A scalar is determined by a single number (or function) which does not change under a change of coordinate system.

How the components of a vector change under the rotation of coordinate system is illustrated in Fig. I. 1. Vector A has the components A_α ($\alpha=1,2,3$) in one Cartesian coordinate system, and components A'_β in another system with the same origin.

If the cosines of the angles between the axes Ox_α and Ox'_β are designated by $a_{\beta'\alpha}$, the relations between the old and new components of vector A take the form

$$A'_\beta=a_{\beta'\alpha}A_\alpha. \tag{I.2}$$

A vector is a quantity which is specified in any coordinate system by the components A_α which under rotation of the coordinate system are transformed into the components A'_β, in accordance with the law given by equation (I.2).

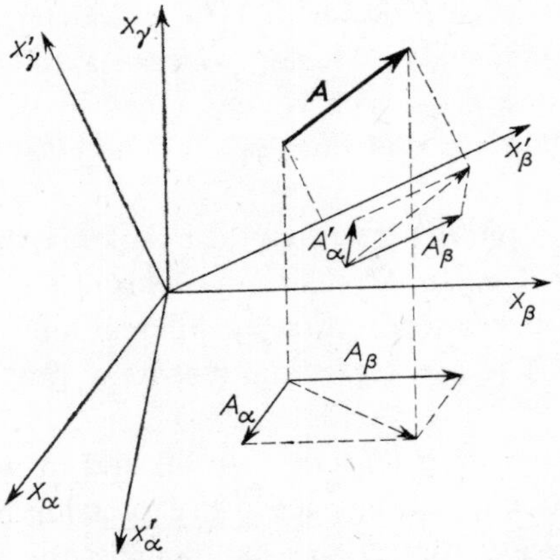

Fig. I.1 Change of vector components under rotation of coordinate system

Tensors of higher rank, e.g. of rank two, are defined in similar fashion. A tensor of rank two is a quantity which is defined in any coordinate system by $3^2=9$ components $A_{\alpha\beta}$ which under rotation of coordinate system are transformed into the components $A'_{\nu\gamma}$ in conformity with the law

$$A'_{\nu\gamma}=a_{\nu'\alpha}a_{\gamma'\beta}A_{\alpha\beta}. \tag{I.3}$$

Tensor Algebra

Only tensors of the same rank can be added, the result being a tensor of the same rank with components equal to the sum of the components of the tensors added, e.g.

$$\mathbf{T}+\mathbf{S}=\mathbf{A}, \quad T_{\alpha\beta}+S_{\alpha\beta}=A_{\alpha\beta}. \tag{I.4}$$

Subtraction of a tensor can be replaced by the addition of the tensor multiplied by the scalar -1.

Multiplication of a tensor by a scalar a yields a tensor of the same rank with components which are a times those of the tensor being multiplied, e.g.

$$a\mathbf{T}=\mathbf{A}, \quad aT_{\alpha\beta}=A_{\alpha\beta}. \tag{I.5}$$

When the order of the indices of tensor $\mathbf{T}$ is changed, the result is the tensor $\tilde{\mathbf{T}}$ conjugate to the original tensor, and

$$T_{\alpha\beta}=\tilde{T}_{\beta\alpha}. \tag{I.6}$$

In matrix notation going over from tensor $\mathbf{T}$ to its conjugate $\tilde{\mathbf{T}}$ entails the interchange of rows and columns:

$$\mathbf{T}=[T_{\alpha\beta}]=\begin{bmatrix} T_{\alpha\alpha} & T_{\alpha\beta} & T_{\alpha\gamma} \\ T_{\beta\alpha} & T_{\beta\beta} & T_{\beta\gamma} \\ T_{\gamma\alpha} & T_{\gamma\beta} & T_{\gamma\gamma} \end{bmatrix}, \tag{I.7}$$

$$\tilde{\mathbf{T}}=[T_{\beta\alpha}]=\begin{bmatrix} T_{\alpha\alpha} & T_{\beta\alpha} & T_{\gamma\alpha} \\ T_{\alpha\beta} & T_{\beta\beta} & T_{\gamma\beta} \\ T_{\alpha\gamma} & T_{\beta\gamma} & T_{\gamma\gamma} \end{bmatrix}. \tag{I.8}$$

A tensor of rank two is symmetric if its components which differ by the interchange of two indices are equal to each other:

$$T_{\alpha\beta}=T_{\beta\alpha}. \tag{I.9}$$

A symmetric tensor is conjugate to itself:

$$\mathbf{T}^{\mathrm{s}}=\tilde{\mathbf{T}}^{\mathrm{s}}. \tag{I.10}$$

A symmetric tensor of rank two has only six independent components:

$$\mathbf{T}^{\mathrm{s}}=[T^{\mathrm{s}}_{\alpha\beta}]=\begin{bmatrix} T_{\alpha\alpha} & T_{\alpha\beta} & T_{\alpha\gamma} \\ T_{\alpha\beta} & T_{\beta\beta} & T_{\beta\gamma} \\ T_{\alpha\gamma} & T_{\beta\gamma} & T_{\gamma\gamma} \end{bmatrix}. \tag{I.11}$$

A tensor of rank two is antisymmetric if its components which differ by the interchange of two indices have the same magnitudes but are of opposite sign:

$$\mathbf{T}^{\mathrm{a}}=-\tilde{\mathbf{T}}^{\mathrm{a}}, \quad T_{\alpha\beta}=-T_{\beta\alpha}. \tag{I.12}$$

The components of an antisymmetric tensor with both indices the same are equal to zero:

$$T_{\alpha\alpha} = -T_{\alpha\alpha} = 0. \tag{I.13}$$

An antisymmetric tensor of the second rank has only three independent components, and hence in three-dimensional spac: is equivalent to the vector:

$$\mathbf{T}^{a} = [T^{a}_{\alpha\beta}] = \begin{bmatrix} 0 & T_{\alpha\beta} & T_{\alpha\gamma} \\ -T_{\alpha\beta} & 0 & T_{\beta\gamma} \\ -T_{\alpha\gamma} & -T_{\beta\gamma} & 0 \end{bmatrix}. \tag{I.14}$$

Every tensor of rank two can be decomposed into a symmetric part

$$\mathbf{T}^{s} = \tfrac{1}{2}(\mathbf{T} + \tilde{\mathbf{T}}) \tag{I.15}$$

and an antisymmetric part

$$\mathbf{T}^{a} = \tfrac{1}{2}(\mathbf{T} - \tilde{\mathbf{T}}) \tag{I.16}$$

so that

$$\mathbf{T} = \mathbf{T}^{s} + \mathbf{T}^{a} = \tfrac{1}{2}(\mathbf{T} + \tilde{\mathbf{T}}) + \tfrac{1}{2}(\mathbf{T} - \tilde{\mathbf{T}}). \tag{I.17}$$

The product of two tensors is a tensor whose components are formed by the successive multiplication of the components of the first tensor by all the components of the second. The rank of the tensor resulting from the multiplication of two tensors is equal to the sum of the ranks of the tensors multiplied. In the general case the product of two tensors is not subject to the commutative law:

$$w_{\alpha} T_{\beta\gamma} = (\boldsymbol{w}\mathbf{T})_{\alpha\beta\gamma},$$

$$T_{\alpha\beta} w_{\gamma} = (\mathbf{T}\boldsymbol{w})_{\alpha\beta\gamma},$$

$$\boldsymbol{w}\mathbf{T} \neq \mathbf{T}\boldsymbol{w}.$$

The product of two vectors is a tensor called a *dyadic*:

$$\mathbf{A} = \boldsymbol{w}\boldsymbol{v} = \begin{bmatrix} w_{\alpha} v_{\alpha} & w_{\alpha} v_{\beta} & w_{\alpha} v_{\gamma} \\ w_{\beta} v_{\alpha} & w_{\beta} v_{\beta} & w_{\beta} v_{\gamma} \\ w_{\gamma} v_{\alpha} & w_{\gamma} v_{\beta} & w_{\gamma} v_{\gamma} \end{bmatrix}, \tag{I.18}$$

$$A_{\alpha\beta} = (\boldsymbol{w}\boldsymbol{v})_{\alpha\beta} = w_{\alpha} v_{\beta}.$$

The scalar or dot product of two tensors arises from the ordinary product of two tensors if the summation is performed over identical (dummy) indices occurring in both tensors being multiplied. As a result of this summation the tensor is contracted once, that is, its rank is reduced by two. Since a tensor of the lowest rank (rank zero) is a scalar, a vector cannot be contracted. The scalar product in index-less notation is indicated by a dot between the tensors under multiplication, and the scalar double product associated with the two-fold contraction of tensors (reduction of rank by four) is denoted by two dots.

Examples of the scalar products of tensors:

$$(\boldsymbol{wv})_{\alpha\alpha} = w_\alpha v_\alpha = \sum_{\alpha=1}^{3} w_\alpha v_\alpha = a \qquad \text{(scalar)}$$

$$\boldsymbol{w}\cdot\boldsymbol{v} = a\,,$$

$$\boldsymbol{w}\cdot\boldsymbol{w} = |\boldsymbol{w}|^2 = w^2\,;$$

$$w_\beta T_{\alpha\beta} = \sum_{\beta=1}^{3} w_\beta T_{\alpha\beta} = v_\alpha \qquad \text{(vector)}$$

$$\boldsymbol{w}\cdot\mathbf{T} = \boldsymbol{v}\,;$$

$$S_{\alpha\gamma} T_{\gamma\beta} = \sum_{\gamma=1}^{3} S_{\alpha\gamma} T_{\gamma\beta} = A_{\alpha\beta} \qquad \text{(tensor of rank two)},$$

$$\mathbf{S}\cdot\mathbf{T} = \mathbf{A}\,.$$

In the general case the scalar product of tensors is not subject to the commutative law:

$$\boldsymbol{w}\cdot\mathbf{T} \neq \mathbf{T}\cdot\boldsymbol{w}\,.$$

The scalar product of vector $\boldsymbol{c}$ by a dyadic $\boldsymbol{wv}$ can be computed in two ways:

$$(\boldsymbol{wv})\cdot\boldsymbol{c} = \boldsymbol{w}(\boldsymbol{v}\cdot\boldsymbol{c})\,,$$

$$\boldsymbol{c}\cdot(\boldsymbol{wv}) = (\boldsymbol{c}\cdot\boldsymbol{w})\,\boldsymbol{v}\,.$$

An example of a scalar double product of tensors:

$$S_{\alpha\beta} T_{\beta\alpha} = \sum_{\alpha,\beta=1}^{3} S_{\alpha\beta} T_{\beta\alpha} = a \qquad \text{(scalar)},$$

$$\mathbf{S} : \mathbf{T} = a\,.$$

Properties analogous to the multiplication of numbers by unity is possessed by the unit tensor $\boldsymbol{\delta}$ of rank two:

$$\boldsymbol{\delta}\cdot\mathbf{T} = \mathbf{T}\cdot\boldsymbol{\delta} = \mathbf{T}\,, \qquad \mathbf{T}\cdot\mathbf{T}^{-1} = \boldsymbol{\delta}\,. \tag{I.19}$$

The unit tensor, called the *Kronecker delta* (or *symbol*), can clearly be written in indicial notation as

$$\delta_{\alpha\beta} = \begin{cases} 1, & \text{if} \quad \alpha = \beta\,, \\ 0, & \text{if} \quad \alpha \neq \beta\,, \end{cases} \tag{I.20}$$

or in matrix notation as

$$\boldsymbol{\delta} = [\delta_{\alpha\beta}] = \begin{bmatrix} 1 & 0 & 0 \\ 0 & 1 & 0 \\ 0 & 0 & 1 \end{bmatrix}. \tag{I.21}$$

The scalar double product of two unit tensors is equal to

$$\boldsymbol{\delta} : \boldsymbol{\delta} = \sum_{\alpha=1}^{3} \delta_{\alpha\alpha} = 1+1+1=3\,. \tag{I.22}$$

The scalar double product of a tensor of rank two and a unit tensor of rank two yields a scalar, called the *trace of the tensor*, equal to the sum of the diagonal components:

$$\begin{aligned} &T_{\alpha\beta}\,\delta_{\alpha\beta} = T_{\alpha\alpha} = \sum_{\alpha=1}^{3} T_{\alpha\alpha} = \operatorname{Tr}\mathbf{T} = T\,,\\ &\mathbf{T} : \boldsymbol{\delta} = \operatorname{Tr}\mathbf{T} = T\,. \end{aligned} \tag{I.23}$$

The trace, as a scalar, does not change under a change of the coordinate systems and hence is called a linear invariant of the tensor.

A tensor whose trace is zero is called a *deviator*:

$$\mathbf{D} : \boldsymbol{\delta} = \operatorname{Tr}\mathbf{D} = 0\,. \tag{I.24}$$

The scalar product of a tensor $\mathbf{T}$ and a vector $\boldsymbol{v}$ is a vector which in the general case differs from $\boldsymbol{v}$ in direction as well as in magnitude. If for a tensor $\mathbf{T}$ there exist vectors $\boldsymbol{v}$ such that scalar multiplication by $\mathbf{T}$ changes only their magnitude and not their directions, that is,

$$T_{\alpha\beta}\,v_\beta = a v_\alpha\,,$$

then these vectors determine directions of the principal tensor axes.

For a spherical tensor arbitrary directions may be directions of the principal tensor axes,

$$[T_{\alpha\beta}] = a \begin{bmatrix} 1 & 0 & 0 \\ 0 & 1 & 0 \\ 0 & 0 & 1 \end{bmatrix} = a\delta_{\alpha\beta}\,. \tag{I.25}$$

Under scalar multiplication by a spherical tensor every vector changes its length but not its direction.

Any tensor is decomposable into a deviator and a spherical tensor:

$$\begin{aligned} &T_{\alpha\beta} = T_{\alpha\beta} - \tfrac{1}{3}T_{\gamma\gamma}\,\delta_{\alpha\beta} + \tfrac{1}{3}T_{\gamma\gamma}\,\delta_{\alpha\beta} = D_{\alpha\beta} + \tfrac{1}{3}T_{\gamma\gamma}\,\delta_{\alpha\beta}\,,\\ &\mathbf{T} = \mathbf{D} + \tfrac{1}{3}\boldsymbol{\delta}\operatorname{Tr}\mathbf{T}\,. \end{aligned} \tag{I.26}$$

The tensor

$$D_{\alpha\beta} = T_{\alpha\beta} - \tfrac{1}{3}T_{\gamma\gamma}\,\delta_{\alpha\beta} \tag{I.27}$$

is a deviator since its trace is zero:

$$D_{\alpha\alpha} = T_{\alpha\alpha} - \tfrac{1}{3}T_{\gamma\gamma}\,\delta_{\alpha\alpha} = T_{\alpha\alpha} - \tfrac{1}{3}T_{\alpha\alpha}\cdot 3 = 0\,.$$

Vector Products

The vector or cross product $\boldsymbol{A}\times\boldsymbol{B}$ of two vectors $\boldsymbol{A}$ and $\boldsymbol{B}$ is a vector $\boldsymbol{C}=\boldsymbol{A}\times\boldsymbol{B}$ which is perpendicular to the plane of $\boldsymbol{A}$ and $\boldsymbol{B}$. If we look along $\boldsymbol{C}$ from its tail to its point, the smaller angle between $\boldsymbol{A}$ and $\boldsymbol{B}$ is in the clockwise direction. The magni-

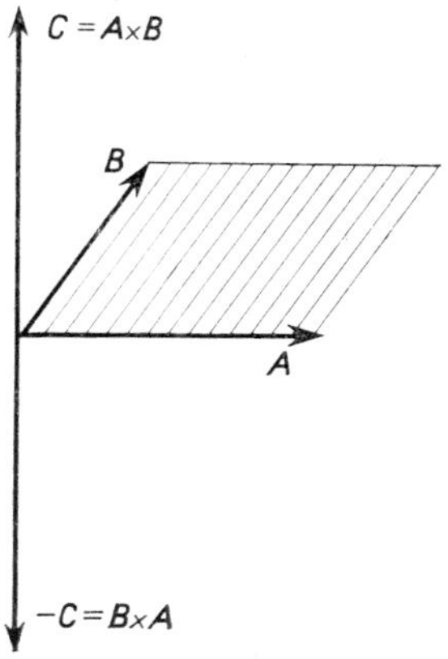

Fig. I.2 Vector product

tude of $\boldsymbol{C}$ is equal to the area of the parallelogram with $\boldsymbol{A}$ and $\boldsymbol{B}$ as its sides (Fig. I.2):

$$|\boldsymbol{C}|=|\boldsymbol{A}\times\boldsymbol{B}|=AB\sin(\boldsymbol{A},\boldsymbol{B}). \qquad \text{(I.28)}$$

In contradistinction to the scalar product of two vectors, the vector product is not commutative but when the order of the factors is changed the sign of the vector product changes:

$$\boldsymbol{A}\times\boldsymbol{B}=-(\boldsymbol{B}\times\boldsymbol{A}). \qquad \text{(I.29)}$$

On the other hand, just as the scalar product of two vectors, the vector product obeys the associative law:

$$\boldsymbol{A}\times(\boldsymbol{B}+\boldsymbol{C})=(\boldsymbol{A}\times\boldsymbol{B})+(\boldsymbol{A}\times\boldsymbol{C}). \qquad \text{(I.30)}$$

The vector product of two parallel vectors is zero.

The components of a vector product are equal to

$$C_\alpha=A_\beta B_\gamma-A_\gamma B_\beta, \qquad \text{(I.31)}$$

where α, β, γ permute cyclically.

A triple scalar product does not change under cyclical permutation of the vectors:

$$(\boldsymbol{A}\times\boldsymbol{B})\cdot\boldsymbol{C}=(\boldsymbol{B}\times\boldsymbol{C})\cdot\boldsymbol{A}=(\boldsymbol{C}\times\boldsymbol{A})\cdot\boldsymbol{B}. \qquad \text{(I.32)}$$

If two vectors in a triple scalar product are identical or parallel, the product is zero:

$$(\boldsymbol{A}\times\boldsymbol{B})\cdot\boldsymbol{A}=(\boldsymbol{A}\times\boldsymbol{B})\cdot\boldsymbol{B}=(\boldsymbol{A}\times\boldsymbol{A})\cdot\boldsymbol{B}=0. \qquad \text{(I.33)}$$

The triple vector product of $\boldsymbol{A}$, $\boldsymbol{B}$, and $\boldsymbol{C}$ is

$$\begin{aligned} \boldsymbol{A}\times(\boldsymbol{B}\times\boldsymbol{C}) &= \boldsymbol{B}(\boldsymbol{A}\cdot\boldsymbol{C})-\boldsymbol{C}(\boldsymbol{A}\cdot\boldsymbol{B}), \\ (\boldsymbol{A}\times\boldsymbol{B})\times\boldsymbol{C} &= \boldsymbol{B}(\boldsymbol{A}\cdot\boldsymbol{C})-\boldsymbol{A}(\boldsymbol{B}\cdot\boldsymbol{C}). \end{aligned} \tag{I.34}$$

Tensor Analysis

Physical quantities of a tensorial nature may vary in time and space, forming non-stationary tensor fields, or may vary only in space, forming stationary tensor fields. A tensor field is a correspondence which associates a strictly specified value of the given tensor with each point in space.

Differentiation of a tensor with respect to a scalar, e.g. time, does not change its rank.

The spatial differentiation of a tensor raises its rank by unity, that is, may be replaced by multiplication by the vector $\boldsymbol{\nabla}$, called the nabla, Hamiltonian operator, or del operator:

$$\boldsymbol{\nabla}=\frac{\partial}{\partial \boldsymbol{x}}\rightarrow\frac{\partial}{\partial x_\alpha}. \tag{I.35}$$

The gradient of a scalar field a,

$$\operatorname{grad} a=\frac{\partial a}{\partial \boldsymbol{x}}=\boldsymbol{\nabla} a \tag{I.36}$$

is a vector with the components

$$\frac{\partial a}{\partial x_\alpha}=a_{,\alpha}, \tag{I.37}$$

where the subscript $_{,\alpha}$ has been used to indicate differentiation with respect to x_α.

The derivative of a scalar a with respect to a vector is a vector whose components are the derivatives of the given scalar quantity with respect to the particular components of the vector.

The derivative of a scalar a in the direction $\boldsymbol{l}$ is equal to the projection of the gradient of the scalar onto that direction:

$$\frac{\mathrm{d}a}{\mathrm{d}\boldsymbol{l}}=\boldsymbol{l}\cdot\boldsymbol{\nabla} a\rightarrow l_\alpha a_{,\alpha}. \tag{I.38}$$

The gradient of a vector field $\boldsymbol{v}$,

$$\operatorname{Grad}\boldsymbol{v}=\frac{\partial \boldsymbol{v}}{\partial \boldsymbol{x}}=\boldsymbol{\nabla}\boldsymbol{v} \tag{I.39}$$

is a tensor of rank two with the components

$$\frac{\partial v_\alpha}{\partial x_\beta}=v_{\alpha,\beta}. \tag{I.40}$$

The derivative of a vector $\boldsymbol{v}$ in the direction $\boldsymbol{l}$ is calculated as follows:

$$\frac{\mathrm{d}\boldsymbol{v}}{\mathrm{d}\boldsymbol{l}}=\boldsymbol{l}\cdot(\nabla\boldsymbol{v})\rightarrow l_\beta v_{\alpha,\beta}\,. \tag{I.41}$$

When contraction is performed once (summation over repeated indices) the divergence is obtained instead of the gradient.

The divergence of a vector field $\boldsymbol{v}$,

$$\operatorname{div}\boldsymbol{v}=\frac{\partial}{\partial\boldsymbol{x}}\cdot\boldsymbol{v}=\nabla\cdot\boldsymbol{v}\,, \tag{I.42}$$

is a scalar, as is evident from the indicial notation

$$\frac{\partial v_\alpha}{\partial x_\alpha}=v_{\alpha,\alpha}\,. \tag{I.43}$$

The divergence of tensor field $\mathbf{T}$,

$$\operatorname{Div}\mathbf{T}=\frac{\partial}{\partial\boldsymbol{x}}\cdot\mathbf{T}=\nabla\cdot\mathbf{T}\,, \tag{I.44}$$

is a vector with the components

$$\frac{\partial T_{\alpha\beta}}{\partial x_\alpha}=T_{\alpha,\beta,\alpha}\,. \tag{I.45}$$

The Laplace operator or Laplacian,

$$\nabla^2=\nabla\cdot\nabla=\operatorname{div}\operatorname{grad}=\frac{\partial}{\partial\boldsymbol{x}}\cdot\frac{\partial}{\partial\boldsymbol{x}}\,, \tag{I.46}$$

is a scalar. In indicial notation the Laplacian of a scalar a is of the form

$$\frac{\partial^2 a}{\partial x_\alpha^2}=a_{,\alpha\alpha}\,. \tag{I.47}$$

The Laplacian of a vector field $\boldsymbol{v}$ can be transformed as follows:

$$\nabla^2\boldsymbol{v}=\nabla(\nabla\cdot\boldsymbol{v})-\nabla\times(\nabla\times\boldsymbol{v})\,. \tag{I.48}$$

The curl of a vector $\boldsymbol{v}$ is defined as

$$\operatorname{curl}\ \boldsymbol{v}=\nabla\times\boldsymbol{v}\rightarrow(\nabla\times\boldsymbol{v})_\gamma=\frac{\partial v_\beta}{\partial x_\alpha}-\frac{\partial v_\alpha}{\partial x_\beta} \tag{I.49}$$

where α, β, γ permute cyclically.

The antisymmetric part of the gradient of vector $\boldsymbol{v}$ is equal to half its curl:

$$(\nabla\boldsymbol{v})^a=\tfrac{1}{2}\ \operatorname{curl}\ \boldsymbol{v}=\tfrac{1}{2}(\nabla\times\boldsymbol{v})\,. \tag{I.50}$$

The expression $(\boldsymbol{v}\cdot\nabla)\boldsymbol{v}$ which appears in the substantial derivative of velocity can be written as

$$(\boldsymbol{v}\cdot\nabla)\,\boldsymbol{v}=\tfrac{1}{2}\nabla(\boldsymbol{v}\cdot\boldsymbol{v})-\boldsymbol{v}\times(\nabla\times\boldsymbol{v})\,. \tag{I.51}$$

The products of vectors are differentiated as follows:

$$\nabla\cdot(a\boldsymbol{v})=\operatorname{div}(a\boldsymbol{v})=\frac{\partial(av_\alpha)}{\partial x_\alpha}+\frac{\partial(av_\beta)}{\partial x_\beta}+\frac{\partial(av_\gamma)}{\partial x_\gamma}$$

$$=a\left(\frac{\partial v_\alpha}{\partial x_\alpha}+\frac{\partial v_\beta}{\partial x_\beta}+\frac{\partial v_\gamma}{\partial x_\gamma}\right)+\left(v_\alpha\frac{\partial a}{\partial x_\alpha}+v_\beta\frac{\partial a}{\partial x_\beta}+v_\gamma\frac{\partial a}{\partial x_\gamma}\right).$$

Finally,

$$\nabla\cdot(a\boldsymbol{v})=a(\nabla\cdot\boldsymbol{v})+\boldsymbol{v}\cdot\nabla a\,, \tag{I.52}$$

$$\operatorname{div}(a\boldsymbol{v})=a\operatorname{div}\boldsymbol{v}+\boldsymbol{v}\cdot\operatorname{grad}a\,.$$

$$\nabla\cdot(\boldsymbol{v}\times\boldsymbol{w})=\boldsymbol{w}\cdot(\nabla\times\boldsymbol{v})-\boldsymbol{v}\cdot(\nabla\times\boldsymbol{w}) \tag{I.53}$$

$$\nabla\times(a\boldsymbol{v})=(\nabla a)\times\boldsymbol{v}+a(\nabla\times\boldsymbol{v})\,. \tag{I.54}$$

In particular,

$$\nabla\times(\nabla a)=\operatorname{curl}(\operatorname{grad}a)=0 \tag{I.55}$$

$$\nabla\cdot(\nabla\times\boldsymbol{v})=\operatorname{div}(\operatorname{curl}\boldsymbol{v})=0\,. \tag{I.56}$$

Nonequilibrium thermodynamics often uses the Gauss–Ostrogradsky theorem: the flux of a vector through a closed surface A is equal to the volume integral of the divergence of the vector $\boldsymbol{v}$ for the entire space of volume V bounded by that surface:

$$\int_A \boldsymbol{v}\cdot\mathrm{d}\boldsymbol{A}=\int_V \operatorname{div}\boldsymbol{v}\,\mathrm{d}V=\int_V \nabla\cdot\boldsymbol{v}\,\mathrm{d}V\,. \tag{I.57}$$

APPENDIX II

STATISTICAL DERIVATION OF THE ONSAGER RECIPROCAL RELATIONS

The state of a system can be specified in terms of independent macroscopic parameters A_i ($i=1, 2, 3, \ldots, n$) which are even functions of the velocities of molecules, e.g. specific energy and concentration, or such parameters B_i ($i=1, 2, 3, \ldots, m$) which are odd functions of the velocities of molecules, e.g. momentum density. The values of these parameters in the equilibrium state will be denoted, respectively, by A_i^0 and B_i^0, and their fluctuations or random deviations from the state of equilibrium by

$$\alpha_i = A_i - A_i^0 \gtrless 0, \quad \beta_i = B_i - B_i^0 \gtrless 0. \tag{II.1}$$

The fluctuations of parameters, with the exception of entropy fluctuations, may take any values. At very low temperatures and with a very rapid change in fluctuations, quantum fluctuations predominate over thermodynamic fluctuations. However, under the conditions considered here the discussion may be confined to thermodynamic fluctuations.

The state of equilibrium, in the strict sense of the word, is attainable only in an isolated system and in approximation, with the assumption of weak interactions, also between system and environment. By the second law of thermodynamics entropy attains a maximum in an equilibrium state, hence entropy fluctuations can assume only negative values, that is, per unit volume

$$\Delta(\rho s) = \Delta S_v = S_v - S_v^0 < 0. \tag{II.2}$$

Since large, spontaneous fluctuations of entropy are rather improbable, it is sufficient to limit considerations to the first three terms of the Taylor series formed by decomposing entropy about its equilibrium value:

$$\begin{aligned} S_v = S_v^0 &+ \sum_{i=1}^{n} \left(\frac{\partial S_v}{\partial \alpha_i}\right)^0 \alpha_i + \frac{1}{2} \sum_{i,k=1}^{n} \left(\frac{\partial^2 S_v}{\partial \alpha_i \partial \alpha_k}\right)^0 \alpha_i \alpha_k \\ &+ \sum_{i=1}^{m} \left(\frac{\partial S_v}{\partial \beta_i}\right)^0 \beta_i + \frac{1}{2} \sum_{i,k=1}^{m} \left(\frac{\partial^2 S_v}{\partial \beta_i \partial \beta_k}\right)^0 \beta_i \beta_k . \end{aligned} \tag{II.3}$$

Since entropy reaches its maximum at equilibrium,

$$\left(\frac{\partial S_v}{\partial \alpha_i}\right)^0 = 0, \quad \left(\frac{\partial S_v}{\partial \beta_i}\right)^0 = 0 \tag{II.4}$$

for any parameters α_i or β_i.

Matrices of the second derivatives of the entropy of a unit volume with respect to the parameter fluctuations will be denoted by

$$\left(\frac{\partial^2 S_v}{\partial \alpha_i \partial \alpha_k}\right)^0 = -g_{ik}\,; \qquad \left(\frac{\partial^2 S_v}{\partial \beta_i \partial \beta_k}\right)^0 = -h_{ik}\,. \tag{II.5}$$

In accordance with the foregoing assumptions and notation, with the assumption of weak interactions with the environment the fluctuation of the entropy for a unit volume can be written as a quadratic form of the fluctuation of the parameters α_i and β_i:

$$\Delta S_v = -\tfrac{1}{2}\sum_{i,k=1}^{n} g_{ik}\,\alpha_i \alpha_k - \tfrac{1}{2}\sum_{i,k=1}^{m} h_{ik}\,\beta_i \beta_k\,. \tag{II.6}$$

If there is no external magnetic field and no rotation of the medium under consideration, the expression above does not have any terms which simultaneously contain fluctuations of α_i and β_i since ΔS_v must be an even function of the velocity of molecules.

Thermodynamic forces will be defined as the partial derivatives of the entropy fluctuations of a unit volume with respect to the fluctuations α_i of the parameters A_i:

$$X_i = \frac{\partial \Delta (S_v)}{\partial \alpha_i} = -\frac{1}{2}\sum_{k=1}^{n} g_{ik}\frac{\partial(\alpha_k \alpha_i)}{\partial \alpha_i}$$

$$= -\frac{1}{2}\sum_{k=1}^{n} g_{ik}\left(\alpha_k + \alpha_i \frac{\partial \alpha_k}{\partial \alpha_i}\right) = -\sum_{k=1}^{n} g_{ik}\,\alpha_k$$

$$(i = 1, 2, 3, \ldots, n), \tag{II.7}$$

where, in view of the independence of the parameter fluctuations,

$$\frac{\partial \alpha_k}{\partial \alpha_i} = \delta_{ki}\,. \tag{II.8}$$

Similarly, for parameters of type B_i thermodynamic forces can be written as

$$Y_i = \frac{\partial \Delta S_v}{\partial \beta_i} = -\sum_{k=1}^{m} h_{ik}\,\beta_k \quad (i = 1, 2, 3, \ldots, m). \tag{II.9}$$

The generalized flows, being the substantial derivatives of the fluctuations of parameters α_i and β_i, are in the first approximation expressed by linear phenomenological equations

$$J_i = \frac{\mathrm{d}\alpha_i}{\mathrm{d}t} = \sum_{k=1}^{n} L_{ik}^{(\alpha\alpha)} X_k + \sum_{k=1}^{m} L_{ik}^{(\alpha\beta)} Y_k \quad (i = 1, 2, 3, \ldots, n), \tag{II.10}$$

$$I_i = \frac{\mathrm{d}\beta_i}{\mathrm{d}t} = \sum_{k=1}^{n} L_{ik}^{(\beta\alpha)} X_k + \sum_{k=1}^{m} L_{ik}^{(\beta\beta)} Y_k \quad (i = 1, 2, 3, \ldots, m). \tag{II.11}$$

The entropy source strength following from occurrence of fluctuations

$$\Phi_s = \frac{d\Delta S_v}{dt} = -\frac{1}{2}\sum_{i,k=1}^{n} g_{ik}\left(\alpha_i \frac{d\alpha_k}{dt} + \frac{d\alpha_i}{dt}\alpha_k\right) - \frac{1}{2}\sum_{i,k=1}^{m} h_{ik}\left(\beta_i \frac{d\beta_k}{dt} + \frac{d\beta_i}{dt}\beta_k\right)$$

$$= -\sum_{i,k=1}^{n} g_{ik}\alpha_k \frac{d\alpha_i}{dt} - \sum_{i,k=1}^{m} h_{ik}\beta_k \frac{d\beta_i}{dt} = \sum_{i=1}^{n} J_i X_i + \sum_{i=1}^{m} I_i Y_i \qquad \text{(II.12)}$$

is a bilinear expression in the thermodynamic forces and generalized flows. The coefficients g_{ik} or h_{ik} are first derivatives of the thermodynamic forces X_i or Y_i, that is, second derivatives of the entropy fluctuations ΔS_v of a unit volume with respect to the independent variables α_i and β_i, and hence are characterized by the symmetry property

$$g_{ik} = g_{ki}, \qquad h_{ik} = h_{ki}. \qquad \text{(II.13)}$$

To simplify the mathematical relations, the considerations will for the moment be limited to thermodynamic forces of type A_i. In conformity with Boltzmann's statistical interpretation, the entropy fluctuations ΔS_v of a unit volume are proportional to the thermodynamic probability $f(\alpha_i)$ of the given state of fluctuations occurring in the system,

$$\Delta S_v \sim k \ln f(\alpha_i), \qquad \text{(II.14)}$$

where k is Boltzmann's constant.

The probability of fluctuation of α_i is described by the Gaussian distribution:

$$f(\alpha_i) = C \exp\left(\frac{\Delta S_v}{k}\right) = C \exp\left(-\frac{1}{2k}\sum_{i,k=1}^{n} g_{ik}\alpha_i\alpha_k\right). \qquad \text{(II.15)}$$

The constant C is found from the normalization condition which requires that the sum of the probabilities of the system occurring in every state characterized by the fluctuations of all the parameters be equal to unity;

$$\int_{-\infty}^{+\infty} f(\alpha_i)\, d\alpha_i = C \int_{-\infty}^{+\infty} \exp\left(\frac{\Delta S_v}{k}\right) d\alpha_i$$

$$= C \int_{-\infty}^{+\infty} \exp\left(-\frac{1}{2k}\sum_{i,k=1}^{n} g_{ik}\alpha_i\alpha_k\right) d\alpha_i = 1$$

$$(i = 1, 2, 3, \ldots, n). \qquad \text{(II.16)}$$

To evaluate this integral the fluctuations of the parameters are subjected to a linear transformation

$$\alpha_i = a_{ik}\alpha'_k, \qquad \text{(II.17)}$$

which takes the quadratic form (II.6) into a sum of squares

$$\sum_{i,k=1}^{n} g_{ik}\alpha_i\alpha_k = \sum_{i=1}^{n} \alpha_i'^2 = \sum_{i,k=1}^{n} \alpha_i'\alpha_k'\delta_{ik}, \tag{II.18}$$

whereby the coefficients of the transformation a_{ik} must satisfy the condition

$$g_{ik}\,a_{il}\,a_{km} = \delta_{lm}. \tag{II.19}$$

This relation can also be written as

$$|g|\,|a|^2 = 1, \tag{II.20}$$

where $|g|$ is the determinant of the matrix of elements g_{ik}, and $|a|$ is the determinant of the matrix of elements a_{il}, whereas the determinant $|\delta_{lm}| = 1$.

When the transformations have been carried out and when it has been noted that the Jacobian

$$\frac{\partial(\alpha_i)}{\partial(\alpha_k')} = |a|, \tag{II.21}$$

a normalization condition is obtained in the form

$$C|a| \int_{-\infty}^{+\infty} \exp\left(-\frac{1}{2k}\alpha_i'^2\right) \mathrm{d}\alpha_i' = 1. \tag{II.22}$$

This integral can now be treated as the product of n integrals which, if relation (II.20) is taken into account, yields

$$C = \sqrt{\frac{|g|}{(2\pi k)^n}}. \tag{II.23}$$

The thermodynamic forces can now be written as

$$X_i = \frac{\partial \Delta S_v}{\partial \alpha_i} = -\sum_{k=1}^{n} g_{ik}\alpha_k = k\frac{\partial \ln f}{\partial \alpha_i} = \frac{k}{f}\frac{\partial f}{\partial \alpha_i}. \tag{II.24}$$

The mean value of the product $\alpha_i X_i$ is

$$\langle \alpha_i X_k \rangle = \int_{-\infty}^{+\infty} \alpha_i X_k f\,\mathrm{d}\alpha_k = k\int_{-\infty}^{+\infty} \alpha_i \frac{\partial f}{\partial \alpha_k}\mathrm{d}\alpha_k = -k\int_{-\infty}^{+\infty} f\frac{\partial \alpha_k}{\partial \alpha_i}\mathrm{d}\alpha_i, \tag{II.25}$$

where the integration by parts has allowed for the fact that in the limits of integration the probability f of an infinitely large positive or negative fluctuation occurring is zero:

$$|\alpha_i f|_{-\infty}^{+\infty} = 0.$$

Since α_i and α_k are independent fluctuations of the parameters, upon taking acount of (II.8) and the normalization condition (II.16), one obtains the mean value

$$\langle \alpha_i X_k \rangle = -k\delta_{ik}, \tag{II.26}$$

and upon using (II.24), one also gets

$$\langle X_i X_k \rangle = -\sum_{j=1}^{n} g_{ij} \langle a_j X_k \rangle = k g_{ik} \tag{II.27}$$

or

$$\langle \alpha_i \alpha_k \rangle = -\sum_{j=1}^{n} g_{ij}^{-1} \langle X_j a_k \rangle = k g_{ik}^{-1}, \tag{II.28}$$

where g_{ik}^{-1} is the inverse to the matrix g_{ki}, that is, $g_{im} g_{mk}^{-1} = \delta_{ik}$.

Relation (II.28) concerns fluctuations $\alpha_i(t)$ and $\alpha_k(t)$ occurring at the same instant of time t,

$$\langle \alpha_i(t) \alpha_k(t) \rangle = \langle \alpha_k(t) \alpha_i(t) \rangle \neq 0,$$

that is to say, there is a correlation between the fluctuations of different parameters at the same place and at the same time. Such a relation is valid for the correlation of fluctuations of the same parameter at different places and at the same time. It remains to consider how the value of the fluctuation $\alpha_i(t)$ of parameter A_i at time t affects the value of the fluctuation $\alpha_j(t+\tau)$ of parameter A_j at time $t+\tau$ ($\tau>0$). The correlation will be considered of the time fluctuations of the parameters determined by the mean value $\langle \alpha_i(t)\, \alpha_j(t+\tau) \rangle$. This dependence is due to the fact not only that the probability density of fluctuations at one point depends on the probability density of fluctuations at another point but also that the probability density of fluctuations at time $t+\tau$ depends on the probability density of fluctuations at some earlier time t.

The expression $\langle \alpha_i(t)\, \alpha_j(t+\tau) \rangle$ is averaged in time by considering the mean value of this expression in a time interval θ which tends to infinity:

$$\langle \alpha_i(t) \alpha_j(t+\tau) \rangle = \lim_{\theta \to \infty} \int_0^{\theta} \alpha_i(t) \alpha_j(t+\tau)\, \mathrm{d}t. \tag{II.29}$$

The principle of microscopic reversibility asserts that the probability of a microscopic process taking place in one direction of time is the same as that in the other direction. If there is no external magnetic field and no rotation of the fluid under consideration, the microscopic equations of motion for the individual molecules are invariant under the time transformation $t \to -t$, that is, if the signs of the time are changed all molecules travel along their paths in the opposite direction. In reference to the correlation of fluctuations in time this means that when averaging, one may first take the value α_i at time t and then the value α_k at time $t+\tau$, or conversely, that is,

$$\langle \alpha_i(t) \alpha_k(t+\tau) \rangle = \langle \alpha_i(t+\tau) \alpha_k(t) \rangle. \tag{II.30}$$

This equality is differentiated with respect to time τ and then τ is set equal to 0, whereupon

$$\left\langle \alpha_i(t)\frac{\mathrm{d}\alpha_k(t)}{\mathrm{d}t}\right\rangle = \left\langle \frac{\mathrm{d}\alpha_i(t)}{\mathrm{d}t}\alpha_k(t)\right\rangle. \tag{II.31}$$

The mean value of changes during fluctuations of parameters is subject to the same linear phenomenological laws as are macroscopic processes

$$\left\langle \frac{\mathrm{d}\alpha_i}{\mathrm{d}t}\right\rangle = \sum_{l=1}^{n} L_{il}^{(\alpha\alpha)} X_l = J_i, \tag{II.32}$$

whereby equations (II.31) and (II.32) yield

$$\langle \alpha_i \sum_{l=1}^{n} L_{kl}^{(\alpha\alpha)} X_l\rangle = \langle \alpha_k \sum_{l=1}^{n} L_{il}^{(\alpha\alpha)} X_l\rangle. \tag{II.33}$$

The fluctuations are such small deviations from the equilibrium state that the phenomenological coefficients may be assumed to be constant, and hence also

$$\sum_{l=1}^{n} L_{kl}^{(\alpha\alpha)} \langle \alpha_i X_l\rangle = \sum_{l=1}^{n} L_{il}^{(\alpha\alpha)} \langle \alpha_k X_l\rangle. \tag{II.34}$$

When relation (II.26) is taken into account, the result is

$$-k\sum_{l=1}^{n} L_{kl}^{(\alpha\alpha)} \delta_{il} = -k\sum_{l=1}^{n} L_{il}^{(\alpha\alpha)} \delta_{kl}. \tag{II.35}$$

Since $\delta_{il} \neq 0$ only if $i=l$, and since $\delta_{kl} \neq 0$ only if $k=l$, finally

$$L_{ki}^{(\alpha\alpha)} = L_{ik}^{(\alpha\alpha)}. \tag{II.36}$$

The Onsager phenomenological coefficients satisfy the symmetry condition.

The same kind of relation is obtained for

$$L_{ki}^{(\beta\beta)} = L_{ik}^{(\beta\beta)}, \tag{II.37}$$

that is, in the case when both quantities β_i and β_k change sign when the sign of time is changed.

If only one fluctuation changes sign and the other does not when the sign of time is changed, then

$$\left\langle \alpha_i\frac{\mathrm{d}\beta_k}{\mathrm{d}t}\right\rangle = -\left\langle \frac{\mathrm{d}\alpha_i}{\mathrm{d}t}\beta_k\right\rangle \tag{II.38}$$

and the phenomenological coefficients possess the antisymmetry property

$$L_{ki}^{(\alpha\beta)} = -L_{ik}^{(\beta\alpha)}. \tag{II.39}$$

These are the Onsager–Casimir reciprocal relations.

REFERENCES

Classical Thermodynamics

[1] J. Werle, *Termodynamika fenomenologiczna* (Phenomenological Thermodynamics), PWN, Warsaw, 1957.

[2] H. B. Callen, *Thermodynamics*, John Wiley and Sons, Inc., New York, 1961.

[3] M. Tribus, *Thermostatics and Thermodynamics*, D. Van Nostrand Co., Inc., Princeton, N. Y., 1961.

[4] S. L. Soo, *Analytical Thermodynamics*, Prentice-Hall, Inc., Englewood Cliffs, N. Y., 1962.

[5] R. Giles, *Mathematical Foundations of Thermodynamics*, Pergamon Press, Oxford, 1964.

[6] G. N. Hatsopoulos and J. H. Keenan, *Principles of General Thermodynamics*, John Wiley and Sons, Inc., New York, 1965.

[7] M. W. Zemansky and H. C. Van Ness, *Basic Engineering Thermodynamics*, McGraw-Hill Book Co., Inc., New York, 1966.

[8] B. Staniszewski, *Termodynamika* (Thermodynamics), PWN, Warsaw, 1969.

Nonequilibrium Thermodynamics

[9] K. G. Denbingh, *The Thermodynamics of the Steady State*, Methuen and Co., Ltd., London, 1951.

[10] K. E. Grew and T. L. Ibbs, *Thermal Diffusion in Gases*, Cambridge University Press, London, 1952.

[11] J. O. Hirschfelder, C. F. Curtiss, and R. B. Bird, *Molecular Theory of Gases and Liquids*, John Wiley and Sons, Inc., New York, 1954.

[12] S. R. de Groot, *Thermodynamics of Irreversible Processes*, North-Holland Publ. Co., Amsterdam, 1952.

[13] W. Finkelnburg and H. Maecker, *Elektrische Bögen und thermisches Plasma*, in: *Handbuch der Physik*, Springer-Verlag, Berlin, 1956.

[14] L. D. Landau and E. M. Lifshitz, *Fluid Mechanics*, Pergamon Press, Oxford, and Addison-Wesley, Reading, Mass., 1959.

[15] J. Meixner and H. G. Reik, *Thermodynamik der irreversiblen Prozesse*, in: *Handbuch der Physik*, Bd. III/2, Springer-Verlag, Berlin, 1959.

[16] R. B. Bird, W. E. Stewart, and E. N. Lightfoot, *Transport Phenomena*, John Wiley and Sons, Inc., New York, 1960.

[17] S. R. de Groot, (ed.), *Termodinamika dei processi irreversibili*, Societa Italiana di Fisica, Bologna, 1960.

[18] J. Kaye and J. A. Welsh (ed.), *Direct Conversion of Heat to Electricity*, John Wiley and Sons, Inc., New York, 1960.

[19] C. Truesdell and R. A. Toupin, *The Classical Field Theories*, in: *Handbuch der Physik*, Bd. III/1, Springer-Verlag, Berlin, 1960.

[20] N. W. Snyder (ed.), *Energy Conversion for Space Power*, Academic Press, Inc., New York, 1961.

[21] D. D. Fitts, *Nonequilibrium Thermodynamics*, McGraw-Hill Book Co., Inc., New York, 1962.

[22] S. R. de Groot and P. Mazur, *Non-Equilibrium Thermodynamics*, North-Holland Publ. Co. Amsterdam, 1962.

[23] K. Gumiński, *Termodynamika procesów nieodwracalnych* (Thermodynamics of Irreversible Processes), PWN, Warsaw, 1962.
[24] S. S. Chang, *Energy Conversion*, Prentice-Hall, Inc., New York, 1963.
[25] R. Haase, *Thermodynamik der irreversiblen Prozesse*, D. Steinkopff Verlag, Darmstadt, 1963.
[26] A. V. Lykov and J. A. Mikhailov, *Teoria teplo- i massoperenosa* (The Theory of Heat and Mass Transfer), Gosenergoizdat, Moscow, 1963.
[27] P. Rysselberghe, *Thermodynamics of Irreversible Processes*, Hermann et Cie, Paris, and Blaisdel Publ., New York, 1963.
[28] A. Katchalsky and P. F. Curran, *Nonequilibrium Thermodynamics in Biophysics*, Harvard University Press, Cambridge, Mass., 1965.
[29] *Non-Equilibrium Thermodynamics, Variational Techniques and Stability*, University of Chicago Press, Chicago, Ill., 1965.
[30] G. W. Sutton and A. Sherman, *Engineering Magnetohydrodynamics*, McGraw-Hill Book Co., Inc., New York, 1965.
[31] C. Truesdell and W. Noll, *The Nonlinear Field Theories of Mechanics*, in: *Handbuch der Physik*, Bd. III/3, Springer-Verlag, Berlin, 1965.
[32] I. Prigogine, *Introduction to Thermodynamics of Irreversible Processes*, John Wiley and Sons, Inc., New York, 1967.
[33] R. S. Schechter, *The Variational Method in Engineering*, McGraw-Hill Book Co., Inc., New York, 1967.
[34] R. J. Tykodi, *Thermodynamics of Steady States*, The MacMillan Co., New York, 1967.
[35] G. Kalitzin, *Thermodynamik irreversibler Prozesse*, VEB Deutscher Verlag f. Grundstoffindustrie, Leipzig, 1968.
[36] S. L. Soo, *Direct Energy Conversion*, Prentice-Hall, Inc., Englewood Cliffs, N. Y., 1968.
[37] C. Truesdell, *The Nonlinear Field Theories in Mechanics*, in: *Topics in Nonlinear Physics*, Springer-Verlag, Berlin, 1968.
[38] S. Wiśniewski, *Analiza termodynamiczna zjawisk wymiany ciepła i masy w ośrodku pochłaniającym promieniowanie* (Thermodynamic Analysis of Heat- and Mass-Exchange Phenomena in a Radiation-Absorbing Medium), WAT, Warsaw, 1968.
[39] I. Gyarmati, *Non-Equilibrium Thermodynamics*, Springer-Verlag, Berlin, 1970.
[40] P. Glansdorff and I. Prigogine, *Thermodynamic Theory of Structure, Stability and Fluctuations*, John Wiley and Sons, Inc., New York, 1971.

INDEX OF SUBJECTS